Advanced Concepts of Wave Processes in Solids

Advanced Concepts of Wave Processes in Solids

Edited by **Ryan White**

Clanrye International

New Jersey

Published by Clanrye International,
55 Van Reypen Street,
Jersey City, NJ 07306, USA
www.clanryeinternational.com

Advanced Concepts of Wave Processes in Solids
Edited by Ryan White

International Standard Book Number: 978-1-63240-018-5 (Hardback)

Printed in the United States of America.

Contents

Preface

An elucidative account on the advanced concepts of wave process in solids has been provided in this profound book. Wave propagation in solids has been extensively analyzed and major developments in this field have been accomplished not only for the enhancement of calculus methodologies, but also for the significant progresses achieved in the elucidation of novel types of materials. This book provides the readers with innovative and existing research studies elucidating developments in both the domains. It provides information regarding propagation of waves in complex materials and in-depth investigations on associated dispersion relations. It also highlights novel applications regarding the analysis of wave processes in conventional solids and lays emphasis on different simulation availabilities in the areas of geomaterials, seismology, damaging, and multi-wave propagation. The book will serve as a valuable resource for engineers, students and veteran scientists who are well-versed about various aspects of wave propagation in solids.

This book is the end result of constructive efforts and intensive research done by experts in this field. The aim of this book is to enlighten the readers with recent information in this area of research. The information provided in this profound book would serve as a valuable reference to students and researchers in this field.

At the end, I would like to thank all the authors for devoting their precious time and providing their valuable contribution to this book. I would also like to express my gratitude to my fellow colleagues who encouraged me throughout the process.

Editor

Section 1

Wave Propagation in Complex Materials

Chapter 1

Nonlinear Acoustic Waves in Fluid-Saturated Porous Rocks – Poro-Acoustoelasticity Theory

Jing Ba, Hong Cao and Qizhen Du

Additional information is available at the end of the chapter

1. Introduction

The two elastic constants of bulk and shear moduli are insufficient to describe the nonlinear acoustic nature of solid materials under higher confining pressure. The theory of third-order elastic constants (nonlinear acoustics) has long been developed to analyze the velocity *vs* stress relationships in theoretical and experimental research for hyperelastic solid materials [1-7].

The 3rd order constants theory in solid material was completed with the papers by Toupin and Bernstein, Jones and Kobett and Truesdell [8-10]. Truesdell used four 3rd-order elastic constants in his general theory for isotropic solid. Brugger [11] gave the thermodynamic definition of higher order elastic constants. In 1973, Green reviewed 3rd-order constants measurements of various crystals, and gave the relations between the 3rd-order constant notations in isotropic solids by different authors [12].

From late 1980s to the present, 3rd-order nonlinear elasticity theory has been applied to rock experiments [13, 14]. For the special case of porous solid materials, Winkler et al. [15] measured third order elastic constants based on pure solid's acoustoelasticity theory in a variety of dry rocks and found that pure solid's third-order elasticity could successfully describe dry rock's velocity-stress relationships. A similar approach performed on water-saturated rocks showed that traditional 3rd-order elasticity theory for the isotropic solid could not fully describe the stress dependence of velocities in water-saturated rocks [16]. For fluid-saturated porous material, as the confining pressure increases, the unrelaxed fluid's effect must be taken into account when probing into the quantitative relationships between two-phase rock's velocities and confining pressures.

For the fluid/solid composite, Biot derived the two-phase wave equations on the basis of the linear elasticity, in which the coupling motion of solid and fluid was first analyzed. Based on the same linear assumption, some new expansions on Biot's theory have included local fluid flow, dynamic permeability and multi-scale heterogeneity in last two decades [17-23]. Around linear poroelasticity, the research interests of recent years are focused on the frequency-dependent P- and S- waves' velocity and attenuation features which are influenced by patchy saturation, pore distribution and rock microstructure.

According to Biot theory, an isotropic poroelastic medium has four independent static moduli (2nd-order elastic constants). Biot [24] developed semilinear mechanics of porous solids, in which total seven physical constants describes the semilinear properties of an isotropic media, four to characterize the linear behavior and three to characterize the nonlinear behavior. The theory is adapted to solid/fluid composite systems by Norris, Sinha and Kosteck in 1994 [25]. Biot [26] also gave eleven-constant elastic potential function for fluid-saturated porous media, in which seven 3rd-order elastic constants describe the nonlinear features of the two-phase system. Drumheller and Bedford chose a Eulerian reference frame to derive nonlinear equations for wave propagation through porous elastic media in 1980 [27], while Berryman and Thigpen chose a Lagrange reference frame in their theory [28].

Donskoy, Khashanah and McKee [29] derived the nonlinear acoustic wave equations for porous media which established a correlation between the measurable effective nonlinear parameter and structural parameters of the porous medium. Their work is based on the semilinear approximation of Biot's poroelasticity theory. The assumption of semilinearity implies a linear relation between the volume change of the solid matrix and the effective stress, therefore a modified strain (*Equ. 7* in [29], *Equ.* 54 in [24]) is used and the effective stress is included, so that the nonlinear acoustoelasticity equations can be simplified and only three independent 3rd-order elastic constants are needed to be considered in application. Dazel and Tournat [30] considered one-dimensional problems following this theoretical approach and derived the solutions for the second harmonic Biot waves. The wave velocity dispersion and dissipation were analyzed in a semi-infinite medium. Zaitsev, Kolpakov and Nazarov [31, 32] theoretically analyzed the experimental results of the propagation of low-frequency video-pulse signals in dry and water-saturated river sand and demonstrated the nonlinearity of such a loose granular media can be non-quadratic in amplitude.

Grinfeld and Norris [33] derived wave speed formulas in closed-pore jacketed and open-pore jacketed configurations to find a complete set of seven 3rd-order elastic moduli for poroelastic medium. Nevertheless, solid and fluid's finite strain is not included in Grinfeld and Norris's work, so that 2nd-order elastic constants will not appear in the velocity square *vs* compression ratio slope relations for both transverse and longitudinal waves. This means there will be no compatibility between their poro-acoustoelasticity theory and the pure solid material's acoustoelasticity theory, within which 2nd-order elastic constants do affect the velocity square *vs* compression ratio slopes (*i.e.* the slope in equation (1a) is $7\lambda+10\mu+6l+4m$, where λ and μ are 2nd-order elastic constants, and l are m the 3rd-order elastic constants.).

In this paper, we derive the wave propagation equations by substituting the 11-term potential function and the Biot kinetic energy function into Lagrange equations. Finite strain is included, and the velocities and quality factors of P- and S-waves are analytically expressed as functions of components' strains, confining pressure, pore fluid pressure and rock's 2nd- and 3rd-order moduli in cases of hydrostatic and uniaxial loading conditions. The reduction of this theory to a pure solid's acoustoelasticity theory is discussed. Numerical examples for velocity and attenuation are given which are corresponding to the open-pore jacketed and closed-pore jacketed rock tests, respectively. In the last section, we perform ultrasonic P-velocity measurements in the "open-pore jacketed" and "close-pore jacketed" tests with hydrostatic loading. The comparisons between theory and rock experimental data are then given.

2. Nonlinear wave equations for poroelasticity

2.1. Review on nonlinear acoustoelasticity for pure solid material

3rd-order elastic constants theory has long been developed to describe the wave propagation and mechanical deformation in pure solid material. The strain energy function for isotropic solid was given with three independent 3rd-order elastic constants (see Table1).

For nonlinear waves propagating in an isotropic solid media, relations between velocity square and confining pressure was derived as [3, 5, 12, 34].

$$\rho_0 V_{LP}^2 = (\lambda+2\mu) - \frac{P}{3\lambda+2\mu}(7\lambda+10\mu+6l+4m), \quad \text{(a)}$$

$$\rho_0 V_{TP}^2 = \mu - \frac{P}{3\lambda+2\mu}(3\lambda+6\mu+3m-\frac{1}{2}n), \quad \text{(b)}$$

$$\rho_0 V_{La}^2 = (\lambda+2\mu) - \frac{P}{3\lambda+2\mu}[\frac{\lambda+\mu}{\mu}(4\lambda+10\mu+4m)+\lambda+2l], \quad \text{(c)}$$

$$\rho_0 V_{Lb}^2 = (\lambda+2\mu) - \frac{P}{3\lambda+2\mu}[2l-\frac{2\lambda}{\mu}(\lambda+2\mu+m)], \quad \text{(d)} \qquad (1)$$

$$\rho_0 V_{Ta}^2 = \mu - \frac{P}{3\lambda+2\mu}\left(4\lambda+4\mu+m+\frac{\lambda n}{4\mu}\right), \quad \text{(e)}$$

$$\rho_0 V_{Tb}^2 = \mu - \frac{P}{3\lambda+2\mu}(m+\frac{\lambda n}{4\mu}+\lambda+2\mu), \quad \text{(f)}$$

$$\rho_0 V_{Tc}^2 = \mu - \frac{P}{3\lambda+2\mu}(m-\frac{\lambda+\mu}{2\mu}n-2\lambda). \quad \text{(g)}$$

Here, λ and μ are 2nd-order elastic constants. l, m and n are 3rd-order elastic constants. P is confining pressure. ρ_0 is density. The subscripts L and T denote longitudinal and transverse waves. The subscript P denotes hydrostatic loading. The subscript a denotes waves propagating parallel to uniaxial pressure. In waves propagating perpendicular to uniaxial pressure, subscript b and c respectively denote waves' vibrating parallel and perpendicular to uniaxial pressure.

Brugger [11]	Truesdell [10]	Toupin & Bernstein [8]	Landau & Lifshitz [6], Goldberg [7]	Murnaghan [2, 3], Hughs & Kelly [5]
$4C_{456}$	$-4\mu+4\alpha_6$	$4\nu_3$	A	n
C_{144}	$-\frac{\mu}{2}\alpha_4$	ν_2	B	$m-\frac{1}{2}n$
$\frac{1}{2}C_{123}$	$\lambda+\mu\alpha_3+\frac{\mu}{2}\alpha_4$	$\frac{1}{2}\nu_1$	C	$l-m+\frac{1}{2}n$

Table 1. Brief relations between 3rd-order elastic constants by different authors [12]

2.2. Review on linear, semilinear and nonlinear mechanics for poroelasticity

In an isotropic medium, the strain energy is a function of four variables (see *equ.* 3.1 in [35]), the three invariants I_1, I_2, I_3 of solid strain components and the fluid displacement divergence ζ .

$$W = W(I_1, I_2, I_3, \zeta), \tag{2}$$

Biot demonstrated that the expansion to the third degree of equation (2) leads to eleven elastic constants, and he only considered the linear relations to derive the quadratic form for W in [35]. The semilinear mechanics for porous media was developed by Biot in [24].

In nonlinear investigations for poroelasticity, perform Taylor expansion on equation (2) to the 3rd order so that we derive the power series of four independent variables I_1 , I_2 I_3 and ζ (see *equ.* 5.9 in [26]).

$$\begin{aligned}2W = M_1 I_1^2 + M_2 I_2 + M_3 \zeta^2 + M_4 \zeta I_1 + M_5 I_1^3 + M_6 I_3 + M_7 \zeta^3 \\ + M_8 I_1 I_2 + M_9 \zeta I_2 + M_{10} I_1^2 \zeta + M_{11} I_1 \zeta^2.\end{aligned} \tag{3}$$

Equation (3) is an elastic potential expression with four 2nd-order elastic constants and seven 3rd-order elastic constants.

We list in table 2 the different notations for 2nd- and 3rd-order elastic constants which have been used in literatures of poroelasticity.

There are 4 2nd-order terms in equation (3). $M_1 I_1^2$ (the dilatation term) and $M_2 I_2$ (the shear term) are mainly dependent on the elastic characteristics of the solid skeleton, $M_3 \zeta^2$ is related to the pore fluid, and $M_4 \zeta I_1$ is associated with the coupling effect between solid and fluid. If pore fluid has a extremely high bulk modulus (fluid's 2nd-order elastic constants) while solid grain has relatively lower bulk modulus for a specific porous material, $M_3 \zeta^2$ may become the most significant term. But as to the natural rocks, solid grain's bulk modulus is much higher than pore fluid (generally solid modulus is one magnitude higher than fluid), therefore, $M_1 I_1^2$ and $M_2 I_2$ are the most significant, $M_4 \zeta I_1$ is the moderate, and $M_3 \zeta^2$ is the least significant.

For a gas-saturated rock (in this case pore fluid has a very low elastic constant), $M_4\zeta I_1$ and $M_3\zeta^2$ tend to zero, and $M_1 I_1^2$ and $M_2 I_2$ dominate the effective elastic nature of the fluid/solid composite. For the rocks saturated with water or light oil, $M_4\zeta I_1$ is generally more significant than $M_3\zeta^2$, nevertheless, $M_4\zeta I_1$'s effects will get weaker with a high-porosity and softer solid skeleton. For the high-porosity poor-consolidated sand/sandstone, the term of $M_3\zeta^2$ may become more significant than $M_4\zeta I_1$.

	Biot [17, 18]; Ba [23, 36]	Biot [26]	Biot [24]; Donskoy, Khashanah & McKee [29]; Dazel & Tournat [30]	Pierce [37]; Norris, Sinha & Kostek [25, 38]	Berryman & Thigpen [28]	Grinfeld & Norris [33]	Bemer, Boutéca, Vincké, Hoteit & Ozanam [39]
2nd	A, N, Q, R	C_1, C_2, C_6, C_8	λ, μ, α, M	A, N_1	a, b, c, d	$\Gamma_{11}, \Gamma_2, \Gamma_{1\zeta}, \Gamma_{\zeta\zeta}$	K_0, K_s, K_{fl}, ϕ_0
3rd	-	$C_3, C_4, C_5, C_7, C_9, C_{10}, C_{11}$	D, F, G	B	e, f, g, m, n	$\Gamma_{111}, \Gamma_{12}, \Gamma_3, \Gamma_{1\zeta\zeta}, \Gamma_{11\zeta}, \Gamma_{2\zeta}, \Gamma_{\zeta\zeta\zeta}$	D, F

Table 2. Different notations of 2nd and 3rd order elastic constants for poroelasticity

	1st	2nd	3rd	4 th	5th
Soft solid skeleton saturated with extremely high-modulus fluid	$M_7\zeta^3$	$M_{11}I_1\zeta^2$	$M_{10}I_1^2\zeta$	$M_9\zeta I_2$	$M_5 I_1^3, M_6 I_3, M_8 I_1 I_2$
Rock saturated with water/oil	$M_5 I_1^3, M_6 I_3, M_8 I_1 I_2$	$M_{10}I_1^2\zeta$	$M_9\zeta I_2$	$M_{11}I_1\zeta^2$	$M_7\zeta^3$
Rock saturated with gas	$M_5 I_1^3, M_6 I_3, M_8 I_1 I_2$	-	-	-	$M_7\zeta^3, M_9\zeta I_2, M_{10}I_1^2\zeta, M_{11}I_1\zeta^2$

(1st, the most significant level; 2nd, the 2nd significant level; 3rd, the 3rd significant level; 4th, the 4th significant level; 5th, the lowest significant level, whose actual effect can be neglected.)

Table 3. The significance of relative-magnitude of the 7 3rd-order terms in the nonlinear elastic potential expression

As to the 7 3rd-order terms in equation (3), $M_5 I_1^3$, $M_6 I_3$ and $M_8 I_1 I_2$ are related to solid skeleton, $M_7\zeta^3$ is related to pore fluid, while $M_9\zeta I_2$, $M_{10}I_1^2\zeta$ and $M_{11}I_1\zeta^2$ are related to the fluid-solid coupling effect. The significance of the relative magnitude of these terms are listed in Table 3. As to the actual rocks in nature, the terms of $M_5 I_1^3$, $M_6 I_3$ and $M_8 I_1 I_2$ are in the most significant level. In gas-saturated rocks, $M_7\zeta^3$, $M_9\zeta I_2$, $M_{10}I_1^2\zeta$ and $M_{11}I_1\zeta^2$ can be neglected in comparison with $M_5I_1^3$, M_6I_3 and $M_8I_1I_2$, which means fluid's effect can be completely neglected. Therefore, the acoustoelasticity theory for a pure solid can be sufficient to describe the nonlinear wave phenomena in gas-saturated or dry rock samples. But as the fluid's elastic modulus increases, the terms of $M_9\zeta I_2$ and $M_{10}I_1^2\zeta$ will play an important role in its nonlinear elastic effects on the whole solid/fluid system. In this case, the pure solid's three 3rd-order elastic constants may be not enough to give a full description for nonlinear phenomena in a water/oil-saturated rock, which has been discussed in the experimental studies by Winkler et al [15, 16].

2.3. Finite strain in solid/fluid system

If the fluid-saturated rock is loaded under high confining pressure, the infinitesimal strain expression is insufficient to describe solid's and fluid's microscale finite deformation. On this occasion, the Lagrangian strain tensor has much higher precision than infinitesimal strain for solid material. It can be written

$$\varepsilon_{ij} = \frac{1}{2}\left(\frac{\partial u_j}{\partial x_i} + \frac{\partial u_i}{\partial x_j} + \frac{\partial u_l}{\partial x_i}\frac{\partial u_l}{\partial x_j}\right), \; i, j, l = 1,2,3 \tag{4}$$

where u_i denotes solid displacement in i direction. The convention of summation over repeated indices is adopted.

Different from Lagrangian description in solid material, the nonlinear problems of the acoustics of fluids are usually formulated in terms of an Eulerian description of wave motion [38, 40, 41]. Kostek, Sinha and Norris [38] gave the explicit relations for the two descriptions in inviscid fluid. In dealing with nonlinear problems involving both fluid and solid, the unified treatment of Lagrangian variables is used in this paper. If an assumption is reasonably made that fluid viscosity's effect on rock's shear deformation can be neglected so that fluid's shear deformation is completely ignored, the fluid's finite strain under high confining pressure can be approximately written

$$\zeta_{ii} = \frac{\partial U_i}{\partial x_i} + \frac{1}{2}\frac{\partial U_i}{\partial x_i}\frac{\partial U_i}{\partial x_i}, \qquad i = 1,2,3 \tag{5}$$

where U_i denotes fluid displacement in i direction. The convention of summation over repeated indices is not adopted in equation (5). Here we only take into account the longitudinal finite deformation for those low viscous fluid like water and gas. For some non-Newtonian fluid such as bitumen and heavy oil, finite shear deformation need be considered.

In this paper, two types of rock experimental configurations, "open-pore jacketed" and "closed-pore jacketed", are considered. The "closed-pore jacketed" system [33] corresponds to constancy of pore fluid mass, which means the fluid/solid composite are closed-pore jacketed with a impervious deformable jacket under confining pressure so that pore fluid cannot flow out solid matrix in experiments. The "closed-pore jacketed" system is the simplest and rather difficult to realize in practice. The "open-pore jacketed" system [33, 42] corresponds to the constancy of fluid pressure, which means the fluid/solid composite are closed-pore jacketed with a thin impermeable jacket and the inside of the jacket is made to communicate with the atmosphere through a tube, so that in experiment pore fluid will flow out under confining pressure. The "open-pore jacketed" configuration is very common in actual rock tests.

Pore fluid will not be finitely deformed like solid if the porous solid skeleton is "open-pore jacketed", especially in case of highly permeable rocks being saturated with low viscous fluid. In the open-pore jacketed porous media, fluid particles would rather flow out the matrix through connected pores and throats when being subject to pressure gradient, rather than get compressed into a finite deformation state. On the other hand, if the porous structure is "closed-pore jacketed" or the whole fluid system could be regarded as a closed system, finite deformation of the inclosed pore fluid will happen.

Several wave speed formulas for specific experimental configurations are derived by Grinfeld and Norris [33]. However, finite strain (equations 4~5) has not been considered in their derivation, therefore their poro-acoustoelasticity velocity expressions can not be reduced to pure solid's acoustoelasticity expressions (equations 1a~1g) if we perform a gedanken experiment by replacing pore fluid with solid in a solid/fluid composite.

2.4. Nonlinear wave equations in fluid-saturated solid structure

In this paper, we use a theoretical approach similar to Biot's [35] and Norris's [33], but in our development finite strain is used instead of infinitesimal strain, so that this newly developed theory will be more appropriate to describe wave phenomena in the finite-deformed solid/fluid composed system.

The dissipation function and the kinetic energy of a unit volume for the isotropic fluid-solid composed system [35] is given by

$$2D = \phi^2 \frac{\eta}{k}((\dot{u}_1 - \dot{U}_1)^2 + (\dot{u}_2 - \dot{U}_2)^2 + (\dot{u}_3 - \dot{U}_3)^2) \tag{6}$$

$$2T = \rho_{11}(\dot{u}_1^2 + \dot{u}_2^2 + \dot{u}_3^2) + 2\rho_{12}(\dot{u}_1\dot{U}_1 + \dot{u}_2\dot{U}_2 + \dot{u}_3\dot{U}_3) + \rho_{22}(\dot{U}_1^2 + \dot{U}_2^2 + \dot{U}_3^2) \tag{7}$$

Mass parameters ρ_{11}, ρ_{12} and ρ_{22} have been defined by [17,18]. ϕ, η and k are the constants of porosity, fluid viscosity and permeability.

By applying Lagrange's equations and take u_i and U_i as generalized coordinates, generalized forces of solid and fluid phase are derived.

$$\begin{aligned} f_i &= \frac{d}{dt}\left(\frac{\partial T}{\partial \dot{u}_i}\right) + \frac{\partial D}{\partial \dot{u}_i}, \qquad &\text{(a)} \\ F_i &= \frac{d}{dt}\left(\frac{\partial T}{\partial \dot{U}_i}\right) + \frac{\partial D}{\partial \dot{U}_i} \qquad &\text{(b)} \end{aligned} \tag{8}$$

where f_i and F_i satisfy

$$f_i = \frac{d}{dx_j}\left(\frac{\partial W}{\partial(\frac{\partial u_i}{\partial x_j})}\right),$$

$$F_i = \frac{d}{dx_j}\left(\frac{\partial W}{\partial(\frac{\partial U_i}{\partial x_j})}\right).$$

Moreover,

$$a_{ij} = \frac{\partial W}{\partial \varepsilon_{ij}},\ b_{ij} = \frac{\partial W}{\partial \zeta_{ij}},\ i,j = 1,2,3 \tag{9}$$

and

$$\mathbf{J} = \begin{bmatrix} 1+u_{1,1} & u_{1,2} & u_{1,3} \\ u_{2,1} & 1+u_{2,2} & u_{2,3} \\ u_{3,1} & u_{3,2} & 1+u_{3,3} \end{bmatrix}^{\mathrm{T}},\ \mathbf{K} = \begin{bmatrix} 1+U_{1,1} & 0 & 0 \\ 0 & 1+U_{2,2} & 0 \\ 0 & 0 & 1+U_{3,3} \end{bmatrix}^{\mathrm{T}} \tag{10}$$

where $u_{i,j}$ designates $\frac{\partial u_i}{\partial x_j}$.

By substituting equations (9~10) into equations (8a~b), nonlinear wave equations in 2-phase medium are written as

$$\begin{aligned} a_{ik,j}\mathbf{J}_{kj} + a_{ik}\mathbf{J}_{kj,j} &= \rho_{11}\ddot{u}_i + \rho_{12}\ddot{U}_i + b(\dot{u}_i - \dot{U}_i) \quad &(a) \\ b_{ik,j}\mathbf{K}_{kj} + b_{ik}\mathbf{K}_{kj,j} &= \rho_{12}\ddot{u}_i + \rho_{22}\ddot{U}_i - b(\dot{u}_i - \dot{U}_i), \quad &(b) \end{aligned} \tag{11}$$

where $b = \phi^2 \frac{\eta}{k}$.

By substituting equations (4~5) into equation (3), substituting equation (3) into equation (9), then solving a_{ij} and b_{ij}, and finally substituting into (11a~b), we derive the nonlinear acoustic wave propagation equations in 2-phase medium in which both solid and fluid's finite strain are considered. The full expansion scheme for equations (11a~b) is very complicated. In this paper, we only give simplified schemes for some particular experimental configurations.

3. Wave speeds for specific experimental configurations

3.1. The 3rd-order elastic constants for solid/fluid composite

There are four 2nd-order and seven 3rd-order elastic constants in poro-acoustoelasticity theory. The determination of four 2nd-order elastic constants has been discussed by

[42]. In this subsection methods of measurement are described for the determination of the seven 3rd-order elastic constants. With the determination of all the 11 elastic constants, the poro-acoustoelasticity theory will be directly applicable to nonlinear solid/fluid systems.

Studies of Winkler et al. [16] showed that traditional 3rd-order elastic constants theory of pure solid material gives much better description for observed velocity-pressure relations in dry rocks than in saturated ones. Therefore we assume the three 3rd-order elastic constants of l_b, m_b and n_b are sufficient for describing nonlinear acoustic features in dry rocks.

Three gedanken experiments have been discussed to determine the four 2nd-order elastic constants for poroelasticity. Similar approach was also adopted on the determination of the six 2nd-order elastic constants in double-porosity models [43]. For the case of poro-acoustoelasticity, the problem comes to much more complicated for determining the total seven 3rd-order elastic constants.

Five gedanken experiments are designed.

1. Because the fluid viscosity effect has been neglected and it will not directly contribute to the whole structure's higher-order distortion, the 3rd-order elastic coefficient of I_3 in equation (3) will satisfy $M_6 = 2n_b$.
2. If the fluid-saturated rock sample surrounded by a flexible rubber is subject to a hydrostatic pressure P_h and the fluid is allowed to squirt out, all the outside pressure will be confined to the solid skeleton because fluid will squirt out when rock sample is squeezed. For the generalized stress tensors for solid and fluid phase a_{ij} and b_{ij}, because the rock sample only responds to frame stiffness and fluid pressure is nearly zero,

$$\begin{aligned} a_{ii} &= P_h = K_b e + l_b e^2 + \frac{1}{9} n_b e^2, && \text{(a)} \\ b_{ii} &= 0, && \text{(b)} \end{aligned} \qquad (12)$$

where $e = \varepsilon_{11} + \varepsilon_{22} + \varepsilon_{33}$. K_b is the bulk modulus of solid matrix.

For isotropic rocks under hydrostatic loading, $\varepsilon_{11} = \varepsilon_{22} = \varepsilon_{33}$. By substituting equations (12a~b) into equations (9), we get one relation between M_5, M_7, M_8, M_9, M_{10}, M_{11} and l_b, m_b, n_b.

3. The fluid-saturated rock sample surrounded by a flexible rubber is subject to a uniaxial pressure P_u along axis 1, fluid being allowed to squirt out.

The solid and fluid pressures are expressed as

$$a_{11} = P_u = (K_b + \frac{4}{3}\mu_b)e - 2\mu_b(\varepsilon_{22} + \varepsilon_{33}) + (l_b + 2m_b)e^2$$
$$- 2m_b[e(\varepsilon_{22} + \varepsilon_{33}) + \varepsilon_{22}\varepsilon_{33} + \varepsilon_{11}\varepsilon_{33} + \varepsilon_{11}\varepsilon_{22}] + n_b\varepsilon_{22}\varepsilon_{33}, \quad \text{(a)}$$
$$a_{22} = a_{33} = 0 = (K_b + \frac{4}{3}\mu_b)e - 2\mu_b(\varepsilon_{11} + \varepsilon_{22}) + (l_b + 2m_b)e^2 \quad (13)$$
$$- 2m_b[e(\varepsilon_{11} + \varepsilon_{33}) + \varepsilon_{22}\varepsilon_{33} + \varepsilon_{11}\varepsilon_{33} + \varepsilon_{11}\varepsilon_{22}] + n_b\varepsilon_{11}\varepsilon_{33}, \quad \text{(b)}$$
$$b_{ii} = 0 \quad \text{(c)}$$

where solid strain in two transverse directions are identical as $\varepsilon_{22} = \varepsilon_{33}$.

By substituting equations (13a~c) into equations (9), we get one more relation.

4. Let the open-pore jacketed porous sample be subjected to a uniform hydrostatic pressure P_f.

In the acoustoelasticity of solid grains and fluid, strains can be solved through

$$P_f = K_s e + K_{s3}e^2, \quad \text{(a)}$$
$$P_f = K_f\zeta + K_{f3}\zeta^2 \quad \text{(b)}$$
$$a_{ii} = (1-\phi)P_f, \quad \text{(c)} \quad (14)$$
$$b_{ii} = \phi P_f \quad \text{(d)}$$

where K_s and K_f are the solid grain and fluid's bulk modulus. K_{s3} and K_{f3} indicate the solid grain and fluid's 3rd-order moduli [38].

ϕ is the porosity.

By substituting equations (14a~d) into equations (9), we get another two relations.

5. The closed-pore jacketed porous rock is subjected to a uniaxial pressure P_u along axis 1, while at the same time the pore pressure is kept with P_p. Let $P_u = P_p$, we derive the last two relations as below.

$$P_p = K_s(\varepsilon'_{11} + \varepsilon'_{22} + \varepsilon'_{33}) + K_{s3}(\varepsilon'_{11} + \varepsilon'_{22} + \varepsilon'_{33})^2, \quad \text{(a)}$$
$$P_p = K_f\zeta + K_{f3}\zeta^2 \quad \text{(b)}$$
$$0 = (K_b + \frac{4}{3}\mu_b)e'' - 2\mu_b(\varepsilon''_{22} + \varepsilon''_{33}) + (l_b + 2m_b)e''^2$$
$$- 2m_b[e''(\varepsilon''_{22} + \varepsilon''_{33}) + \varepsilon''_{22}\varepsilon''_{33} + \varepsilon''_{11}\varepsilon''_{33} + \varepsilon''_{11}\varepsilon''_{22}] + n_b\varepsilon''_{22}\varepsilon''_{33}, \quad \text{(c)}$$
$$-P_p = (K_b + \frac{4}{3}\mu_b)e'' - 2\mu_b(\varepsilon''_{11} + \varepsilon''_{22}) + (l_b + 2m_b)e''^2 \quad (15)$$
$$- 2m_b[e''(\varepsilon''_{11} + \varepsilon''_{33}) + \varepsilon''_{22}\varepsilon''_{33} + \varepsilon''_{11}\varepsilon''_{33} + \varepsilon''_{11}\varepsilon''_{22}] + n_b\varepsilon''_{11}\varepsilon''_{33}. \quad \text{(d)}$$
$$a_{11} = (1-\phi)P_p \quad \text{(e)}$$
$$a_{22} = a_{33} = -\phi P_p \quad \text{(f)}$$
$$b_{ii} = \phi P_p \quad \text{(g)}$$

The total strain in solid frame is then calculated with $\varepsilon_{11} = \varepsilon'_{11} + \varepsilon''_{11}$, $\varepsilon_{22} = \varepsilon'_{22} + \varepsilon''_{22}$, $\varepsilon_{33} = \varepsilon'_{33} + \varepsilon''_{33}$, where ε'_{ij} and ε''_{ij} denote solid strains induced by pore pressure and effective pressure respectively.

By substituting equations (15a~g) into equations (9), we get the last two relations.

The total seven relations in gedanken experiments (1~5) are well-posed to solve out the seven unknown 3rd-order elastic constants.

3.2. Wave velocity expressions under hydrostatic confining pressure

Based on poro-acoustoelasitcity theory, relations between acoustic wave velocity and loading pressure will be derived, which can be directly applicable to the actual measurements for fluid-saturated porous solids, especially in rock experiments.

In this paper we only consider the case of rocks being subject to hydrostatic or uniaxial loading in experiments. No transverse stress is loaded on samples therefore both shear deformation of solid and fluid can be neglected.

For simplicity let us consider the plane P-waves propagating along the axis "1". Wave equations (11a~b) are reduced to

$$\begin{aligned} a_{11,1} + a_{11,1}u_{1,1} + a_{11}u_{1,11} &= \rho_{11}\ddot{u}_1 + \rho_{12}\ddot{U}_1 + b(\dot{u}_1 - \dot{U}_1), \quad &\text{(a)} \\ b_{11,1} + b_{11,1}U_{1,1} + b_{11}U_{1,11} &= \rho_{12}\ddot{u}_1 + \rho_{22}\ddot{U}_1 - b(\dot{u}_1 - \dot{U}_1) \quad &\text{(b)} \end{aligned} \qquad (16)$$

For hydrostatic loading, the "large" part of the static strain in "small-on-large" constitutive relations induced by static confining pressure is expressed as

$$\varepsilon_{ij} = \begin{pmatrix} \alpha & 0 & 0 \\ 0 & \alpha & 0 \\ 0 & 0 & \alpha \end{pmatrix}, \; \zeta_{ij} = \begin{pmatrix} \beta & 0 & 0 \\ 0 & \beta & 0 \\ 0 & 0 & \beta \end{pmatrix} \qquad (17)$$

where α is a component of the solid's deformation in the 3D space under hydrostatic loading, while β is a component of the fluid's deformation.

The small dynamic strain induced by the plane P-waves' vibration along axis 1 is

$$\varepsilon_{ij} = \begin{pmatrix} Se^{i[\omega t - kx_1(1+\alpha)]} & 0 & 0 \\ 0 & 0 & 0 \\ 0 & 0 & 0 \end{pmatrix}, \; \zeta_{ij} = \begin{pmatrix} S'e^{i[\omega t - kx_1(1+\beta)]} & 0 & 0 \\ 0 & 0 & 0 \\ 0 & 0 & 0 \end{pmatrix} \qquad (18)$$

where S and S' are wave amplitudes in solid and fluid phases. ω and k denote angle frequency and wave number.

By substituting equations (17~18) into equations (16a~b), neglecting all terms whose power orders are higher than 1 and eliminating S and S', equations (16a~b) will be simplified to

$$\begin{vmatrix} \Upsilon_1 k^2 - \rho_{11}\omega^2 + ib\omega & (\Upsilon_2 + M_4\beta)k^2 - \rho_{12}\omega^2 - ib\omega \\ (\Upsilon_2 + M_4\alpha)k^2 - \rho_{12}\omega^2 - ib\omega & \Upsilon_3 k^2 - \rho_{22}\omega^2 + ib\omega \end{vmatrix} = 0, \quad (19)$$

where

$$\Upsilon_1 = M_1 + (7M_1 + M_2 + 9M_5 + 2M_8)\alpha + (\frac{3}{2}M_4 + 3M_{10})\beta,$$

$$\Upsilon_2 = \frac{1}{2}M_4 + (\frac{1}{2}M_4 + M_9 + 3M_{10})\alpha + (3M_{11} + \frac{1}{2}M_4)\beta,$$

and

$$\Upsilon_3 = M_3 + (\frac{3}{2}M_4 + 3M_{11})\alpha + (9M_7 + 7M_3)\beta.$$

Frequency-dependent P velocity can be solved through $V = k / \omega$. The fast and slow P-wave velocities respectively correspond to the two solutions of the quadratic equation.

Similar to the reduction process from Biot equations to Gassmann equations [44] in zero frequency limit, if we neglect the relative motion between solid matrix and pore fluid, which means fluid particles have the same vibration amplitudes as solid in wave propagation (that is, $S = S'$), equation (19) will be reduced to

$$\rho V^2 = \Upsilon_1 + 2\Upsilon_2 + \Upsilon_3 + M_4(\alpha + \beta). \quad (20)$$

where $\rho = \rho_{11} + 2\rho_{12} + \rho_{22}$ denotes the average density [35].

Equation (20) is also a nonlinear extension of Gassmann theory. The difference between solid and fluid's "small" dynamic strain induced by wave vibration is neglected, while solid and fluid's "large" static strain induced by loading are still different with $\alpha \neq \beta$.

Another necessary condition must be satisfied to reduce equation (20) to equation (1a). By assuming pore fluid have the same static deformation as solid skeleton under hydrostatic loading, which means solid and fluid share the same constitutive relation, both in "small" dynamic and "large" static part, and then substituting $\alpha = \beta$ in equation (20),

$$\begin{aligned} \rho V^2 &= M_1 + M_3 + M_4 \\ &+ [7(M_1 + M_3 + M_4) + M_2 + 9(M_5 + M_7 + M_{10} + M_{11}) + 2(M_8 + M_9))]\alpha \end{aligned} \quad (21)$$

Equation (21) is identical to equation (1a) if we take the average body deformation of $-\dfrac{P}{3\lambda + 2\mu}$ for α. The constant relations between different studies are listed in table 4 (As to the elastic potential W in equation 3, a similar notation in equation 5.9 in [26] has been defined as a function of the relative variation of fluid content m, $m = \phi(\zeta - I_1)$. Since [26] has used a different invariant for fluid phase, the constant relations between [26] and this paper are very complicated. If we assume ζ is also used in equation 5.9 in [26] by replacing m with ζ, the constants can be listed in the column 4 of table 4.)

This paper	Biot [17, 18]	Murnaghan [2] Hughs & Kelly [5]	Biot [26]	Grinfeld & Norris [33]
M_1	$A+2N$, P	-	$2C_1$	$2\Gamma_{11}$
M_3	R	-	$2C_6$	$2\Gamma_{\zeta\zeta}$
M_4	$2Q$	-	$2C_8$	$2\Gamma_{1\zeta}$
$M_1+M_3+M_4$	$A+2N+2Q+R$	$\lambda+2\mu$	-	-
$\frac{M_2}{4}$	$-N$	$-\mu$	$\frac{C_2}{2}$	$\frac{\Gamma_2}{2}$
M_5	-	-	$2C_3$	$2\Gamma_{111}$
M_6	-	$2n$	$2C_5$	$2\Gamma_3$
M_7	-	-	$2C_7$	$2\Gamma_{\zeta\zeta\zeta}$
M_8	-	-	$2C_4$	$2\Gamma_{12}$
M_9	-	-	$2C_{11}$	$2\Gamma_{2\zeta}$
M_{10}	-	-	$2C_9$	$2\Gamma_{11\zeta}$
M_{11}	-	-	$2C_{10}$	$2\Gamma_{1\zeta\zeta}$
$\frac{M_5+M_7+M_{10}+M_{11}}{2}$	-	$\frac{l+2m}{3}$	-	-
$\frac{M_8+M_9}{2}$	-	$-2m$	-	-

Table 4. The relations between elastic constants as used in different studies. (The relation between column 1 and column 3 holds when poro-acoustoelasticity theory is reduced to acoustoelasticity theory.)

For plane S-waves propagating along 2 axis and vibrating along 1 axis,

$$\begin{aligned} &a_{12,2}+a_{12,2}u_{1,1}+a_{22}u_{1,22}=\rho_{11}\ddot{u}_1+\rho_{12}\ddot{U}_1+b(\dot{u}_1-\dot{U}_1) && \text{(a)} \\ &0=\rho_{12}\ddot{u}_1+\rho_{22}\ddot{U}_1-b(\dot{u}_1-\dot{U}_1) && \text{(b)} \end{aligned} \qquad (22)$$

The small dynamic strain induced by S-waves is

$$\varepsilon_{ij}=\begin{pmatrix} Se^{i[\omega t-kx_2(1+\alpha)]} & 0 & 0 \\ 0 & 0 & 0 \\ 0 & 0 & 0 \end{pmatrix},\ \zeta_{ij}=\begin{pmatrix} S'e^{i[\omega t-kx_2(1+\beta)]} & 0 & 0 \\ 0 & 0 & 0 \\ 0 & 0 & 0 \end{pmatrix} \qquad (23)$$

By substituting equation (17) and equation (23) into equations (22a~b),

$$\begin{vmatrix} \Psi k^2 - \rho_{11}\omega^2 + ib\omega & -\rho_{12}\omega^2 - ib\omega \\ -\rho_{12}\omega^2 - ib\omega & -\rho_{22}\omega^2 + ib\omega \end{vmatrix} = 0 \tag{24}$$

where

$$\Psi = -\frac{M_2}{4} + (3M_1 - \frac{M_6}{4} - \frac{3}{4}M_8)\alpha + (\frac{3}{2}M_4 - \frac{3}{4}M_9)\beta .$$

Frequency-dependent S-velocity can be solved through equation (24).

By assuming pore fluid has the same static deformation as solid skeleton, equation (24) comes to

$$\rho V^2 = -\frac{M_2}{4} + (3M_1 + \frac{3}{2}M_4 - \frac{M_6}{4} - \frac{3}{4}M_8 - \frac{3}{4}M_9)\alpha. \tag{25}$$

Equation (25) can not be directly reduced to equation (1b), because low viscous fluid's shear constitutive functions has been neglected in Biot theory. Return to pure solid's nonlinear acoustic theory needs to regard pore fluid as solid grains with the same shear constitutive behaviors. Therefore, neglecting solid/fluid relative deformation and adding the neglected term of fluid's shear deformation, S-velocity expression is reduced to

$$\rho V^2 = -\frac{M_2}{4} + (3M_1 + 3M_4 + 3M_3 - \frac{M_6}{4} - \frac{3}{4}M_8 - \frac{3}{4}M_9)\alpha, \tag{26}$$

which is identical to equation (1b).

3.3. Velocity expressions under Uniaxial confining pressure

For uniaxial loading along axis 1, the "large" static strain in "small-on-large" constitutive relations induced by confining pressure is expressed as

$$\varepsilon_{ij} = \begin{pmatrix} \alpha & 0 & 0 \\ 0 & \alpha' & 0 \\ 0 & 0 & \alpha' \end{pmatrix}, \ \zeta_{ij} = \begin{pmatrix} \beta & 0 & 0 \\ 0 & \beta & 0 \\ 0 & 0 & \beta \end{pmatrix} \tag{27}$$

The small dynamic strain induced by plane P-waves vibration along axis 1 is

$$\varepsilon_{ij} = \begin{pmatrix} Se^{i[\omega t - kx_1(1+\alpha)]} & 0 & 0 \\ 0 & 0 & 0 \\ 0 & 0 & 0 \end{pmatrix}, \ \zeta_{ij} = \begin{pmatrix} S'e^{i[\omega t - kx_1(1+\beta)]} & 0 & 0 \\ 0 & 0 & 0 \\ 0 & 0 & 0 \end{pmatrix} \tag{28}$$

By substituting equations (27~28) into equations (16a~b),

$$\begin{vmatrix} \Upsilon_1 k^2 - \rho_{11}\omega^2 + ib\omega & (\Upsilon_2 + M_4\beta)k^2 - \rho_{12}\omega^2 - ib\omega \\ (\Upsilon_2 + M_4\alpha)k^2 - \rho_{12}\omega^2 - ib\omega & \Upsilon_3 k^2 - \rho_{22}\omega^2 + ib\omega \end{vmatrix} = 0, \tag{29}$$

where

$$\Upsilon_1 = M_1 + (5M_1 + 3M_5)\alpha + (2M_1 + M_2 + 6M_5 + 2M_8)\alpha' + (\frac{3}{2}M_4 + 3M_{10})\beta,$$

$$\Upsilon_2 = \frac{1}{2}M_4 + (\frac{1}{2}M_4 + M_{10})\alpha + (M_9 + 2M_{10})\alpha' + (3M_{11} + \frac{1}{2}M_4)\beta$$

and

$$\Upsilon_3 = M_3 + (\frac{1}{2}M_4 + M_{11})\alpha + (M_4 + 2M_{11})\alpha' + (9M_7 + 7M_3)\beta.$$

The two frequency-dependent P velocities are solved out from equation (29). The equation will be reduced to equation (19) in hydrostatic loading configuration if $\alpha = \alpha'$ is satisfied.

If the difference between solid and fluid deformation is neglected for uniaxial loading,

$$\rho V^2 = \Upsilon_1' + 2\Upsilon_2' + \Upsilon_3' + M_4(\alpha + \beta). \tag{30}$$

Equation (30) can not be directly reduced to equation (1c) like the reduction process from (19) to (1a) in hydrostatic case. In equation (30), if we replace the fluid strain β with the solid strain α along axis 1 and replace β with α' along axis 2 and axis 3 (the two orthogonal directions perpendicular to axis 1), we can derive an analytical scheme which is completely compatible with equation (1c).

$$\begin{aligned} \rho V^2 = {} & M_1 + M_3 + M_4 \\ & + [5(M_1 + M_3 + M_4) + 3(M_5 + M_7 + M_{10} + M_{11})]\alpha \\ & + [2(M_1 + M_3 + M_4) + M_2 + 6(M_5 + M_7 + M_{10} + M_{11}) + 2(M_8 + M_9)]\alpha' \end{aligned} \tag{31}$$

where

$$\alpha \approx -\frac{(\lambda + \mu)P_u}{\mu(3\lambda + 2\mu)}, \quad \alpha' \approx \frac{\lambda P_u}{2\mu(3\lambda + 2\mu)}.$$

With the same "large" static strain, for transverse waves propagating perpendicular to uniaxial loading and vibrating along loading,

$$\varepsilon_{ij} = \begin{pmatrix} Se^{i[\omega t - kx_2(1+\alpha')]} & 0 & 0 \\ 0 & 0 & 0 \\ 0 & 0 & 0 \end{pmatrix}, \; \zeta_{ij} = \begin{pmatrix} S'e^{i[\omega t - kx_2(1+\beta)]} & 0 & 0 \\ 0 & 0 & 0 \\ 0 & 0 & 0 \end{pmatrix}, \tag{32}$$

Then

$$\begin{vmatrix} \Psi k^2 - \rho_{11}\omega^2 + ib\omega & -\rho_{12}\omega^2 - ib\omega \\ -\rho_{12}\omega^2 - ib\omega & -\rho_{22}\omega^2 + ib\omega \end{vmatrix} = 0, \tag{33}$$

where

$$\Psi = -\frac{M_2}{4} + (M_1 - \frac{1}{4}M_8)\alpha + (2M_1 - \frac{1}{2}M_8 - \frac{M_6}{4})\alpha' + (\frac{3}{2}M_4 - \frac{3}{4}M_9)\beta .$$

By replacing β with α and α' and include fluid's shear constitutive relations, equation (33) will be reduced to an identical scheme of equation (1f).

$$\rho V^2 = -\frac{M_2}{4} + [(M_1 + M_3 + M_4)(\alpha + 2\alpha') - \frac{M_6}{4}\alpha' - \frac{1}{4}(M_8 + M_9)(\alpha + 2\alpha')] \tag{34}$$

For longitudinal waves propagating perpendicular to uniaxial loading, assume rock sample is subject to a uniaxial pressure along axis 2 and the P-waves transmit along axis 1. The "large" strains are expressed as

$$\varepsilon_{ij} = \begin{pmatrix} \alpha' & 0 & 0 \\ 0 & \alpha & 0 \\ 0 & 0 & \alpha' \end{pmatrix}, \ \zeta_{ij} = \begin{pmatrix} \beta & 0 & 0 \\ 0 & \beta & 0 \\ 0 & 0 & \beta \end{pmatrix} \tag{35}$$

The small dynamic strain induced by plane P-waves propagating along axis 1 is

$$\varepsilon_{ij} = \begin{pmatrix} Se^{i[\omega t - kx_1(1+\alpha')]} & 0 & 0 \\ 0 & 0 & 0 \\ 0 & 0 & 0 \end{pmatrix}, \ \zeta_{ij} = \begin{pmatrix} S'e^{i[\omega t - kx_1(1+\beta)]} & 0 & 0 \\ 0 & 0 & 0 \\ 0 & 0 & 0 \end{pmatrix} \tag{36}$$

Therefore,

$$\begin{vmatrix} \Upsilon_1 k^2 - \rho_{11}\omega^2 + ib\omega & (\Upsilon_2 + M_4\beta)k^2 - \rho_{12}\omega^2 - ib\omega \\ (\Upsilon_2 + M_4\alpha')k^2 - \rho_{12}\omega^2 - ib\omega & \Upsilon_3 k^2 - \rho_{22}\omega^2 + ib\omega \end{vmatrix} = 0 \tag{37}$$

where

$$\Upsilon_1 = M_1 + (M_1 + \frac{1}{2}M_2 + 3M_5 + M_8)\alpha + (6M_1 + \frac{1}{2}M_2 + 6M_5 + M_8)\alpha' + (\frac{3}{2}M_4 + 3M_{10})\beta ,$$

$$\Upsilon_2 = \frac{1}{2}M_4 + (\frac{1}{2}M_9 + M_{10})\alpha + (\frac{1}{2}M_4 + \frac{1}{2}M_9 + 2M_{10})\alpha' + (3M_{11} + \frac{1}{2}M_4)\beta$$

and

$$\Upsilon_3 = M_3 + (\frac{1}{2}M_4 + M_{11})\alpha + (M_4 + 2M_{11})\alpha' + (9M_7 + 7M_3)\beta .$$

The two frequency-dependent P velocities can be solved from equation (37).

If solid/fluid relative motion is neglected, replace fluid strain β with α and α',

$$\begin{aligned}\rho V^2 &= M_1 + M_3 + M_4 \\ &+ [(M_1 + M_3 + M_4) + \frac{1}{2}M_2 + 3(M_5 + M_7 + M_{10} + M_{11}) + (M_8 + M_9)]\alpha \\ &+ [6(M_1 + M_3 + M_4) + \frac{1}{2}M_2 + 6(M_5 + M_7 + M_{10} + M_{11}) + (M_8 + M_9)]\alpha'.\end{aligned} \tag{38}$$

Equation (38) is identical to equation (1d).

For transverse waves propagating along uniaxial loading and vibrating perpendicular to loading, the static strain is expressed as equation (35) and the dynamic strain is

$$\varepsilon_{ij} = \begin{pmatrix} Se^{i[\omega t - kx_2(1+\alpha)]} & 0 & 0 \\ 0 & 0 & 0 \\ 0 & 0 & 0 \end{pmatrix}, \ \zeta_{ij} = \begin{pmatrix} S'e^{i[\omega t - kx_2(1+\beta)]} & 0 & 0 \\ 0 & 0 & 0 \\ 0 & 0 & 0 \end{pmatrix} \tag{39}$$

Then

$$\begin{vmatrix} \Psi k^2 - \rho_{11}\omega^2 + ib\omega & -\rho_{12}\omega^2 - ib\omega \\ -\rho_{12}\omega^2 - ib\omega & -\rho_{22}\omega^2 + ib\omega \end{vmatrix} = 0, \tag{40}$$

where

$$\Psi = -\frac{M_2}{4} + (M_1 - \frac{M_2}{2} - \frac{1}{4}M_8)\alpha + (2M_1 + \frac{M_2}{2} - \frac{1}{2}M_8 - \frac{M_6}{4})\alpha' + (\frac{3}{2}M_4 - \frac{3}{4}M_9)\beta.$$

A similar operation for the reduction to pure solid's acoustoelasticity leads to

$$\rho V^2 = -\frac{M_2}{4} + [(M_1 + M_3 + M_4)(\alpha + 2\alpha') - \frac{M_2}{2}(\alpha - \alpha') - \frac{M_6}{4}\alpha' - \frac{1}{4}(M_8 + M_9)(\alpha + 2\alpha')] \tag{41}$$

which is identical to equation (1e).

For transverse waves propagating perpendicular to uniaxial loading and vibrating perpendicular to uniaxial loading, the dynamic strains are expressed as

$$\varepsilon_{ij} = \begin{pmatrix} Se^{i[\omega t - kx_3(1+\alpha')]} & 0 & 0 \\ 0 & 0 & 0 \\ 0 & 0 & 0 \end{pmatrix}, \ \zeta_{ij} = \begin{pmatrix} S'e^{i[\omega t - kx_3(1+\beta)]} & 0 & 0 \\ 0 & 0 & 0 \\ 0 & 0 & 0 \end{pmatrix} \tag{42}$$

Therefore

$$\begin{vmatrix} \Psi k^2 - \rho_{11}\omega^2 + ib\omega & -\rho_{12}\omega^2 - ib\omega \\ -\rho_{12}\omega^2 - ib\omega & -\rho_{22}\omega^2 + ib\omega \end{vmatrix} = 0, \tag{43}$$

where

$$\Psi = -\frac{M_2}{4} + (M_1 + \frac{M_2}{2} - \frac{1}{4}M_8 - \frac{M_6}{4})\alpha + (2M_1 - \frac{M_2}{2} - \frac{1}{2}M_8)\alpha' + (\frac{3}{2}M_4 - \frac{3}{4}M_9)\beta .$$

The reduction operation leads to

$$\rho V^2 = -\frac{M_2}{4} + [(M_1 + M_3 + M_4)(\alpha + 2\alpha') + \frac{M_2}{2}(\alpha - \alpha') - \frac{M_6}{4}\alpha' - \frac{1}{4}(M_8 + M_9)(\alpha + 2\alpha')] \quad (44)$$

which is identical to equation (1g).

Equations (19, 24, 29, 33, 37, 40, 43) are the central result of this paper, which are applicable in rock tests.

4. Numerical tests for poro-acoustoelasticity

The elastic and density coefficients of a virtual water-saturated porous rock sample are listed in table 5. The 2nd-order elastic constants and the density coefficients come from the actual sample of water-saturated Coldlake sandstone [45]. The seven 3rd-order elastic constants are assumed for numerical tests.

4.1. Wave speeds for sample under hydrostatic loading

If the rock sample is open-pore jacketed and subject to a hydrostatic pressure P_h, and the fluid pressure is hold on one atmosphere (can be zero in approximation), we have the Biot's constitutive relations as $P_h = Ae + \frac{2}{3}Ne + Q\zeta$ and $0 = Qe + R\zeta$. In the virtual tests, the hydrostatic pressure changes from 0 to 30MPa. The solid and fluid's strains are directly solved. By substituting $\alpha = e/3$ and $\beta = \zeta/3$ into equation (19) and equation (24), open-pore jacketed wave speeds are calculated.

For closed-pore jacketed rock sample in hydrostatic loading, the Biot's constitutive relations come to $P_s = Ae + \frac{2}{3}Ne + Q\zeta$, $P_f = Qe + R\zeta$, $P_h = P_s + P_f$ and $P_f / \phi = K_f\zeta$. Hydrostatic loading changes from 0 to 30MPa. By solving solid and fluid's strain and substituting into equation (19) and equation (24), "closed pore" wave speeds are calculated.

The waves center frequency ranges from 100 to 1000 KHz. The numerical results for open- and closed-pore jacketed rock sample under hydrostatic loading are shown in figures 1~2.

Figure 1 shows the relations between wave velocity and attenuation and confining pressure. Fast P wave velocity, Fast P wave inverse quality factor and S wave velocity are sensitive to the change in hydrostatic pressure. Fast P wave attenuation is significantly increased in loading process. Slow P wave's velocity in open-pore jacketed test reaches a maximum value around 13MPa. Moreover, S wave attenuation is entirely not affected by hydrostatic pressure (it holds on 2.1569×10^{-4} at 1MHz), and slow P attenuation has very small change in relation to loading, therefore both have not been drawn in figure 1.

M_1	8.566 GPa	ρ_{11}	2111 Kg/ m^3
M_2	-11.704 GPa	ρ_{12}	-348 Kg/ m^3
M_3	0.7285 GPa	ρ_{22}	697 Kg/ m^3
M_4	2.7 GPa	ϕ	0.335
M_5	244.3 GPa	k	1×10^{-12} m^2
M_6	131.2 GPa	K_b	2.5 GPa
M_7	7.0 GPa		
M_8	-449.8 GPa		
M_9	-149.9 GPa		
M_{10}	65.1 GPa		
M_{11}	24.1 GPa		

Table 5. The coefficients of the virtual rock sample.

Comparing with open-pore jacketed test, closed-pore jacketed test has lower fast P velocity (2380~2580m/s) and higher S and slow P velocity (1177~1284m/s and 52.5~61.7m/s respectively). Fast P inverse quality factor changes from 0.5×10^{-6} to 4.6×10^{-6}, which are much lower than open-pore jacketed tests (0.5×10^{-6} ~4.5×10^{-5}). The sign of the predicted $1/Q$ is positive. The $1/Q$ value is below 10^{-4}, which agrees with the description of dissipation of traditional Biot theory [17, 18] and is obviously lower than the level of the attenuation which can be caused by local fluid flow mechanism. Local fluid flow is not considered in this study, and according to Dvorkin et al. 1994 [46], the magnitude of $1/Q$ produced by local fluid flow may be 1~2 orders magnitude higher than the one produced by Biot friction. Pressure is mainly confined on solid skeleton in open-pore jacketed configuration, so solid's nonlinear acoustic wave feature dominates the whole rock's speeds. Pore water is more likely to be compressed in closed-pore jacketed test and its finite deformation undertakes part of pressure, therefore water's acoustoelasticity should be considered for closed-pore jacketed case. Actually, water's velocity- pressure slope is much lower than solid, therefore, closed-pore jacketed test will have lower fast P velocity than open-pore jacketed test since part of pressure is consumed by pore water instead of by solid matrix. The higher slow P velocity and lower inverse quality factors physically means solid/fluid coupling effect is strengthened, while Biot dissipation is weakened.

Velocities and inverse quality factors *vs* pressure and attenuation are drawn in figure 2. For both cases, it is obvious that fast P wave and S wave are more sensitive to the changes in confining pressure than the changes in wave frequencies, while slow P waves are opposite. As to inverse quality factors, only fast P wave is sensitive to both the confining pressure and the wave frequency. It reaches peak value at the highest center frequency and the largest pressure. S wave's inverse quality factor slightly increases when center frequency increases and shows no response to the loading variations.

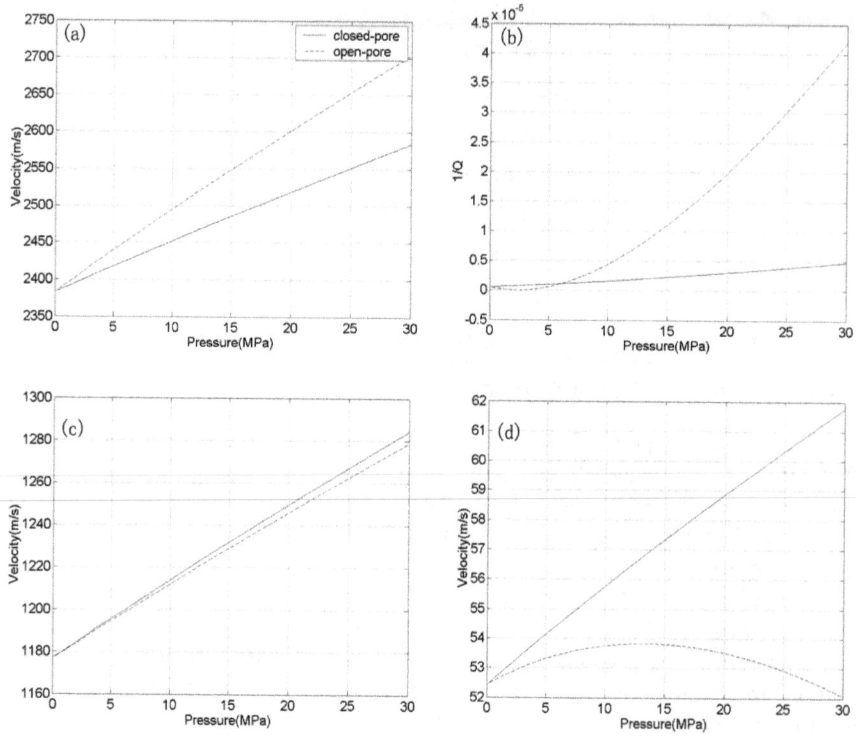

Figure 1. The results in hydrostatic loading with central frequency at 1M Hz. (a) Fast P wave velocity (dashed line for open-pore jacketed, solid line for closed-pore jacketed), (b) Fast P wave inverse quality factor, (c) S wave velocity, (d) Slow P wave velocity.

Comparing open-pore jacketed results with closed-pore jacketed results shows that S wave inverse quality factors of the two cases are entirely equal and not affected by loading. Slow P inverse quality factors of the two cases have very small difference. S and slow P velocities of closed-pore jacketed test are always slightly higher than open-pore jacketed test. Fast P velocity of open-pore jacketed test is significantly higher than closed-pore jacketed test as confining pressure increases. Fast P inverse quality factor in open-pore jacketed test is lower than closed-pore jacketed test when confining pressure is below 5MPa, but gets much higher than closed-pore jacketed test when pressure reaches 30MPa.

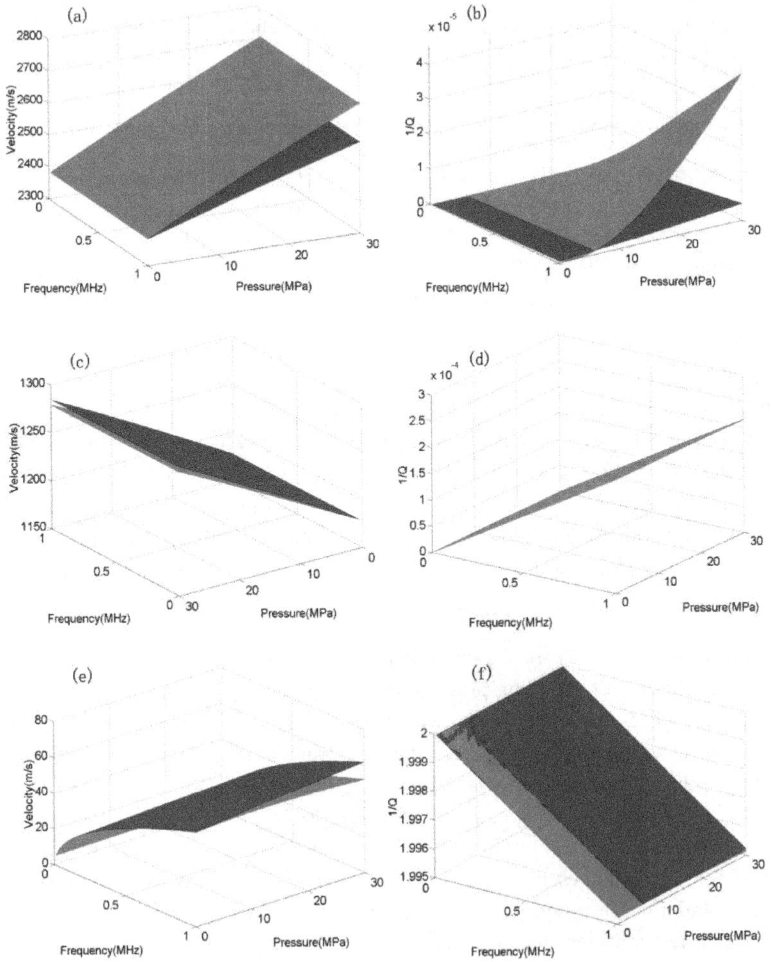

Figure 2. The velocities and inverse quality factors *vs* pressure and centre frequency in hydrostatic loading (red faces for open-pore jacketed, blue faces for closed-pore jacketed). (a) Fast P wave velocity, (b) Fast P wave inverse quality factor, (c) S wave velocity, (d) S wave inverse quality factor, (e) Slow P wave velocity, (f) Slow P wave inverse quality factor.

Mechanisms of "local fluid flow" have not been taken into account in this work. The pores in the rock are assumed to be equant when we derive the nonlinear acoustical wave equations for solid/fluid composite, which is also a basic assumption in Biot's theory [35]. In real rocks which contain both equant pores and soft pores, the contributions of wave-induced local fluid flow may be important for the explanation of the observed attenuation in rock tests. The produced attenuation in this study is mainly associated to the traditional Biot friction and may be underestimated in comparison with an actual laboratory measurement. In the limitation of Biot's theory, the prediction of dissipation can be appropriate only in the case that local fluid flow can be neglected.

4.2. Wave speeds for open-pore jacketed sample under uniaxial loading

In this subsection a virtual test is analyzed with uniaxial loading configuration. The uniaxial pressure is P_u. The fluid pressure is hold around zero. Biot's constitutive relations are written as $P_u = Ae + 2Ne_{11} + Q\zeta$, $0 = Ae + 2Ne_{22} + Q\zeta$, and $0 = Qe + R\zeta$. Let uniaxial pressure change from 0 to 30MPa, e_{11}, e_{22} and ζ are solved out. Substitute $\alpha = e_{11}$, $\alpha' = e_{22}$ and $\beta = \zeta / 3$ into equations (29, 33, 37, 40, 43). Wave center frequency changes from 100 to 1000 KHz.

Figure 3a shows that fast P velocity changes from 2380m/s to 2690m/s along loading direction and changes from 2383m/s to 2389m/s perpendicular to loading direction, while figure 3c shows slow P wave velocity increases slightly in transverse direction and decreases slightly in parallel direction in the loading process. On the same loading level, fast P inverse quality factor is higher in parallel direction than in transverse direction when loading pressure exceeds 5MPa, and both are much lower than in hydrostatic test. The $1/Q$ *vs* pressure curve seems to have a trough around 6MPa along loading and around 8MPa perpendicular to loading. The sign of the predicted $1/Q$ is positive and the value is below 10^{-4}.

Numerical results for S waves are shown in figure 3d. All S velocities in three directions are sensitive to the changes of confining pressure. S wave transmitting along loading direction has the highest speed comparing with two other directions, changing from 1177m/s to 1228m/s. S wave whose propagating and vibrating direction are both perpendicular to loading has relatively lower speed changing from 1177m/s to 1224m/s. S wave vibrating along uniaxial loading has the lowest speed changing from 1177m/s to 1184m/s. All three S velocities in uniaxial configuration are lower than the open-pore jacketed hydrostatic test, where S speed changes from 1177m/s to 1280m/s. For all the three directions in uniaxial test and for the unique direction in hydrostatic test, S waves share the same inverse quality factor independent to the pressure. It only changes with wave centre frequency.

As has been shown in figure 1b and figure 3b, the predicted attenuation increases with confining pressure in average, which disagrees with the experimental results in [47]. Local fluid flow, which dominates the wave attenuation and dissipation phenomena in the rocks under low confining pressure, is depressed since the flat throats and soft pores tend to close under higher confining pressure, so that the attenuation in actual experiment generally decreases with effective stress. The mechanism of local fluid flow and its relation to the confining pressure has not been analyzed in this study. The numerical results are reasonable in condition that the rock pores are assumed to be equant and the local fluid flow mechanism can be neglected.

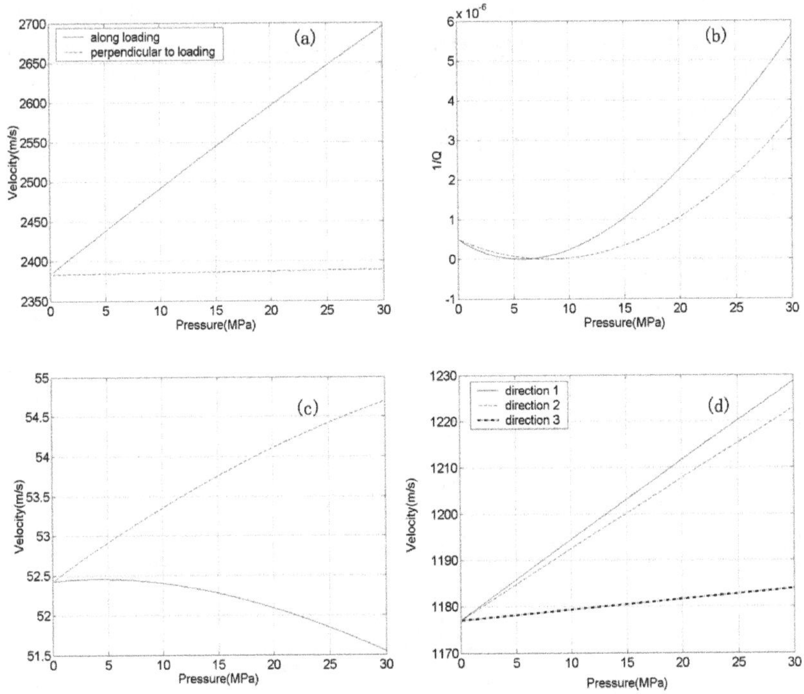

Figure 3. The open-pore jacketed results under uniaxial loading with central frequency at 1M Hz. (a) Fast P wave velocity (dashed line for along loading, solid line for perpendicular to loading), (b) Fast P wave inverse quality factor, (c) Slow P wave velocity, (d) S wave velocity (direction 1: $\parallel P_u$ propagating, $\perp P_u$ vibrating; direction 2: $\perp P_u$ propagating, $\parallel P_u$ vibrating; direction 3: $\perp P_u$ propagating, $\perp P_u$ vibrating).

4.3. Wave speeds for closed-pore jacketed sample under uniaxial loading

If water-saturated rock sample is closed-pore jacketed and subject to uniaxial pressure. In process of loading, because compressed pore water is not allowed to flow out from rubber jacket, the fluid phase will impose an extra stress increment on solid matrix in transverse directions. The six poroelastic constitutive relations are $P_{su} = Ae + 2Ne_{11} + Q\zeta$, $P_f = Qe + R\zeta$, $P_u = P_{su} + P_f$, $P_{sv} = Ae + 2Ne_{22} + Q\zeta$, $0 = P_{sv} + P_f$ and $P_f / \phi = K_f \zeta$, where P_{su} and P_{sv} respectively denote solid stress components along the loading direction and perpendicular to loading.

Numerical results for P waves are figured in figures 4a~4c. First particular feature in this configuration is that fast P speed in vertical direction decrease slightly (from 2385 to 2345 m/s) when confining pressure increases, while all results in former three configurations show opposite trends. Fast P velocity changes from 2380 to 2660m/s along loading, slightly lower than the open-pore jacketed uniaxial test. As is shown in figure 4b, another particular feature is that fast P inverse quality factors along uniaxial loading decreases with rise of pressure. Fast P inverse quality factor in vertical direction increases with loading and is much higher than in parallel direction ($0.5\text{~}5.5 \times 10^{-6}$ *vs* $0.5\text{~}3.5 \times 10^{-6}$), while in open-pore jacketed test fast P inverse quality factor in vertical direction is lower than in parallel direction ($5.2\text{~}2.4 \times 10^{-7}$ *vs* $0.5\text{~}3.2 \times 10^{-6}$).

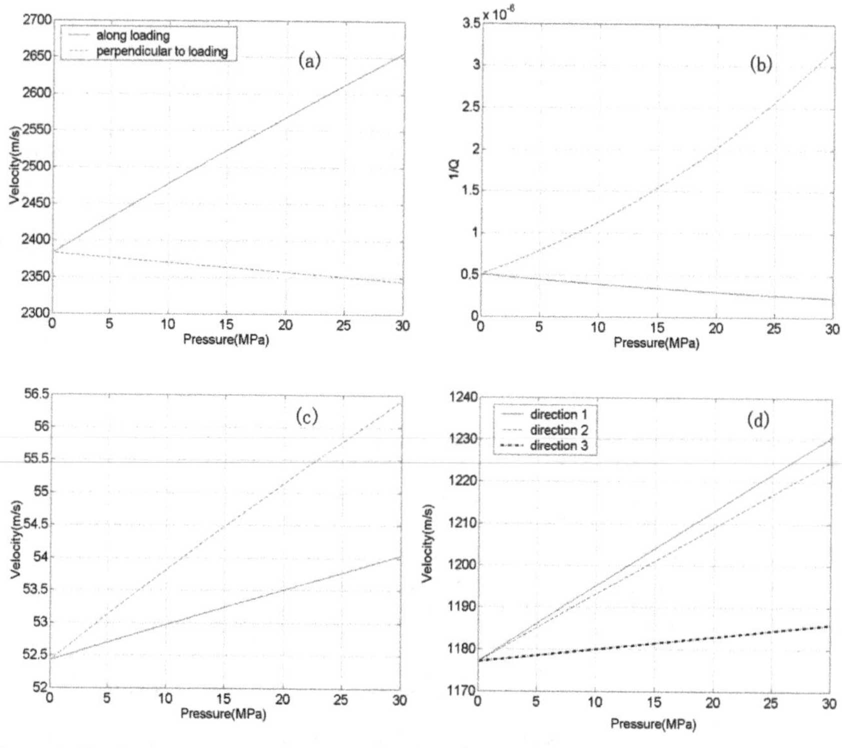

Figure 4. The closed-pore jacketed results under uniaxial loading with central frequency at 1M Hz. (a) Fast P wave velocity (dashed line for along loading, solid line for perpendicular to loading), (b) Fast P wave inverse quality factor, (c) Slow P wave velocity, (d) S wave velocity (direction 1: $\| P_u$ propagating, $\perp P_u$ vibrating; direction 2: $\perp P_u$ propagating, $\| P_u$ vibrating; direction 3: $\perp P_u$ propagating, $\perp P_u$ vibrating).

Numerical results for S waves in closed-pore jacketed sample under uniaxial loading are shown in figure 4d. S wave transmitting along loading has the highest speed ranging from 1177m/s to 1231m/s. S velocities in two vertical directions respectively range from 1177m/s to 1225m/s and from 1177m/s to 1186m/s. S-velocities in all three directions increase as loading rises. S velocities of closed-pore jacketed uniaxial test are very close to open-pore jacketed uniaxial tests. In all test configurations of this paper, S inverse quality factors seem to share the same value which is only slightly dependent to wave centre frequency.

5. Experimental data

5.1. Physical setup

Two hydrostatic loading tests (one for "open-pore jacketed" configuration, another for "closed-pore jacketed" configuration) are performed on a sandstone sample with moderate porosity (13.26 percent) and low permeability (1.21mD). The sample is collected from a gas reservoir in southwest China. It is from the depth of around 2000 meters from surface. The sandstone sample is mainly constructed by quartz and feldspar. Minor clays and rock fragments reside inside pores and grains. It is moderately sorted. The grain size ranges from below 0.1mm to around 1 mm. The pore size ranges from below 0.1 mm to around 0.4 mm. Most grains contact well, so as to form a rigid solid skeleton. The average grain density is 2.659g/cm^3. The grain bulk modulus is 39.0 GPa.

The experimental setup consists of a digital oscilloscope and a pulse generator. In the test, the rock sample is jacketed with a rubber tubing to isolate it from the confining pressure. The receiving transducer is connected to the digitizing board in the PC through a signal amplifier. A pore fluid inlet in the endplate allows passage of pore fluid through the sample and can help to control the pore pressure inside rocks in experiments. We give a small modification on the original inlet instrument by adding a valve (it is closely connected to the inlet), so that the "closed-pore jacketed" configuration can be realized. The open-pore and closed-pore tests are performed respectively. In each test, the rock is full saturated with water, and both confining pressure and pore pressure is raised to 10 MPa before the P-wave speed measurements. In the "closed-pore jacketed" test, we keep the valve closed and raises the confining pressure from 10 to 62 MPa with an interval of around 4 MPa. The P-wave speed is measured and recorded in each pressure level. In the "open-pore jacketed" test, the valve is kept open in measuring process. The pore pressure is close to atmospheric pressure when confining pressure is not so high (it can be neglected comparing with the confining loading).

5.2. Discussions on the theory and experimental data

The relationship between the measured velocities of P waves and the confining pressure in the "open-pore jacketed" and "closed-pore jacketed" tests are shown in figure 5. In the both cases, measured P velocity increases as the confining pressure gets higher. Comparing with the closed-pore test, the velocity-pressure relationship in the open-pore test has a higher slope in a range of 10-35 MPa if it is fitted with a liner correlation, and the difference of measured velocities between open-pore and closed-pore tests increases as the confining pressure rises. However, when confining pressure increases in a higher range of 40-62 MPa, this difference decreases as the confining pressure rises. The observed velocity-pressure results of open-pore test will approach to the results of closed-pore test.

Based on the poro-acoustoelasticity theory, 11 elastic constants have to be determined if we want to predict the velocity-pressure relationships in actual rock. The four 2nd order elastic constants can be theoretically calculated with a Biot-Gassmann theory [48]. The dry bulk modulus can be estimated through Pride's frame moduli expression for consolidated sandstone [49] as $K_d = K_s(1-\phi)/(1+c\phi)$, where K_d, K_s and ϕ are dry rock modulus, grain bulk modulus and porosity. c are the consolidation parameter. We choose $c = 8$ since it is a moderate porosity and good consolidated sandstone. According to Table 3, the 4 2nd order elastic constants can be calculated as $M_1 = 37.4GPa$, $M_2 = \text{-}53.6GPa$, $M_3 = 0.273GPa$ and $M_4 = 1.84GPa$. As to the 7 3rd-order elastic constants, more prior information about each component's higher-order elastic modulus should be fully realized if we want to give an applicable theoretical estimate. Moreover, some uniaxial loading tests and S velocity measurements should also be performed in the "open/closed-pore jacketed" configurations, so that all the 7 3rd-order elastic constants can be precisely determined and discussed. Nevertheless, these tests also require further improvements on our current experimental instruments. We will leave the issue for a future consideration.

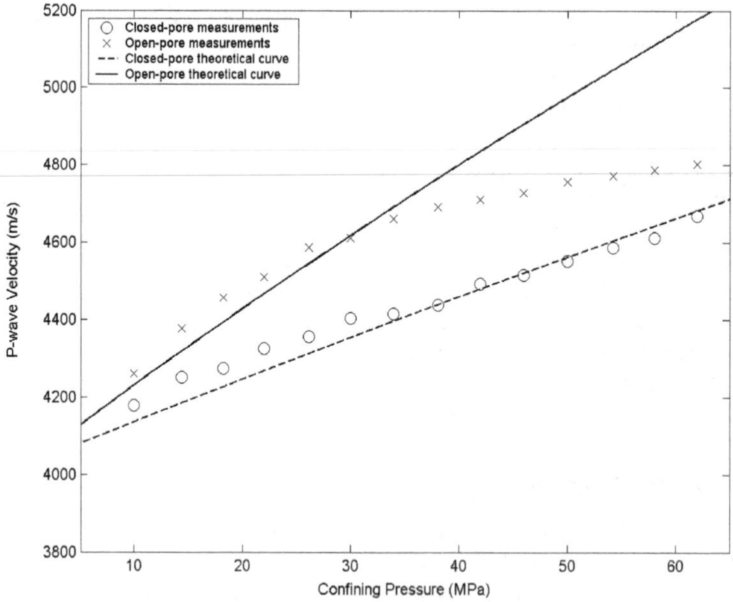

Figure 5. Comparison between experimental data and theoretical curves.

In this paper, we use 4 3rd order elastic constants (M_5, M_6, M_8 and M_{10}) to fit the measured results in the open-pore and closed-pore jacketed tests, and neglect other 3 terms, since these four terms are considered as the most important for actual rocks as has been discussed in section 2.

Based on the numerical approach in Section 4, we perform numerical calculation and draw the velocity-pressure curves in figure 5. They are both in good agreement with open-pore and closed-pore measurements in a low effective pressure range. We choose $M_5 = 1800GPa$, $M_6 = -2500GPa$, M_8=2300GPa and M_{10}=100GPa in calculation. M_5, M_6 and M_8 are mostly associated with the solid skeleton's higher order elastic characteristics. The order of magnitude of these constants (KGPa) agrees well with the measurement of 3rd-order elastic constants on rocks performed by Winkler et. al. [15, 16]. As a complement to the pure solid's 3rd-order elastic constant theory, the new 3rd elastic constant M_{10} is related to the higher-order coupling effect between solid and fluid. Its order of magnitude is around 100GPa. It effects on the wave speed expressions for fluid-saturated rocks, and holds a lower magnitude than M_5, M_6 and M_8, which has been discussed in Section 2. A higher value of M_{10} will produce unreasonable results for wave speeds.

The observed differences between "open-pore jacketed" and "closed-pore jacketed" tests decrease as we still increase the confining pressure in a high pressure range (higher than 40 MPa), which do not agree with our theoretical analysis. Actually, the "open-pore jacketed" measured speeds approach to the "closed-pore jacketed" results in high pressure end. In high confining pressure range, the theoretical results fit well with the "closed-pore" measurement, while give an estimate higher than the actually measured speeds in the "open-pore" tests. As confining pressure is getting higher for an open-pore system, the stress difference between solid skeleton and pore fluid is consequently increased. The effective pressure is also increased. Pore fluid continues to flow out and the soft pores such as grain cracks will tend to close. In a high effective pressure range, the pore system of rock in a "open-pore" test will tend to act like a theoretical "closed-pore" system, where soft pores are closed and fluid stops flowing out. The basic assumption of a theoretical "open-pore" configuration breaks down in this case.

In a real sandstone, at higher pressure (above 50 MPa), the increase in the values of wave velocities essentially saturates [50, 51]. This saturation effect is caused exclusively by gradual closing of the soft crack-like fraction of the rock-frame porosity. This crack-like fraction strongly dominates the nonlinearity of the material in the pressure range of several tens MPa, whereas visible pores are less affected. The compressibility of crack-like pores strongly depends on the fact whether the filling fluid is confined inside or can be squeezed out. Thus below 50MPa the difference between "open-pore jacketed" and "closed-pore jacketed" tests is obvious until the most part of the cracks become closed. The application of this theory is limited to the pressure range lower than 50MPa. Especially for the "open-pore jacketed" case, the theoretical predictions will be unreliable when the effective pressure exceeds the limitation.

6. Conclusions

The nonlinear acoustic wave propagation equations are derived in this paper. For hydrostatic and uniaxial loading configurations of rock tests, the analytical deformation-dependent P- and S- speed expressions are presented. We discuss the reductions from current theoretical approach to the traditional works of acoustoelasticity in pure solid material. Neglecting solid/fluid differences both in a large static part of strain induced by confining pressure and in a small dynamic part of strain induced by wave's propagation and vibration, the velocity equations in this paper (equations 21, 26, 31, 34, 38, 41 & 44) are completely compatible with the expressions of acoustoelasticity theory in pure solid material (equations 1a~g).

We design a virtual rock sample with seven assumed 3rd-order elastic constants. We perform four numerical tests on this virtual rock sample with different considerations: open-pore jacketed sample under hydrostatic loading, open-pore jacketed sample under uniaxial loading, closed-pore jacketed sample under hydrostatic loading and closed-pore jacketed sample under uniaxial loading. Numerical results show that fast P wave's inverse quality factor is both sensitive to the confining pressure and wave centre frequency, which agrees with the laboratory observations of our former studies [47] in the aspect that P wave's inverse quality factors are more sensitive to pressure than S wave's. For the virtual rock sample, velocities of fast P waves and S waves seem to be more sensitive to pressure change than to changes in wave centre frequency, while slow P wave shows an opposite feature. In closed-pore jacketed uniaxial loading configuration, fast P wave's velocity in transverse direction is in decreasing trend with the rise of loading pressure, which is completely contrary to other three configurations. In all test configurations of this paper, S inverse quality factors share the same value which is independent to confining loading. S inverse quality factor can be influenced by wave centre frequency.

We perform the "open-pore jacketed" and "closed-pore jacketed" P-velocity measurements on a sandstone sample under hydrostatic loading. The observed P-velocity differences between the "open-pore jacketed" and "closed-pore jacketed" tests are getting larger as the confining pressure is increased in the range of 10~35 MPa, which agrees well with theoretical prediction, while in a higher confining loading range (>40MPa), the tested "open-pore jacketed" velocity approaches to the "closed-pore jacketed" velocity. When hydrostatic confining pressure is increased and pore pressure keeps constant in an "open-pore" experiment, fluid will continue to flow out and microscopic connections between pores tend to close. When effective pressure reaches a high level, the rock in an "open-pore" test will approach to the theoretical configuration of the "closed-pore" system, where the pore fluid will actually stop flowing out under loading. The basic assumptions of the theoretical "open-pore jacketed" configuration is violated. A rock under high effective pressure in an "open-pore jacketed" test is actually "closed". In a high confining loading, the crack-like fraction will become gradually closed while the visible pores remain almost affected. Mostly the equant pores remain survived for which the compressibility weakly depend on the presence of the fluid (does not matter "closed-" or "open-pore"). The theory presented in this paper can give reasonable predictions for wave speeds in a confining pressure range lower than 50 MPa.

Moreover, in this work, the mechanism of "local fluid flow" have been neglected when we theoretically derive the nonlinear wave equations. The heterogeneity in pore structures and fluid distributions will induce local fluid flow if elastic waves squeeze the porous rock, which surely will effect on the waves' propagation velocities and intrinsic attenuation [23, 36, 50, 52]. However, to include the mechanism of local fluid flow in a poro-acoustoelasticity theory will produce too complicated expressions to be handled on the present stage, and there probably will be several additional 2nd order terms [52, 53] and much more 3rd order terms in the strain energy expression. The present paper is limited to the area of applicability of the Biot approach, whereas rocks in nature usually contain the stiff and the soft pores. It may be important and essential that the local fluid flow mechanism should be analyzed in a poro-acoustoelasticity context, since the effects of confining pressure on pore shapes are also believed to be relevant to the local fluid flow and will consequently change the observed waves' velocities and attenuation in laboratory, which can be described in detail by introducing additional 2nd and 3rd order terms. This extension will be a future study.

Author details

Jing Ba and Hong Cao
Geophysical Department, Research Institute of Petroleum Exploration and Development, PetroChina, Beijing, China
Key Laboratory of Geophysics, PetroChina, Beijing, China

Qizhen Du
School of Geosciences, China University of Petroleum (East China), Qingdao, China

Acknowledgement

Discussions with Winkler K. W. were helpful. Wu X. Y. and Hao Z. B. helped in designing the experimental setup for a "closed-pore jacket" rock test and performing the measurements. This research was sponsored by the Major State Basic Research Development Program of China (973 Program, 2007CB209505), the CNPC 12-5 Basic Research Plan (2011A-3601) the Natural Science Foundation of China (41074087, 41104066) and the RIPED Youth Innovation Foundation (2010-A-26-01).

7. References

[1] Truesdell C (1965) Continuum Mechanics IV: Problems of Non-linear Elasticity. Gordon & Breach, New York.
[2] Murnaghan F D (1937) Finite deformations of an elastic solid. Amer. J. Math., 59: 235-260.
[3] Murnaghan F D (1951) Finite deformation of an elastic solid. John Wiley & Sons, Inc., New York.
[4] Hearmon R F S (1953) 'Third-order' elastic coefficients. Acta Crystallographica, 6: 331-340.
[5] Hughs D S, Kelly J L (1953) Second-order elastic deformation of solids. Phys. Rev., 92: 1145-1149.
[6] Landau L D, Lifshitz E M (1959) Theory of Elasticity. Pergamon Oxford, London.

[7] Goldberg Z A (1961) Interaction of plane longitudinal and transverse elastic waves. Akusticheskii Zhurnal., 6(3): 307-310.
[8] Toupin P A, Bernstein B (1961) Sound waves in deformed perfectly elastic materials. Acoustoelastic effect. J. Acoust. Soc. Am., 33: 216-225.
[9] Jones G L, Kobett D (1960) Interaction of elastic waves in an isotropic solid. J. Acoust. Soc. Am., 35: 5-10.
[10] Truesdell C (1961) General and exact theory of waves in finite elastic strain. Arch. Ration. Mech. Anal., 8: 263-296.
[11] Brugger K (1964) Thermodynamic definition of higher order elastic coefficients. Phys. Rev., 133: A1611-A1612.
[12] Green R E (1973) Ultrasonic investigation of Mechanical Properties. Treatise on Materials Science and Technology, Vol. 3. Academic Press, New York.
[13] Johnson P A, Shankland T J (1989) Nonlinear generation of elastic waves in granite and sandstone: continuous wave and travel time observations. J. Geophys. Res., 94: 17729.
[14] Meegan G D, Johnson P A, Guyer R A, McCall K R (1993) Observations on nonlinear elastic wave behaviour in sandstone. J. Acoust. Soc. Am., 94: 3387-3391.
[15] Winkler K W, Liu X (1996) Measurements of third-order elastic constants in rocks. J. Acoust. Soc. Am., 100: 1392-1398.
[16] Winkler K W, McGowan L (2004) Nonlinear acoustoelastic constants of dry and saturated rocks. J. Geophys. Res., 109: B10204.
[17] Biot M A (1956a) Theory of propagation of elastic waves in a fluid-saturated porous solid. I. Low-frequency range. J. Acoust. Soc. Am., 28: 168-178.
[18] Biot M A (1956b) Theory of propagation of elastic waves in a fluid-saturated porous solid. II. Higher frequency range. J. Acoust. Soc. Am., 28: 179-191.
[19] Dvorkin J, Nur A (1993) Dynamic poroelasticity: A unified model with the squirt flow and the Biot mechanisms. Geophysics, 58(4): 524-533.
[20] Johnson D L (2001) Theory of frequency dependent acoustics in patchy-saturated porous media. J. Acoust. Soc. Am., 110: 682-694.
[21] Carcione J M, Picotti S (2006) P-wave seismic attenuation by slow wave diffusion: Effects of inhomogeneous properties. Geophysics, 71: O1-O8.
[22] Pride S R, Masson Y J (2006) Acoustic attenuation in self-affine porous structures. Phys. Rev. Lett., 97: 184301.
[23]Ba J, Nie J X, Cao H, Yang H Z (2008) Mesoscopic fluid flow simulation in double-porosity rocks. Geophys. Res. Lett., 35: L04303.
[24] Biot M A (1973) Nonlinear and semilinear rheology of porous solids. J. Geophys. Res., 78: 4924-4937.
[25] Norris A N, Sinha B K, Kostek S (1994) Acoustoelasticity of solid/fluid composite systems. Geophys. J. Int., 118: 439-446.
[26] Biot M A (1972) Theory of finite deformations of porous solids. Indiana University Mathematics Journal, 21: 597-620.
[27] Drumheller D S, Bedford A (1980) A thermomechanical theory for reacting immiscible mixtures. Arch. Rational Mech. Anal. 73: 257-284.

[28] Berryman J G, Thigpen L (1985) Nonlinear and semilinear dynamic poroelasticity with microstructure. J. Mech. Phys. Solids, 33: 97-116.

[29] Donskoy D M, Khashanah K, Mckee T G (1997) Nonlinear acoustic waves in porous media in the context of Biot's theory. J. Acoust. Soc. Am., 102(5): 2521-2528.

[30] Dazel O, Tournat V (2010) Nonlinear Biot waves in porous media with application to unconsolidated granular media. J. Acoust. Soc. Am., 127(2): 692-702.

[31] Zaitsev V Y, Kolpakov A B, Nazarov V E (1999a) Detection of acoustic pulses in river sand: experiment. Acoustical Physics, 45(2): 235-241.

[32] Zaitsev V Y, Kolpakov A B, Nazarov V E (1999b) Detection of acoustic pulses in river sand. Theory. Acoustical Physics, 45(3): 347-353.

[33] Grinfeld M A, Norris A N (1996) Acoustoelasticity theory and applications for fluid-saturated porous media. J. Acoust. Soc. Am., 100: 1368-1374.

[34] Pao Y H, Sachse W, Fukuoka H (1984) Acoustoelasticity and ultrasonic measurement of residual stresses. In: Physical Acoustics, Vol. XVII. Orlando.

[35] Biot M A (1962) Generalized theory of acoustic propagation in porous dissipative media. J. Acoust. Soc. Am., 34(9): 1254-1264.

[36] Ba J, Cao H, Yao F C, Nie J X, Yang H Z (2008) Double-porosity rock model and squirt flow in the laboratory frequency band. Applied Geophysics, 5(4): 261-276.

[37] Pierce A D (1981) Acoustics: An introduction to its physical principles and applications. Acoustical Society of America, Woodbury, New York.

[38] Kostek S, Sinha B K, Norris A N (1993) Third-order elastic constants for an inviscid fluid. J. Acoust. Soc. Am., 94(5): 3014-3017.

[39] Bemer E, Boutéca M, Vincké O, Hoteit N, Ozanam O (2001) Poromechanics: From linear to nonlinear poroelasticity and poroviscoelasticity. Oil & Gas Science and Technology-Rev. IFP, 56(6): 531-544.

[40] Beyer R T (1960) Parameter of nonlinearity in fluids. J. Acoust. Soc. Am., 32: 719-721.

[41] Beyer R T (1984) Nonlinear Acoustics in Fluids. Van Nostrand Reinhold, New York.

[42] Biot M A, Willis D G (1957) The elastic coefficients of the theory of consolidation. J. Appl. Mech., 24: 594-601.

[43] Ba J, Yang H Z, Xie G Q (2008) AGILD seismic modelling for double-porosity media. The 23rd PIERS 2008 in Hangzhou, Hangzhou, China.

[44] Gassmann F (1951) Uber die elastizitat, poroser medien. Vier. Der Natur. Gesellschaft, 96: 1-23.

[45] Dai N, Vafidis A, Kanasewich E R (1995) Wave propagation in heterogeneous, porous media: A velocity-stress, finite-difference method. Geophysics, 60(2): 327-340.

[46] Dvorkin J, Nolen-Hoeksema R, Nur A (1994) The squirt-flow mechanism: Macroscopic description. Geophysics, 59(3): 428-438.

[47] Guo M Q, Fu L Y, Ba J (2009) Comparison of stress-associated coda attenuation and intrinsic attenuation from ultrasonic measurements. Geophys. J. Int., doi: 10.1111/j.1365-246X.2009.04159.x.

[48] Johnson D L (1986) Recent developments in the acoustic properties of porous media. In: Sette D, editor. Frontiers in Physical Acoustics XCIII. North Holland Elsevier, New York: 255-290.

[49] Pride S R, Berryman J G, Harris J M (2004) Seisimic attenuation due to wave-induced flow. J. Geophys. Res., 109, B01201, doi:10.1029/2003JB002639.
[50] Mavko G A, Jizba D (1994) The relation between seismic P- and S-wave velocity dispersion in saturated rocks. Geophysics, 59(1): 87-92.
[51] Zaitsev V, Sas P (2004) Effect of high-compiant porosity on variations of P- and S- wave velocities in dry and saturated rocks: comparison between theory and experiment. Physical Mesomechanics, 7(1-2): 37-46.
[52] Ba J, Carcione J M, Nie J X (2011) Biot-Rayleigh theory of wave propagation in double-porosity media. J. Geophys. Res. solid earth, 116: B06202, doi: 10.1029/2010JB008185..
[53] Ba J, Carcione J M, Cao H, Du Q Z, Yuan Z Y, Lu M H (2012) Velocity dispersion and attenuation of P waves in partially-saturated rocks: Wave propagation equations in double-porosity medium. Chinese J. Geophys. (in Chinese), 55(1): 219-231.

Chapter 2

Linear Wave Motions in Continua with Nano-Pores

Pasquale Giovine

Additional information is available at the end of the chapter

1. Introduction

The first attempts to describe the behaviour of porous materials with the use of an additional degree of kinematical freedom, in order to refine the Cauchy's theory, are due principally to Nunziato and Cowin [1, 2] and co-workers. Nevertheless, their voids theory can be considered as a particular case of a general theory of continua with microstructure [3] and so, when we have to consider more complex media with nano-pores, we need to use suggestions of this last theory [4]. In fact, a nano-pore in a thermoelastic solid is roughly ellipsoidal, unlike small lacunae finely dispersed in the solid matrix that can be supposed all spherical and for which the volume fraction suffices to describe the microdeformation (see, also, [5, 6]). Cowin itself remarked the importance of the shape of the holes in the description of lacunae containing osteocytes or of bone canaliculi [7, 8]: in the human bone, *e.g.*, the lacunae are almost ellipsoidal with mean values along the axes of about 4 μm, 9 μm and 22 μm. And, as a matter of fact, the voids theory does not predict size effects in torsion of bars in an isotropic material, while they occur both in torsion and in bending, as observed for bones and polymer foam materials in [9]. Even if some problem of physical concreteness or of mathematical hardness could arise [10, 11], a better improvement of the voids theory, within a microstructured scheme, is necessary in order to characterize the more complex structure.

A direct way to proceed is to consider the thermoelastic solid with nano-pores as a continuum with an ellipsoidal microstructure (see [4, 12]) which describes media whose each material element contains a large cavity, that does not diffuse through the skeleton, filled by an elastic inclusion, or an inviscid fluid, both of negligible mass (*e.g.*, composite materials reinforced with chopped elastic fibers, porous media with elastic granular inclusions, real ceramics, *etc.*): this cavity is able to have a microstretch different from, and independent of, the local affine deformation deriving from the macromotion and so can allow distinct microstrains along the principal axes of microdeformation, in absence of microrotations . The "tortuosity" matrix, a macroscopic geometrical symmetric tensor that expresses the effects of the geometry of the microscopic pores' surface, was previously presented in [13], but in [14] the model of a microstretched medium has been firstly used to study materials with distributions of aligned ellipsoidal vacuous pores and explicit computations have been carried out for matrix materials subjected to axisymmetric and plane strain loading conditions; then the model has been used to formulate more general multiphase theories of microstructured media [15]. Furthermore, some tension tests and numerical results have been presented in [16] for a similar model of a microstretched medium. The quoted model [4] is surely complementary to the use of the Cosserat theory [9], when microrotations are of interest in the analysis, but no the microdeformations: merely, we wish to observe that our theory contains naturally the voids theory by constraining the microstrain to be spherical. In particular, in [12] it has been observed that, during quasi-static homogeneous motion, the porous solid behaves like an isotropic simple material with fading memory in the linear range and it reduces to a viscoelastic medium when the microstructural variable remains spherical as in [2]. More generally, for complete microdeformations, the framework of media with affine structure better depicts macro- and micro-motion (see [3, 17]). Finally, our model [4] was used to analyze nonlinear wave propagation in constrained porous media [18] and to examine adsorption and diffusion of pollutants in soils, viewed as an immiscible mixture of materials with, and without, microstructure [19, 20].

In this chapter, we extend the linear theory [12] of elastic solids with nano-pores to the thermoelastic case and include a rate effect in the holes' response, which results in internal dissipation from experimental evidence [21]; after we make a complete study of the propagation of linear waves. In particular, in §2 we apply the general theory of continua with microstructure to the ellipsoidal case and furnish balance equations and jump conditions; in §3 we present constitutive equations for kinetic energy and co-energy density and for dependent constitutive fields and, after, we use thermodinamic restrictions; in §4 we define small thermoelastic deformations from the reference placement and obtain the linear field balance equations; in §5 we study linear micro-vibrations for which we obtain three admissible modes; in §6 we analyze the propagation of harmonic plane waves and comment the secular equations governing the eight solutions: two shear optical micro-elastic modes, two coupled transverse elastic waves and four coupled longitudinal thermo-elastic waves; in §7 we get the propagation conditions of the macro-acceleration waves for either a heat-conducting or non-conducting isotropic thermoelastic material with nano-pores (corresponding, respectively, to *homothermal* and *homentropic* waves), as well as for *generalized transverse* waves; in §8 we gain the growth equations which govern the propagation of the macro-acceleration waves and discuss the couplings between the higher order discontinuities.

2. Balance laws and jump conditions

We identify the continuous material body with nano-pores $\mathcal{B}$ with a fixed homogeneous and free of residual stresses region of the three dimensional Euclidean space $\mathcal{G}$, called the "natural" reference placement $\mathcal{B}_*$ (see, *e.g.*, §83 of [22]). We suppose that each material element of the continuum contains a nano-pore which is capable to have a microstretch different from, and independent of, the local affine deformation ensuing from the macromotion. Therefore, if we denote the generic material element of $\mathcal{B}_*$ by $\mathbf{x}_*$, the thermomechanical behaviour of $\mathcal{B}$ is described by three smooth mappings on $\mathcal{B}_* \times \Re$ ($\Re$ is the set of real numbers): the spatial position $\mathbf{x} \in \mathcal{G}$, at time τ, of the material point which occupied the position $\mathbf{x}_*$ in the reference placement $\mathcal{B}_*$, the left Cauchy–Green tensor of the micro-deformation $\mathbf{U} \in \mathrm{Sym}^+$, at time τ, of the associated nano-pore (Sym^+ being the collection of second-order symmetric and positive definite tensor fields) and the absolute positive temperature $\theta > 0$.

The spatial position $\mathbf{x}(\mathbf{x}_*, \tau)$ is a one-to-one correspondence, for each τ, between the reference placement $\mathcal{B}_*$ and the current placement $\mathcal{B}_\tau = \mathbf{x}(\mathcal{B}_*, \tau)$ of the body $\mathcal{B}$ and, so, the deformation gradient $\mathbf{F} := \nabla\mathbf{x}(\mathbf{x}_*, \tau) \;\; (= \partial\mathbf{x}/\partial\mathbf{x}_*)$ is a second order tensor with positive determinant. Through the inverse mapping $\mathbf{x}_*(\mathbf{x}, \tau)$ of $\mathbf{x}$, we can consider all the relevant fields in the theory as defined over the current placement $\mathcal{B}_\tau = \mathbf{x}(\mathcal{B}_*, \tau)$ as well as over the reference placement $\mathcal{B}_*$ of the body $\mathcal{B}$.

Hence, a body with nano-pores is like a medium with ellipsoidal microstructure [4] and a rotation $\mathbf{Q} = \mathrm{e}^{-\mathcal{E}\mathbf{s}}$ of the observer of characteristic vector $\mathbf{s}$, where $\mathcal{E}$ is Ricci's permutation tensor and e the basis of natural logarithms, causes the symmetric tensor $\mathbf{U}$ to change into $\mathbf{U}_s = \mathbf{Q}\mathbf{U}\mathbf{Q}^T$; moreover, the infinitesimal generator $\mathcal{A}$ of the group of rotations on the microstructure in Sym$^+$, *i.e.*, the operator describing the effect of a rotation of the observer on the value $\mathbf{U}_s$ of the microstructure to the first order in $\mathbf{s}$ (see §3 of [3]), is given by

$$\mathcal{A}(\mathbf{U}) := \left.\frac{d\mathbf{U}_s}{d\mathbf{s}}\right|_{\mathbf{s}=\mathbf{0}}, \tag{1}$$

that is, in components:

$$\mathcal{A}_{ijk} = U_{il}\mathcal{E}_{ljk} + \mathcal{E}_{ikl}U_{lj}. \tag{2}$$

$\mathcal{A}$ is a third-order tensor, symmetric and positive definite in the first two indices, that is $\mathcal{A}\mathbf{c} \in$ Sym$^+$ for all vectors $\mathbf{c}$.

The expression of the kinetic energy density per unit mass of microstructured bodies is the sum of two terms, the classical one $\frac{1}{2}\dot{\mathbf{x}}^2$ due to the translational inertia and the microstructured one $\kappa(\mathbf{U}, \dot{\mathbf{U}})$ due to the inertia related to the admissible expansional micromotions of the pores' boundaries (the superposed dot denotes material time derivative). This additional term is a non-negative scalar function, homogeneous in $\mathbf{U}$, such that $\kappa(\mathbf{U}, \mathbf{0}) = 0$ and $\frac{\partial^2\kappa}{\partial\dot{\mathbf{U}}^2} \neq \mathbf{0}$, and it is related to the kinetic co-energy density $\chi(\mathbf{U}, \dot{\mathbf{U}})$ by the Legendre transform

$$\frac{\partial\chi}{\partial\dot{\mathbf{U}}} \cdot \dot{\mathbf{U}} - \chi = \kappa \tag{3}$$

(see, also, [23]). The kinetic co-energy χ, as κ, must have the same value for all observers at rest, *i.e.*, it must be invariant under the Galilean group and hence satisfy the condition

$$\dot{\mathcal{A}}^* \frac{\partial\chi}{\partial\dot{\mathbf{U}}} = -\mathcal{A}^* \frac{\partial\chi}{\partial\mathbf{U}}, \tag{4}$$

where the third-order tensor $\mathcal{A}^*$ is defined through the relation $(\mathcal{A}^*\mathbf{C}) \cdot \mathbf{c} := \mathbf{C} \cdot (\mathcal{A}\mathbf{c})$, for all second-order tensors $\mathbf{C}$ and all vectors $\mathbf{c}$. The use of Eq. (2) into Eq. (4) and the multiplication of both sides by the Ricci's tensor $\mathcal{E}$ gives the following kinematic compatibility relation

$$\operatorname{skw}\left[\dot{\mathbf{U}}\frac{\partial\chi}{\partial\dot{\mathbf{U}}} + \mathbf{U}\frac{\partial\chi}{\partial\mathbf{U}}\right] = \mathbf{0}, \tag{5}$$

where 'skw' denotes the skew part of a second–order tensor: $\operatorname{skw}(\cdot) := \frac{1}{2}\left[(\cdot) - (\cdot)^T\right]$ $\left(\text{the symmetric one being } \operatorname{sym}(\cdot) := \frac{1}{2}\left[(\cdot) + (\cdot)^T\right]\right)$.

All the admissible thermo-kinetic processes for porous solids with large irregular voids are governed by the following general system of balance equations proposed in [4]; they are the mass conservation, the Cauchy equation, the micromomentum and moment of momentum balances, the Neumann energy equation and the entropy inequality in the Lagrangian description, respectively:

$$\rho_* = \rho \det \mathbf{F}, \tag{6}$$

$$\rho_* \ddot{\mathbf{x}} = \rho_* \mathbf{f} + \operatorname{Div} \mathbf{P}, \tag{7}$$

$$\rho_* \left[\frac{d}{d\tau} \left(\frac{\partial \chi}{\partial \dot{\mathbf{U}}} \right) - \frac{\partial \chi}{\partial \mathbf{U}} \right] = \rho_* \mathbf{B} - \mathbf{Y} + \operatorname{Div} \mathbf{\Lambda}, \tag{8}$$

$$\mathcal{E} \left(\mathbf{P} \mathbf{F}^T \right) = \mathcal{A}^* \mathbf{Y} + \left(\nabla \mathcal{A}^* \right) \mathbf{\Lambda}, \tag{9}$$

$$\rho_* \dot{\varepsilon} = \mathbf{P} \cdot \dot{\mathbf{F}} + \mathbf{Y} \cdot \dot{\mathbf{U}} + \mathbf{\Lambda} \cdot \nabla \dot{\mathbf{U}} + \rho_* \lambda - \operatorname{Div} \mathbf{h}, \tag{10}$$

$$\rho_* \theta \dot{\eta} \geq \rho_* \lambda - \operatorname{Div} \mathbf{h} + \theta^{-1} \mathbf{h} \cdot \nabla \theta, \tag{11}$$

where ρ is the mass density and ρ_* its referential value; Div means the trace of the nabla: $\operatorname{Div}(\cdot) := \operatorname{tr}(\nabla(\cdot))$; $\mathbf{f}$ is the vector body force, $\mathbf{P}$ the Piola-Kirchhoff stress tensor, ε the specific internal energy density per unit mass, λ the rate of heat generation due to irradiation or heating supply, $\mathbf{h}$ the referential heating flux, η the density of entropy.

Moreover, on the left hand side of the balance of micro-momentum (8) the Lagrangian derivative of the kinetic co-energy χ appears, while, on the right hand side, $\rho_* \mathbf{B}$ and $-\mathbf{Y}$ are the resultant second-order symmetric tensor densities of external and internal microactions, respectively: the first one is interpreted as a controlled pore pressure and the other one includes interactive forces between the gross and fine structures as well as internal dissipative contributions due to the stir of the pores' surface. Finally, $\mathbf{\Lambda}$ is the referential microstress third-order tensor, symmetric in the first two indices, which is related to the capability of recognizing boundary microtractions, even if, in some cases, it expresses weakly non-local internal effects due to the impossibility of defining a physically significant connection on the manifold of the microstructural kinetic parameter $\mathbf{U}$ (see [24]).

The balance of moment of momentum (9) assumes a more significant expression when we use the representation (2) for $\mathcal{A}$; in fact we have

$$\operatorname{skw} \left(\mathbf{P} \mathbf{F}^T \right) = 2 \operatorname{skw} \left[\mathbf{U} \mathbf{Y} + \nabla \mathbf{U} \odot \mathbf{\Lambda} \right], \tag{12}$$

where the tensor product $\odot$ between third-order tensors is so defined:

$$\left(\nabla \mathbf{U} \odot \Lambda \right)_{ij} := \mathrm{U}_{ih,L} \Lambda_{jhL}. \tag{13}$$

Remark. We observe that the voids theories [1, 5] are immediately recovered when the microstretch $\mathbf{U}$ is constrained to be spherical (see, also, §5 of [4]).

In addition to balances (6-8) and (10-12), we need the balance equations at a surface of discontinuity, namely a propagating wave Σ. As it is customary, we assume that the smooth movable surface Σ, that traverses the body $\mathcal{B}$, is oriented and we denote by $\mathbf{n}$ the unit normal vector to Σ in the reference placement $\mathcal{B}_*$ and by v_n the corresponding non-zero normal speed of displacement of Σ at point $(\mathbf{x}_*, \tau)$ in the reference placement. We further assume that some field related to the motion of $\mathcal{B}$ (excepting $\mathbf{x}$, $\mathbf{U}$ and θ) suffers jump discontinuity across Σ and so we employ the usual notation $[\![\cdot]\!]$ for jumps, so that

$$[\![f]\!] = f^{+} - f^{-}, \tag{14}$$

where f^{+} or f^{-} refers to the limit of f as the wave is approached from the right or left, respectively.

Therefore, we can write classical Kotchine's equations, as modified in order to take into account microstructural effects, and a relation that restricts the jump of micromomentum (see [25]-[27]) as it follows:

$$[\![\rho_*(v_n - \dot{\mathbf{x}} \cdot \mathbf{n})]\!] = 0, \quad [\![\rho_* v_n \dot{\mathbf{x}} + \mathbf{P}\,\mathbf{n}]\!] = 0, \quad \left[\!\!\left[\rho_* v_n \frac{\partial \kappa}{\partial \dot{\mathbf{U}}}(\mathbf{U}, \dot{\mathbf{U}}) + \boldsymbol{\Lambda}\,\mathbf{n}\right]\!\!\right] = 0, \tag{15}$$

$$\left[\!\!\left[\rho_* v_n \left(\epsilon + \frac{1}{2}\dot{\mathbf{x}}^2 + \kappa(\mathbf{U}, \dot{\mathbf{U}})\right)\right]\!\!\right] = \left[\!\!\left[\left(\mathbf{h} - \mathbf{P}^T \dot{\mathbf{x}} - \boldsymbol{\Lambda}^* \dot{\mathbf{U}}\right) \cdot \mathbf{n}\right]\!\!\right], \; [\![\rho_* v_n \theta \eta]\!] \geq [\![\mathbf{h} \cdot \mathbf{n}]\!]. \tag{16}$$

The form of the jumps across a propagating wave of higher order time derivatives of principal fields can be obtained from the balance equations (6-8) and (10).

3. Thermodynamic restrictions on the constitutive assumptions

The first proposal is of kinematic character and leads to an expression for the microstructural kinetic co-energy χ for $\mathcal{B}$. We are interested in the linear theory from the next section, so the simplest assumption is to assume the function χ to be quadratic in $\dot{\mathbf{U}}$

$$\chi = \frac{1}{2}(\dot{\mathbf{U}}\mathbf{J}_*) \cdot \dot{\mathbf{U}}, \tag{17}$$

where the second-order referential microinertia tensor field $\mathbf{J}_* \in \mathrm{Sym}^+$ is constant (see [3, 12, 28]). As a consequence of this definition, the left-hand side of Eq. (8) reduces to $(\rho_* \ddot{\mathbf{U}}\mathbf{J}_*)$ and we meet also the requirements of positivity and quadratic form for κ: in fact from (3) we have

$$\kappa = \chi = \frac{1}{2}(\dot{\mathbf{U}}\mathbf{J}_*) \cdot \dot{\mathbf{U}}. \tag{18}$$

In order to study thermodynamic restrictions for thermoelastic materials with nano-pores we must define the Helmholtz free energy density per unit mass $\psi := \epsilon - \theta\eta$ and insert it in the axiom of dissipation (11); after, by using Eq. (10), we obtain the reduced version of the entropy inequality:

$$\rho_* \left(\dot{\psi} + \dot{\theta}\eta\right) \leq \mathbf{P} \cdot \dot{\mathbf{F}} + \mathbf{Y} \cdot \dot{\mathbf{U}} + \boldsymbol{\Lambda} \cdot \nabla \dot{\mathbf{U}} - \theta^{-1}\mathbf{h} \cdot \nabla \theta. \tag{19}$$

Here we generalize constitutive prescriptions presented in [12, 26] by considering both conduction of heat and inelastic surface effects associated to changes in the deformation of nano-pores in the vicinity of the hole boundaries. Therefore, let us call the array $\mathcal{S} := \{\mathbf{F}, \mathbf{U}, \nabla\mathbf{U}, \theta\}$ of independent variables the elastic state of the material with nano–pores and, assuming the equipresence principle, let us postulate the following constitutive relations of thermoelastic kind:

$$\{\psi, \eta, \mathbf{P}, \mathbf{Y}, \boldsymbol{\Lambda}, \mathbf{h}\} = \{\tilde{\psi}, \tilde{\eta}, \tilde{\mathbf{P}}, \tilde{\mathbf{Y}}, \tilde{\boldsymbol{\Lambda}}, \tilde{\mathbf{h}}\}(\mathcal{S}, \dot{\mathbf{U}}, \nabla\theta). \tag{20}$$

Now we have to check the compatibility of these prescriptions with the Clausius-Duhem inequality in its reduced version (19) that must be valid for any choice of the fields in the set $(\mathcal{S}, \dot{\mathbf{U}}, \nabla\theta)$. Consequently, by using the chain rule of differentiation, when the terms are appropriately ordered, the inequality reads:

$$\left(\rho_* \frac{\partial \psi}{\partial \mathbf{F}} - \mathbf{P}\right) \cdot \dot{\mathbf{F}} + \left(\rho_* \frac{\partial \psi}{\partial \mathbf{U}} - \mathbf{Y}\right) \cdot \dot{\mathbf{U}} + \left(\rho_* \frac{\partial \psi}{\partial (\nabla \mathbf{U})} - \mathbf{\Lambda}\right) \cdot \nabla \dot{\mathbf{U}} +$$
$$+\rho_* \left(\eta + \frac{\partial \psi}{\partial \theta}\right) \dot{\theta} + \rho_* \frac{\partial \psi}{\partial \dot{\mathbf{U}}} \cdot \ddot{\mathbf{U}} + \rho_* \frac{\partial \psi}{\partial (\nabla \theta)} \cdot \nabla \dot{\theta} + \frac{1}{\theta} \mathbf{h} \cdot \nabla \theta \leq 0. \tag{21}$$

The left-hand member of 21 is linear with respect to $\dot{\mathbf{F}}, \nabla \dot{\mathbf{U}}, \dot{\theta}, \ddot{\mathbf{U}}$ and $\nabla \dot{\theta}$, quantities that take up arbitrary values; thus the respective coefficients in the linear expression must vanish, and hence:

$$\psi = \tilde{\psi}(\mathcal{S}), \quad \mathbf{P} = \rho_* \frac{\partial \psi}{\partial \mathbf{F}}, \quad \mathbf{\Lambda} = \rho_* \frac{\partial \psi}{\partial (\nabla \mathbf{U})}, \quad \eta = -\frac{\partial \psi}{\partial \theta}. \tag{22}$$

These relations mean that the Helmholtz free energy ψ, the Piola-Kirchhoff stress tensor $\mathbf{P}$, the microstress $\mathbf{\Lambda}$ and the entropy η depend upon the elastic state of the material only; moreover, $\mathbf{P}$, $\mathbf{\Lambda}$ and η are determined as soon as the constitutive equation for ψ is known.

The residual inequality defines the dissipation $\mathcal{D}$ of the thermo-kinetic process

$$\mathcal{D} := \mathbf{H} \cdot \dot{\mathbf{U}} + \frac{1}{\theta} \mathbf{h} \cdot \nabla \theta \leq 0, \tag{23}$$

where $\mathbf{H} := \rho_* \frac{\partial \tilde{\psi}}{\partial \mathbf{U}}(\mathcal{S}) - \tilde{\mathbf{Y}}(\mathcal{S}, \dot{\mathbf{U}}, \nabla \theta)$ is the dissipative part of internal microactions, a symmetric second-order tensor.

For a thermally isotropic porous material which is a *definite heat conductor* [29], Fourier's law gives

$$\mathbf{h} = -\xi(\mathcal{S}) \nabla \theta, \quad \text{with} \quad \xi \geq 0; \tag{24}$$

thus, by using in the energy Eq. (10) relations (22), (24) and $\epsilon = \psi + \theta\eta$, we have that, for a thermoelastic medium with nano–pores, it becomes

$$\rho_* \theta \frac{d\eta}{d\tau} + \mathbf{H} \cdot \dot{\mathbf{U}} = \rho_* \lambda + \mathrm{Div}\, (\xi \nabla \theta). \tag{25}$$

4. Linear field equations

Now we need to introduce the displacement field $\mathbf{u}$ of a material element of the body: it is

$$\mathbf{u}(\mathbf{x}_*, \tau) := \mathbf{x}(\mathbf{x}_*, \tau) - \mathbf{x}_*; \tag{26}$$

therefore, the deformation gradient $\mathbf{F}$ is expressed by:

$$\mathbf{F}(\mathbf{x}_*, \tau) = \mathbf{I} + \nabla \mathbf{u} \tag{27}$$

The natural reference placement $\mathcal{B}_*$ is homogeneous, free of residual macro- and micro-stresses, so $\mathcal{S}_* = \{\mathbf{I}, \mathbf{I}, \mathbf{O}, \theta*\}$ and $\mathbf{P}_*$, $\mathbf{Y}_*$, $\mathbf{\Lambda}_*$, ψ_* all vanish in $\mathcal{B}_*$ ($\mathbf{I} := (\delta_{ik})$ is the identity tensor); moreover, the reference microinertia tensor field $\mathbf{J}_*$ has spherical value:

$\mathbf{J}_* = \kappa_* \mathbf{I}$, where $\kappa_* \geq 0$ is the non-negative microinertia coefficient depending on the reference geometric features of the pores.

Besides, we may take the displacement field $\mathbf{u}$, the infinitesimal strain tensor $\mathbf{E}$, the microstrain tensor $\mathbf{V}$, with its reference gradient $\nabla\mathbf{V}$, and the temperature ϑ and mass ϱ variations from the reference placement as measures of "small" thermoelastic deformations from the reference placement $\mathcal{B}_*$; they are defined by Eq. (26) and as it follows:

$$\mathbf{E} := \mathrm{sym}\,(\nabla\mathbf{u}), \quad \mathbf{V} := \mathbf{U} - \mathbf{I}, \quad \nabla\,\mathbf{V} = \nabla\,\mathbf{U}, \quad \vartheta := \theta - \theta_* \quad \text{and} \quad \varrho := \rho - \rho_*. \tag{28}$$

Hence, in the linear theory we can change the choice of variables of the elastic state $\mathcal{S}$, the new ones being the following $\tilde{\mathcal{S}} := \{\mathbf{E}, \mathbf{V}, \nabla\mathbf{V}, \vartheta\}$; then, in the natural reference placement $\mathcal{B}_*$, it is $\tilde{\mathcal{S}}_* = \{\mathbf{0}, \mathbf{0}, \mathbf{O}, 0\}$.

As observed in the previous section, the free energy ψ determines much of the behaviour of the nano-porous material, thus we suppose that the reference placement $\mathcal{B}_*$ of the body is also a placement of minimum for the free energy and, therefore, we choose the most general homogeneous, quadratic and positive definite form for the free energy ψ valid for a centrosymmetric isotropic linear thermoelastic solids with nano-pores (see [12, 30, 31]):

$$\begin{aligned}\psi(\tilde{\mathcal{S}}) = {} & \frac{(v_l^2 - 2v_t^2)}{2}\,(\mathrm{tr}\,\mathbf{E})^2 + v_t^2\,\mathbf{E}\cdot\mathbf{E} + \frac{\kappa_*\lambda_3}{2}\,(\mathrm{tr}\,\mathbf{V})^2 + \kappa_*\lambda_4\,\mathbf{V}\cdot\mathbf{V} + \kappa_*\lambda_5\,(\mathrm{tr}\,\mathbf{E})\,(\mathrm{tr}\,\mathbf{V}) + \\ & + \kappa_*\lambda_6\,\mathbf{E}\cdot\mathbf{V} + \kappa_*(v_{tm}^2 - v_{sm}^2)\,\mathrm{Div}\,\mathbf{V}\cdot\mathrm{Div}\,\mathbf{V} + \frac{\kappa_*}{2}v_{sm}^2\,\nabla\mathbf{V}\cdot\nabla\mathbf{V} + \\ & + \kappa_*\lambda_1\nabla\,(\mathrm{tr}\,\mathbf{V})\cdot\mathrm{Div}\,\mathbf{V} + \frac{\kappa_*\lambda_2}{2}\,\nabla\,(\mathrm{tr}\,\mathbf{V})\cdot\nabla\,(\mathrm{tr}\,\mathbf{V}) + \frac{\gamma_1}{2}\,\vartheta^2 + \gamma_2\,\vartheta\,\mathrm{tr}\,\mathbf{E} + \kappa_*\gamma_3\,\vartheta\,\mathrm{tr}\,\mathbf{V},\end{aligned} \tag{29}$$

where $\mathrm{tr}\,(\cdot)$ denotes the trace of a tensor, *i.e.*, $\mathrm{tr}\,\mathbf{E} := \mathbf{E}\cdot\mathbf{I}$.

The positiveness of the expression in (29) assures us that the thirteen constant thermoelastic coefficients v_l, v_t, λ_i $(i = 1,\ldots,6)$, γ_j $(j = 1,2,3)$, v_{tm} and v_{sm} must resolve the following system of inequalities:

$$\begin{aligned} & 3v_l^2 > 4v_t^2 > 0, \quad (3v_l^2 - 4v_t^2)(3\,\lambda_3 + 2\,\lambda_4) > \kappa_*(3\,\lambda_5 + \lambda_6)^2, \quad 4\,\lambda_4 v_t^2 > \kappa_*\lambda_6^2, \\ & v_{tm}^2 > v_{sm}^2 > 0, \quad v_{sm}^2\,(6\lambda_1 + 9\lambda_2 + 2v_{tm}^2 + v_{sm}^2) > 4(v_{tm}^2 - v_{sm}^2)^2, \quad \gamma_1 > 0, \\ & \gamma_1\left[\rho_*\kappa_*^{-1}(3v_l^2 - 4v_t^2)(3\,\lambda_3 + 2\,\lambda_4) - (3\,\lambda_5 + \lambda_6)^2\right] + 6\gamma_2\gamma_3(3\,\lambda_5 + \lambda_6) > \\ & \qquad > 3\gamma_2^2\kappa_*^{-1}(3\,\lambda_3 + 2\,\lambda_4) + 3\gamma_3^2(3v_l^2 - 4v_t^2), \quad \gamma_1(3\lambda_3 + 2\lambda_4) > 3\kappa_*\gamma_3^2;\end{aligned} \tag{30}$$

alternatively to the last inequality of relation (30), the following one holds:

$$\gamma_1(3v_l^2 - 4v_t^2) > 3\gamma_2^2. \tag{31}$$

Finally, we need to express the dissipative part $\mathbf{H}$ of internal microactions $\mathbf{Y}$ and the referential heating flux $\mathbf{h}$ within the same linear approximation as the other constitutive terms in the balance equations, thus, since $\mathbf{H}$ must vanish whenever $\dot{\mathbf{V}} = \mathbf{0}$, we take

$$\mathbf{H} = -\rho_*\kappa_*\left[\omega(\mathrm{tr}\,\dot{\mathbf{V}})\mathbf{I} + 2\,\sigma\,\dot{\mathbf{V}}\right] \quad \text{and} \quad \mathbf{h} = -\,\xi_*\nabla\,\theta, \tag{32}$$

with ω and σ inelastic constants and $\xi_* := \xi(\tilde{\mathcal{S}}_*)$. By inserting relations (32) in the dissipation imbalance (23), we obtain in the linear approximation

$$\rho_*\kappa_* \left[\omega(\mathrm{tr}\,\dot{\mathbf{V}})^2 + 2\sigma\,\dot{\mathbf{V}}^2\right] + \frac{\xi_*}{\theta}\,(\nabla\,\theta)^2 \geq 0, \tag{33}$$

which is verified if and only if

$$3\omega + 2\sigma \geq 0, \quad \sigma \geq 0, \quad \text{and} \quad \xi_* \geq 0. \tag{34}$$

Therefore, we derive from constitutive equations (22) and (32) and the definition of $\mathbf{H}$ (see, also, [12]) the following linear constitutive expressions for the dependent fields:

$$\begin{aligned}
\mathbf{P} &= \rho_* \left\{ \left[(v_l^2 - 2v_t^2)\,\mathrm{tr}\,\mathbf{E} + \kappa_*\lambda_5\,\mathrm{tr}\,\mathbf{V} + \gamma_2\vartheta\right]\mathbf{I} + 2\,v_t^2\,\mathbf{E} + \kappa_*\lambda_6\mathbf{V}\right\},\\
\mathbf{Y} &= \rho_*\kappa_* \left\{\left[\mathrm{tr}\,(\lambda_3\mathbf{V} + \lambda_5\mathbf{E} + \omega\dot{\mathbf{V}}) + \gamma_3\,\vartheta\right]\mathbf{I} + 2\,\lambda_4\,\mathbf{V} + \lambda_6\,\mathbf{E} + 2\,\sigma\,\dot{\mathbf{V}}\right\},\\
\mathbf{\Lambda} &= \rho_*\kappa_*\{2\,(v_{tm}^2 - v_{sm}^2)\,\mathrm{syml}\,(\mathrm{Div}\,\mathbf{V}\otimes\mathbf{I}) + v_{sm}^2\nabla\,\mathbf{V} +\\
&\qquad + \lambda_1\left[\mathbf{I}\otimes\mathrm{Div}\,\mathbf{V} + \mathrm{syml}\,(\nabla\,(\mathrm{tr}\,\mathbf{V})\otimes\mathbf{I})\right] + \lambda_2\,\mathbf{I}\otimes\nabla\,(\mathrm{tr}\,\mathbf{V})\},\\
\eta &= -\gamma_1\vartheta - \gamma_2\,\mathrm{tr}\,\mathbf{E} - \kappa_*\gamma_3\,\mathrm{tr}\,\mathbf{V},
\end{aligned} \tag{35}$$

where the left-symmetric part "syml" of a third-order tensor $\mathbf{\Omega}$ means:

$$(\mathrm{syml}\,\mathbf{\Omega})_{ijl} := \frac{1}{2}\left(\Omega_{ijl} + \Omega_{jil}\right), \quad \forall\, i,j,k = 1,2,3. \tag{36}$$

In addition we need the expression of the determinant of the deformation gradient $\mathbf{F}$ in the linear theory:

$$\det\mathbf{F} = \det(\mathbf{I} + \nabla\mathbf{u}) \simeq 1 + \mathrm{Div}\,\mathbf{u} \qquad \left[\Rightarrow (\det\mathbf{F})^{-1} \simeq 1 - \mathrm{Div}\,\mathbf{u}\right]. \tag{37}$$

Remark. When the pores are absent, v_l and v_t reduce to be the usual propagation speeds of dilatational and distortional waves in the linear isothermal elasticity, respectively.

The balance equations for a thermoelastic solid with nano-pores, governing the mass density ρ, the displacement field $\mathbf{u}$, the microstrain tensor $\mathbf{V}$ and the temperature change ϑ, are obtained by substituting constitutive relations (17) and (35) and Eq. $(37)_2$ into the Eqs. (6)-(8) and after by using $(28)_1$ and the fact that $\mathbf{J}_* = \kappa_*\mathbf{I}$; besides, last equation is get by Eq. (25) when linearized relations $(28)_4$, (32) and $(35)_4$ are applied:

$$\varrho = -\rho_*\mathrm{Div}\,\mathbf{u}, \tag{38}$$

$$\ddot{\mathbf{u}} = v_t^2\Delta\mathbf{u} + \nabla\left[(v_l^2 - v_t^2)\mathrm{Div}\,\mathbf{u} + \kappa_*\lambda_5\,\mathrm{tr}\,\mathbf{V} + \gamma_2\vartheta\right] + \kappa_*\lambda_6\,\mathrm{Div}\,\mathbf{V} + \mathbf{f}, \tag{39}$$

$$\begin{aligned}
\ddot{\mathbf{V}} &= v_{sm}^2\,\Delta\mathbf{V} + 2(v_{tm}^2 - v_{sm}^2)\,\mathrm{sym}\,[\nabla\,(\mathrm{Div}\,\mathbf{V})] + \lambda_1\nabla^2(\mathrm{tr}\,\mathbf{V}) +\\
&+ \left[\lambda_1\mathrm{Div}\,(\mathrm{Div}\,\mathbf{V}) + \lambda_2\Delta(\mathrm{tr}\,\mathbf{V}) - \mathrm{tr}\,(\lambda_3\mathbf{V} + \omega\dot{\mathbf{V}}) - \lambda_5\mathrm{Div}\,\mathbf{u} - \gamma_3\vartheta\right]\mathbf{I} -\\
&\qquad - \lambda_6\,\mathrm{sym}\,(\nabla\,\mathbf{u}) - 2\lambda_4\mathbf{V} - 2\sigma\dot{\mathbf{V}} + \kappa_*^{-1}\mathbf{B} \qquad \text{and}
\end{aligned} \tag{40}$$

$$0 = \gamma\Delta\vartheta + \gamma_1\dot{\vartheta} + \gamma_2\mathrm{Div}\,\dot{\mathbf{u}} + \kappa_*\gamma_3\mathrm{tr}\,\dot{\mathbf{V}} + \theta_*^{-1}\lambda, \tag{41}$$

with $\gamma := \xi_*(\rho_*\theta_*)^{-1} \geq 0$.

Moreover, we observe that the balance of moment of momentum (12) in the linear approximation assumes the following more simple expression: $\mathrm{skw}\,(\mathbf{P}-2\mathbf{Y}) = \mathbf{0}$, which is satisfied by the symmetric tensors $(35)_{1,2}$ identically.

At the end we can write the linear equations of jumps by inserting constitutive relations (35) in Eqs.(15) and (16) and by ignoring terms of higher order: therefore they are the following ones:

$$[\![\rho_*(v_n - \dot{\mathbf{x}}\cdot\mathbf{n})]\!] = 0, \tag{42}$$

$$\begin{aligned}&\Big[\!\!\Big[v_n\dot{\mathbf{u}} + \left[(v_l^2 - 2v_t^2)\,\mathrm{Div}\,\mathbf{u} + \kappa_*\lambda_5\,\mathrm{tr}\,\mathbf{V} + \gamma_2\vartheta\right]\mathbf{n}\Big]\!\!\Big] + \\ &\qquad + \Big[\!\!\Big[v_t^2\,[\mathrm{Div}\,(\mathbf{u}\otimes\mathbf{n}) + \nabla\,(\mathbf{u}\cdot\mathbf{n})] + \kappa_*\lambda_6\mathbf{V}\mathbf{n}\Big]\!\!\Big] = \mathbf{0},\end{aligned} \tag{43}$$

$$\begin{aligned}&\Big[\!\!\Big[v_n\dot{\mathbf{V}} + 2\,(v_{tm}^2 - v_{sm}^2)\,\mathrm{sym}\,(\mathrm{Div}\,\mathbf{V}\otimes\mathbf{n}) + v_{sm}^2\,(\nabla\,\mathbf{V})\,\mathbf{n}\Big]\!\!\Big] + \\ &\qquad + [\![\lambda_1\,\{(\mathbf{n}\cdot\mathrm{Div}\,\mathbf{V})\mathbf{I} + \mathrm{sym}\,[\nabla\,(\mathrm{tr}\,\mathbf{V})\otimes\mathbf{n}]\} + \lambda_2\,[\mathbf{n}\cdot\nabla\,(\mathrm{tr}\,\mathbf{V})]\,\mathbf{I}]\!] = \mathbf{O},\end{aligned} \tag{44}$$

$$[\![v_n\,(\gamma_1\vartheta + \gamma_2\,\mathrm{Div}\,\mathbf{u} + \kappa_*\gamma_3\,\mathrm{tr}\,\mathbf{V}) - \gamma\,\nabla\,\vartheta\cdot\mathbf{n}]\!] = 0, \tag{45}$$

last imbalance being satisfied identically by Eq. (45).

Now we can uncouple the spherical and deviatoric components of the linear balance of micromomentum (39) to obtain, respectively:

$$\begin{aligned}\ddot{\nu} = &\left(\frac{1}{3}v_{sm}^2 + \frac{2}{3}v_{tm}^2 + 2\lambda_1 + 3\lambda_2\right)\Delta\nu + \left[2(v_{tm}^2 - v_{sm}^2) + 3\lambda_1\right]\mathrm{Div}\,(\mathrm{Div}\,\mathbf{V}^D) - \\ &-(3\lambda_3 + 2\lambda_4)\,\nu - (3\omega + 2\sigma)\dot{\nu} - (3\lambda_5 + \lambda_6)\mathrm{Div}\,\mathbf{u} - 3\gamma_3\vartheta + \iota,\end{aligned} \tag{46}$$

$$\begin{aligned}\ddot{\mathbf{V}}^D = &\, v_{sm}^2\,\Delta\mathbf{V}^D + 2(v_{tm}^2 - v_{sm}^2)\left\{\mathrm{sym}\left[\nabla\,(\mathrm{Div}\,\mathbf{V}^D)\right]\right\}^D - 2\lambda_4\mathbf{V}^D - 2\sigma\dot{\mathbf{V}}^D + \\ &+ \left[\frac{2}{3}(v_{tm}^2 - v_{sm}^2) + \lambda_1\right](\nabla^2\nu)^D - \lambda_6\,[\mathrm{sym}\,(\nabla\,\mathbf{u})]^D + \kappa_*^{-1}\mathbf{B}^D,\end{aligned} \tag{47}$$

where ν and ι are the traces of $\mathbf{V}$ and $\kappa_*^{-1}\mathbf{B}$, respectively, while the deviatoric part is defined by: $\mathbf{A}^D := \mathbf{A} - 3^{-1}(\mathrm{tr}\,\mathbf{A})\mathbf{I}$, for each symmetric second-order tensor $\mathbf{A}$.

With the same procedure, the respective jump (44) is splitted in the following spherical and deviatoric relations, respectively:

$$\begin{aligned}&\Big[\!\!\Big[v_n\dot{\nu} + \left[\tfrac{1}{3}(v_{sm}^2 + 2v_{tm}^2) + 2\,\lambda_1 + 3\,\lambda_2\right]\nabla\,\nu\cdot\mathbf{n} + [2\,(v_{tm}^2 - v_{sm}^2) + 3\,\lambda_1]\mathrm{Div}\,\mathbf{V}^D\cdot\mathbf{n}\Big]\!\!\Big] = 0, \\ &\Big[\!\!\Big[v_n\dot{\mathbf{V}}^D + v_{sm}^2\,(\nabla\,\mathbf{V}^D)\,\mathbf{n} + \\ &\quad + \left\{\mathrm{sym}\left\{\left[\lambda_1\nabla\,\nu + 2\,(v_{tm}^2 - v_{sm}^2)\left(\mathrm{Div}\,\mathbf{V}^D + \tfrac{1}{3}\nabla\,\nu\right)\right]\otimes\mathbf{n}\right\}\right\}^D\Big]\!\!\Big] = \mathbf{O}.\end{aligned} \tag{48}$$

5. Micro-vibrations in solids with nano-pores

Micro-vibrations, produced during various operations from railway and/or roads to foot traffic and propagated from one medium to another, are one of the main factor for fatigue in structures; moreover, they could also cause serious damages in producing micro and nano scale equipments, other than errors during experiments in high-precision laboratories equipped with lasers, sensors or microscopes.

To study their propagation, let us assume that external volume contributions are null, *i.e.* $\mathbf{f} = \mathbf{0}$, $\mathbf{B} = \mathbf{O}$ and $\lambda = 0$; moreover, there are no macro-displacements in the system, *i.e.* $\mathbf{u} = \mathbf{0}$ and $\varrho = 0$, then let us consider solutions of Eqs. (39), (46), (47) and (41) of the form of thermal micro-vibrations in absence of dissipation as the following ones:

$$\nu = \hat{\nu}\, e^{ib\tau}, \; \mathbf{V}^D = \hat{\mathbf{V}}\, e^{ib\tau}, \; \vartheta = \hat{\vartheta}\, e^{ib\tau}, \; \omega = \sigma = 0, \tag{49}$$

where $\hat{\nu}$, $\hat{\mathbf{V}}$ and $\hat{\vartheta}$ are constant amplitudes, b is the frequency and i is the imaginary unit. Then Eq. (39) is satisfied identically, while the other equations become

$$\left(b^2 - 3\lambda_3 - 2\lambda_4\right)\hat{\nu} = 3\gamma_3\hat{\vartheta}, \quad \left(b^2 - \lambda_4\right)\hat{\mathbf{V}}^D = \mathbf{O}, \quad \gamma_1\hat{\vartheta} + \kappa_*\gamma_3\hat{\nu} = 0. \tag{50}$$

Therefore we have the following admissible results:

o) *dilatational mode:*

$$\begin{aligned} b_d &= \sqrt{3\lambda_3 + 2\lambda_4 - 3\kappa_*\gamma_1^{-1}\gamma_3^2}, \\ \hat{\mathbf{V}}_{11} = \hat{\mathbf{V}}_{22} = \hat{\mathbf{V}}_{33} \; (\Rightarrow \hat{\nu} = 3\,\hat{\mathbf{V}}_{11}), \quad & \hat{\mathbf{V}}_{ij} = 0, \; \forall\, i \neq j, \quad \hat{\vartheta} = -\kappa_*\gamma_1^{-1}\gamma_3\hat{\nu}; \end{aligned} \tag{51}$$

we observe that the frequency b_d of this spatio-thermal oscillation is real for the restriction $(30)_8$ of the free energy density ψ to be a positive definite form;

o) *extensional modes with a constant volume:*

$$b_e = \sqrt{\lambda_4}, \quad \hat{\nu} = \hat{\vartheta} = 0 \; (\Rightarrow \hat{\mathbf{V}}_{33} = -\hat{\mathbf{V}}_{11} - \hat{\mathbf{V}}_{22}), \quad \hat{\mathbf{V}}_{ij} = 0, \; \forall\, i \neq j; \tag{52}$$

also in these modes the frequency b_e of the micro-oscillations is real for the restriction $(30)_3$, while no thermal vibrations are present;

o) *shear modes:*

$$b_s = \sqrt{\lambda_4}, \quad \hat{\mathbf{V}}_{ij} \neq 0, \; \forall\, i \neq j, \quad \hat{\mathbf{V}}_{ii} = 0, \; \forall\, i, \; \Rightarrow \hat{\nu} = \hat{\vartheta} = 0; \tag{53}$$

their frequency b_s coincides with the real frequency b_e of the extensional modes and neither here there are thermal vibrations.

Remark. When we neglect thermic phenomena, our oscillating solutions recover three of the mechanical micro-vibrations obtained for general microstructure in [17].

6. Dispersion relations for plane waves

Now we draw here some results on the propagation of plane wave motion in a linear thermoelastic solids with big pores. We seek solutions of the system of linear balance Eqs. (38), (39), (46), (47) and (41) in the form of traveling harmonic waves (see, also, [32]):

$$\mathbf{u} = \phi(\mathbf{x}_*, \tau)\,\mathbf{w}, \; \nu = \mu\,\phi(\mathbf{x}_*, \tau), \; \mathbf{V}^D = \phi(\mathbf{x}_*, \tau)\,\mathbf{S}, \; \vartheta = \bar{\vartheta}\,\phi(\mathbf{x}_*, \tau), \; \varrho = \bar{\varrho}\,\phi(\mathbf{x}_*, \tau) \tag{54}$$

where $\mathbf{w}$, μ, $\mathbf{S}$ $\bar{\vartheta}$ and $\bar{\varrho}$ are constants which represent the wave amplitude of, respectively, the macro-displacement vector, the micro-strain trace, the deviatoric part of the micro-strain tensor and the temperature and mass fluctuation; besides, the wave function ϕ can be generally represented by the real (or the imaginary) part of a complex function

$$\phi(\mathbf{x}_*, \tau) = \exp(ib\tau - \delta\, \mathbf{n} \cdot \mathbf{x}_*) \quad \text{with} \quad \delta := a + ib/c, \tag{55}$$

where δ is the wave number; $b > 0$ is the frequency; $a(b) > 0$ and $c(b)$ are the wave attenuation and the wave speed, respectively; $\mathbf{n}$ is the unit vector representing the direction of wave propagation, while the unit vector $\hat{\mathbf{w}}$ defines the direction of motion. The specific loss l is defined by $l := 4\pi\left|\frac{ac}{b}\right|$.

Again we suppose that all external sources are zero, *i.e.* $\mathbf{f} = \mathbf{0}$, $\mathbf{B} = \mathbf{O}$ and $\lambda = 0$, hence, by substituting Eqs. (54) and (55) in the linear system (38), (39), (46), (47) and (41), we obtain the following relations:

$$(b^2 + v_t^2\delta^2)\mathbf{w} + (v_l^2 - v_t^2)\delta^2(\mathbf{w}\cdot\mathbf{n})\mathbf{n} - \kappa_*\delta\left[\lambda_6\mathbf{S}\mathbf{n} + \left(\frac{\lambda_6}{3} + \lambda_5\right)\mu\,\mathbf{n}\right] - \gamma_2\delta\bar{\vartheta}\mathbf{n} = \mathbf{0}, \tag{56}$$

$$\left[b^2 + \left(\frac{1}{3}v_{sm}^2 + \frac{2}{3}v_{tm}^2 + 2\lambda_1 + 3\lambda_2\right)\delta^2 - 2\lambda_4 - 3\lambda_3 - ib(2\sigma + 3\omega)\right]\mu +$$
$$+\left[2(v_{tm}^2 - v_{sm}^2) + 3\lambda_1\right]\delta^2(\mathbf{S}\mathbf{n})\cdot\,\mathbf{n} + (\lambda_6 + 3\lambda_5)\,\delta(\mathbf{w}\cdot\mathbf{n}) - 3\gamma_3\bar{\vartheta} = 0, \tag{57}$$

$$\left(b^2 + v_{sm}^2\delta^2 - 2\lambda_4 - 2i\sigma b\right)\mathbf{S} + 2(v_{tm}^2 - v_{sm}^2)\delta^2\left[\text{sym}\,(\mathbf{S}\mathbf{n}\otimes\mathbf{n})\right]^D +$$
$$+\lambda_6\delta\left[\text{sym}\,(\mathbf{w}\otimes\mathbf{n})\right]^D + \left[\frac{2}{3}(v_{tm}^2 - v_{sm}^2) + \lambda_1\right]\delta^2\mu(\mathbf{n}\otimes\mathbf{n})^D = \mathbf{O}, \tag{58}$$

$$(i\gamma\delta^2 - b\gamma_1)\bar{\vartheta} + \gamma_2 b\delta\mathbf{w}\cdot\mathbf{n} - \kappa_*\gamma_3 b\nu = 0 \quad \text{and} \quad \bar{\varrho} = \rho_*\psi h_n. \tag{59}$$

This algebraic system of eleven equations may be combined into five independent systems through linear combinations of those equations: two uncoupled relations, two coupled systems of two equations each and one coupled system of five equations. The study of all five systems needs the introduction of two unit vectors, $\mathbf{e}$ and $\mathbf{f}$, in the plane orthogonal to the direction of propagation $\mathbf{n}$ and such that $\mathbf{e}\cdot\mathbf{f} = 0$. Therefore, we have the following particular occurrences.

6.1. Shear optical waves

From the deviatoric Eq. (58), we obtain two independent dispersion relations relating frequencies b and wave numbers δ; they are two different shear optical micro-waves :

$$\left(b^2 + v_{sm}^2\delta^2 - 2\lambda_4 - 2i\sigma b\right)S_{ef} = 0 \quad \text{and} \tag{60}$$

$$\left(b^2 + v_{sm}^2\delta^2 - 2\lambda_4 - 2i\sigma b\right)(S_{ee} - S_{ff}) = 0, \tag{61}$$

where the subscripts indicates tensor components.

These shear optical micro-modes propagate with attenuation $a_s = \frac{\sigma c}{v_{sm}^2}$ and velocity given by $c_s^2 = \frac{v_{sm}^2}{2\sigma^2}\left[2\lambda_4 - b^2 + \sqrt{(2\lambda_4 - b^2)^2 + 4\sigma^2 b^2}\right]$ without modifying the thermo-elastic features of the matrix material of the porous medium; then the specific loss is $l_s = 4\pi \left| \frac{2\lambda_4 - b^2}{2b\sigma} + \sqrt{1 + \left(\frac{2\lambda_4 - b^2}{2b\sigma} \right)^2} \right|$.

For high frequencies all quantities grow with b, while for low frequencies, the speed and the attenuation approach $v_{sm}\sqrt{2\lambda_4}/\sigma$ and $\sqrt{2\lambda_4}/v_{sm}$, respectively, while l_s is big. Moreover, it is also possible a static solution with attenuation $a_s = \frac{\sqrt{2\lambda_4}}{v_{sm}}$.

6.2. Transverse waves

We get also two different systems of transverse waves, for $j = e, f$, from Eqs. (56) and (58):

$$(b^2 + v_t^2\delta^2)\, w_j - \kappa_* \lambda_6 \delta\, S_{nj} = 0, \tag{62}$$

$$\lambda_6 \delta\, w_j + 2\left(b^2 + v_{tm}^2\delta^2 - 2\lambda_4 - 2i\sigma b\right) S_{nj} = 0. \tag{63}$$

This homogeneous system has a nontrivial solution for the amplitudes w_j and S_{nj} if and only if the following dispersion relation is satisfied by δ:

$$2\left(v_t^2\delta^2 + b^2\right)\left(v_{tm}^2\delta^2 + b^2 - 2\lambda_4 - 2i\sigma b\right) + \kappa_* \lambda_6^2 \delta^2 = 0. \tag{64}$$

Eq. (64) is similar to the dispersion relation for plane thermoelastic waves studied in [33] and our analysis will use results there obtained. The first transverse solution of (64) is associated predominantly with the elastic properties of the material (v_t) and denoted by δ_t; the second one, δ_{tm}, with the properties governing elastic and dissipative changes in porosity (v_{tm}, λ_4, λ_6 and σ).

The analytical solutions of the dispersion relation (64) are quite cryptic and they are summarized in Table 1 of [34] (*modulo* some innocuous identification in notations); in this work we only report their physical interpretation without big difficulties. The coupling of motion Eqs. (56) and (58) of linear macro- and micro-momentum does the wave of dispersive kind, while the presence of big voids adds a dissipative mechanism associated with nano-pores which yields both waves to attenuate. If the dissipation coefficient σ is null, then we recover the presence of the resonance.

When frequencies are low, the elastic wave propagates with speed $v_t\sqrt{1-\zeta}$, where $0 \leq \zeta \left(:= \frac{\kappa_* \lambda_6^2}{4\lambda_4 v_t^2} \right) < 1$ for the inequality $(30)_3$, while the attenuation coefficient and the specific loss remain very small and approach zero with the frequency itself. The predominantly micro-transverse wave propagates with constant speed $\frac{v_{tm}}{\sigma}\sqrt{2\lambda_4(1-\zeta)}$ and with constant attenuation $\frac{\sqrt{2\lambda_4(1-\zeta)}}{v_{tm}}$; nevertheless, its specific loss $l_{tm} = \frac{2\lambda_4(1-\zeta)}{\sigma b}$ is large and in inverse proportion to the frequency b.

At high frequencies, the predominantly elastic transverse wave propagates with the classical speed v_t and, as the frequency approaches infinity, the attenuation coefficient a_{te} and the specific loss l_{te} are very small and approach zero, as for low frequencies. Instead the predominantly micro-transverse wave propagates with attenuation σ/v_{tm} and with constant speed v_{tm}, but with a small specific loss which approach zero when the frequency approaches infinity.

The amplitude ratio R of the micro-wave to the macro-wave is obtained from (62):

$$R = \frac{S_{nj}}{w_j} = \frac{(b^2 + v_t^2\delta^2)}{\kappa_* \lambda_6 \delta}. \tag{65}$$

For the solution predominantly elastic δ_t at large frequencies, the ratio R_t is a constant; at small frequencies it approaches zero with the frequency itself. Instead, the micro-mode δ_{tm} at large frequencies gives a ratio R_{tm} very big, while at low frequencies it is constant.

At the end, we can also have a static solution with attenuation $a_{tm} = \frac{\sqrt{2\lambda_4(1-\zeta)}}{v_{tm}}$ in which the amplitudes are related by $w_j = \frac{\kappa_*\lambda_6}{v_t^2 a} S_{nj}$, for $j = e, f$.

6.3. Longitudinal waves

The remaining equations of the system (56)-(59) furnish the solutions for the longitudinal waves, the only ones which present thermal effects:

$$(b^2 + v_l^2\delta^2) w_n - \kappa_*\delta \left[\lambda_6 S_{nn} + \left(\frac{\lambda_6}{3} + \lambda_5\right)\mu\right] - \gamma_2 \delta \bar{\vartheta} = 0, \tag{66}$$

$$\begin{aligned} &\left[b^2 + \left(\frac{1}{3}v_{sm}^2 + \frac{2}{3}v_{tm}^2 + 2\lambda_1 + 3\lambda_2\right)\delta^2 - 2\lambda_4 - 3\lambda_3 - ib(2\sigma + 3\omega)\right]\mu + \\ &\quad + (\lambda_6 + 3\lambda_5)\,\delta w_n + \left[\frac{2}{3}(v_{tm}^2 - v_{sm}^2) + \lambda_1\right]\delta^2 S_{nn} - 3\gamma_3\bar{\vartheta} = 0, \end{aligned} \tag{67}$$

$$\begin{aligned} &\frac{2}{3}\lambda_6\delta\, w_n + \left[b^2 + \frac{1}{3}\left(4v_{tm}^2 - v_{sm}^2\right)\delta^2 - 2\lambda_4 - 2i\sigma b\right] S_{nn} + \\ &\qquad + \frac{2}{3}\left[\frac{2}{3}(v_{tm}^2 - v_{sm}^2) + \lambda_1\right]\delta^2\mu = 0, \end{aligned} \tag{68}$$

$$\gamma_2 b\delta\, w_n - \kappa_*\gamma_3 b\mu + (i\gamma\delta^2 - b\gamma_1)\bar{\vartheta} = 0 \quad \text{and} \quad \bar{\varrho} = \rho_*\psi w_n. \tag{69}$$

Last equation rules the propagation of mass wave and it relates the mass amplitude $\bar{\varrho}$ directly with that of normal displacement w_n, so when this is calculated, the first one is get by Eq. $(69)_2$.

For the residual amplitudes w_n, μ, S_n and $\bar{\vartheta}$, we must pose the determinant of their coefficients equal to zero in order to have a nontrivial solution of the system (66)-$(69)_1$; therefore, we get a 4th-order equation in δ^2, which can be resolved with the Ferrari-Cardano derivation of the quartic formula, after the application of the Tchirnhaus transformation by mean of numerical techniques (see [32] and [35]). By the way, an exact analytical solution of the dispersion relation for longitudinal waves is very complicated and without interest to be reported here explicitly: instead we are concerned to summarize the behaviour of all wave numbers $\delta_e^2, \delta_d^2, \delta_v^2$ and δ_{th}^2, which are dominated by displacement, deviatoric and spherical parts of microstrain and thermal fields, respectively. Hence, there exist four coupled longitudinal waves: the first one δ_e^2 is predominantly an elastic wave of dilatation, the second one δ_d^2 is associated with an equivoluminal microelastic wave, the third one δ_v^2 is predominantly a volume fraction wave of pure dilatation, the last one corresponding to δ_{th}^2 is similar in character to a thermal wave; from numerical outcomes, the two micro-waves result to be slower than elastic and thermal waves, the elastic one being the fastest: this is in accordance with experimental evidence.

By disregarding micro-rotation and thermal effects, we are able to observe that the micro-wave solution above can be acknowledged in some developments of §8 of [17] for elastic plates and, peculiarly, the velocity of the elastic wave is less than that which would be calculated for classical elasticity v_l due to the phenomenon of the compliance of pores. In addition, if we neglect non-spherical contributions to the microstrain in constitutive equations (35), we find solutions of voids theory [36].

The longitudinal waves are all dispersive in character, because of the coupling of Eqs. (66)-(69)$_1$, and suffer attenuation (the thermal mode with a large coefficient) due to the thermal coupling and to the presence of voids. Furthermore, if the two dissipation coefficient σ and ω are zero, we can observe the phenomenon of the resonance.

For low frequencies, there is no damping effect in either of the four modes. The only significant wave is the δ_e-one, because the other three modes almost do not exist and their attenuation coefficients remain very small and approach zero with the frequency itself; instead the velocity v_l of the δ_e-wave is increased by a small amount due to the thermomechanical coupling, but decreased significantly because of nano-porosity effects; the attenuation is a quite small constant.

If frequencies are high, the δ_d micro-wave, of speed $v_d^2 = 2v_{tm}^2 - v_{sm}^2 + \lambda_1 > 0$ by inspection, and the δ_v one, which travels with velocity $v_v^2 = \frac{5}{3}v_{sm}^2 - \frac{2}{3}v_{tm}^2 + 3\lambda_2 > 0$ for (30)$_{4,5}$, are not accompanied by elastic or thermal modes, which are instead coupled, and vice versa; attenuation coefficients for micro-modes remains small but constant. Elastic mode propagates with the classical speed v_l and attenuation coefficient which approaches zero slowly with the frequency itself; instead the propagation velocity and the attenuation coefficient of the thermal wave δ_{th} sharply increase with the frequency itself, being diffusive in nature. We notice that the high frequency limits of the two micro-elastic waves correspond to the velocities of acceleration waves in the same material, undeformed and at rest, obtained in [26].

Finally, it can be observed numerically that the specific loss l is significantly large when the wave velocity has quite small value in some regions of frequency. The loss due to energy dissipation is comparatively high in case of δ_e and δ_{th}-modes and moderate for predominantly dilatation micro-elastic modes.

7. Macro-acceleration waves

We are now in a position to study acceleration wave propagation. We shall consider the surfaces Σ that are *weak singularities*, defined as those carrying only jumps of the derivatives of order 2 of the macro- and micro-displacement vectors and of order 1 of the thermal variables; these singular surfaces of order 2 are called *macro-acceleration waves* and all external forces and supplies, *i.e.* $\mathbf{f}$, $\mathbf{B}$ and λ, are supposed continuous across them with all the derivatives (see, also, [1, 37]).

Therefore, these peculiar discontinuity surfaces suffer 2nd-order derivative jumps of the displacement $\mathbf{u}$ and 1st-order derivative jumps of the temperature variation ϑ and of the microstrain tensor **V**, which is, in our context, directly related to the left Cauchy-Green tensor **U** of the micro-deformation.

Remark: It is noteworthy that our definition of macro-acceleration wave differs from usual definitions of acceleration waves in, *e.g.*, [27, 38, 39] because they examine singularities of order 2 for both macro- and micro-structural kinematic variables **u** and **V**. Moreover, our study of standard acceleration waves in [26] shows that, in the linear theory, these jumps are in general uncoupled unless the instantaneous acoustic macro- and micro-tensor have some eigenvalue coincident (see, also, the comments about the non-linear theory in §8 of [39]).

The normal velocity of displacement v_n of the macro-acceleration wave is continuous everywhere in the body and, hence, the following Hugoniot-Hadamard compatibility condition for the jump across Σ of the derivatives of an arbitrary field $\mathbf{\Upsilon}$ in $\mathcal{B}_*$ holds (see, *e.g.*, $(2.7)_2$ of [40]):

$$[\![\nabla \mathbf{\Upsilon}]\!] = \nabla [\![\mathbf{\Upsilon}]\!] - v_n^{-1} [\![\dot{\mathbf{\Upsilon}}]\!] \otimes \mathbf{n}. \tag{70}$$

In particular, if the field $\mathbf{\Upsilon}$ is continuous in $\mathcal{B}_*$, Eq. (70) reduces to

$$v_n [\![\nabla \mathbf{\Upsilon}]\!] = -[\![\dot{\mathbf{\Upsilon}}]\!] \otimes \mathbf{n}. \tag{71}$$

7.1. Homothermal case

In the linear approximation, jump Eqs. (42) and (43) along a macro-acceleration wave Σ are identically satisfied; instead the jump balance of energy (45) reduces to the Fourier condition

$$-\gamma [\![\nabla \vartheta]\!] \cdot \mathbf{n} = \gamma v_n^{-1} [\![\dot{\vartheta}]\!] = 0, \tag{72}$$

where Eq. (71) were used. Thus, *in linear porous thermoelasticity*, the first derivatives of ϑ are continuous and *every macro-acceleration wave is homothermal.*

The last jump condition (44) gives the following relation

$$\begin{aligned} & v_n [\![\dot{\mathbf{V}}]\!] + v_{sm}^2 [\![\nabla \mathbf{V}]\!] \mathbf{n} + 2(v_{tm}^2 - v_{sm}^2) \,\mathrm{sym}\, ([\![\mathrm{Div}\, \mathbf{V}]\!] \otimes \mathbf{n}) + \\ & + \lambda_1 \left[(\mathbf{n} \cdot [\![\mathrm{div}\, \mathbf{V}]\!]) \mathbf{I} + \mathrm{sym}\, ([\![\nabla \nu]\!] \otimes \mathbf{n}) \right] + \lambda_2 \, (\mathbf{n} \cdot [\![\nabla \nu]\!]) \mathbf{I} = \mathbf{O}, \end{aligned} \tag{73}$$

while the jumps of the balance laws (39) and (41) and of the derivative of Eq. (38) furnish these other ones, when Eq. (72) is also used:

$$[\![\ddot{\mathbf{u}}]\!] = v_t^2 [\![\Delta \mathbf{u}]\!] + (v_l^2 - v_t^2) [\![\nabla \, (\mathrm{Div}\, \mathbf{u})]\!] + \kappa_* \, (\lambda_5 [\![\nabla \nu]\!] + \lambda_6 [\![\mathrm{Div}\, \mathbf{V}]\!]) \,, \tag{74}$$

$$\gamma \, [\![\Delta \vartheta]\!] + \gamma_2 [\![\mathrm{Div}\, \dot{\mathbf{u}}]\!] + \kappa_* \gamma_3 [\![\dot{\nu}]\!] = 0, \quad [\![\dot{\rho}]\!] + \rho_* [\![\mathrm{Div}\, \dot{\mathbf{u}}]\!] = 0. \tag{75}$$

Now we can use the Hugoniot-Hadamard condition (71) to get a system of algebraic equations where the amplitudes of the discontinuities $[\![\dot{\rho}]\!]$, $[\![\ddot{\mathbf{u}}]\!]$, $[\![\dot{\mathbf{V}}]\!]$ and $[\![\ddot{\vartheta}]\!]$ are the unknown quantities. With this end in view, we employ the following definitions of instantaneous *homothermal acoustic macro-tensor* $\mathcal{U}(\mathbf{n} \otimes \mathbf{n})$ and *micro-tensor* $\mathcal{C}(\mathbf{n} \otimes \mathbf{n})$ (see, for analogy, [26, 39]):

$$\mathcal{U}(\mathbf{n}\otimes\mathbf{n}) := \left[v_t^2\,\mathbf{I} + (v_l^2 - v_t^2)\mathbf{n}\otimes\mathbf{n}\right], \tag{76}$$

$$\mathcal{C}(\mathbf{n}\otimes\mathbf{n}) := [v_{sm}^2\,\mathcal{I} + (v_{tm}^2 - v_{sm}^2)(\mathbf{\Phi}\otimes\mathbf{n} + \mathbf{n}\otimes\mathbf{I}\otimes\mathbf{n}) + \\ +\lambda_2\,\mathbf{I}\otimes\mathbf{I} + \lambda_1\,(\mathbf{I}\otimes\mathbf{n}\otimes\mathbf{n} + \mathbf{n}\otimes\mathbf{n}\otimes\mathbf{I})], \tag{77}$$

where we introduced the fourth-order tensor $\mathcal{I}$ and third-order tensor $\mathbf{\Phi}$ of components, respectively: $\mathcal{I}_{ijkl} := \delta_{ik}\delta_{jl}$ and $\Phi_{ijk} := \delta_{ik}n_j$.

Therefore, from Eqs. (73)-(75), we are able to write the system of equations for the unknown amplitudes in the following manner (see, also, [41]):

$$v_n[\![\dot{\rho}]\!] = \rho_*\,[\![\ddot{\mathbf{u}}]\!]\cdot\mathbf{n}, \tag{78}$$

$$\left[v_n^2\mathbf{I} - \mathcal{U}(\mathbf{n}\otimes\mathbf{n})\right][\![\ddot{\mathbf{u}}]\!] = -\kappa_* v_n(\lambda_5\,\mathbf{n}\otimes\mathbf{I} + \lambda_6\,\mathbf{I}\otimes\mathbf{n})[\![\dot{\mathbf{V}}]\!], \tag{79}$$

$$\left[v_n^2\mathcal{I} - \mathcal{C}(\mathbf{n}\otimes\mathbf{n})\right][\![\dot{\mathbf{V}}]\!] = \mathbf{0}, \tag{80}$$

$$\gamma\,[\![\ddot{\vartheta}]\!] = \kappa_*\gamma_3 v_n^2\,\mathbf{I}\cdot[\![\dot{\mathbf{V}}]\!] - \gamma_2 v_n[\![\ddot{\mathbf{u}}]\!]\cdot\mathbf{n}. \tag{81}$$

In conclusion, the jump of macro-acceleration $[\![\ddot{\mathbf{u}}]\!]$, ruled by Eq. (79), is in general coupled to the discontinuities in the microstructural variable $[\![\dot{\mathbf{V}}]\!]$, unlike the purely acceleration waves studied in [26], as observed in the previous remark.

Hence to classify possible macro-acceleration waves Σ we must analyze in advance Eq. (80) that gives three possible speed of displacement v_n for the surface Σ related to:

- o) *Two shear optical micro-waves,* whose speed propagation is $v_n = v_{sm}$; the amplitude of the discontinuity is $[\![\dot{\mathbf{V}}]\!]_{sm} = \alpha(\mathbf{e}\otimes\mathbf{e} - \mathbf{f}\otimes\mathbf{f}) + \beta(\mathbf{e}\otimes\mathbf{f} + \mathbf{f}\otimes\mathbf{e})$, where α and β are the scalar components of the wave amplitude and, as before in the plane waves study, $\mathbf{e}$ and $\mathbf{f}$ are the unit vectors in the plane orthogonal to $\mathbf{n}$ such that $\mathbf{e}\cdot\mathbf{f} = 0$.

  By inserting this solution in the other three Eqs. (78), (79) and (81), we obtain, in general, that $[\![\ddot{\mathbf{u}}]\!]_{sm} = \mathbf{0}$, $[\![\dot{\rho}]\!]_{sm} = 0$ and $[\![\ddot{\vartheta}]\!]_{sm} = 0$, that is, in essence, the waves carry predominantly a change in the nano-pore structure without altering the thermoelastic features of the matrix material.

  Observation. Instead, in the particular case in which one eigenvalue of $\mathcal{U}$ coincides with that of $\mathcal{C}$, *i.e.*, $v_t = v_{sm}$, (or $v_l = v_{sm}$), there could be also an associated transverse (or longitudinal, respectively) macro-wave of free amplitude which not alters (or which alters, respectively) mass and temperature fields; in this last case we have $[\![\dot{\rho}]\!]_{sm} = \rho_* v_{sm}^{-1}[\![\ddot{u}_n]\!]_{sm}$ and $[\![\ddot{\vartheta}]\!]_{sm} = -\gamma^{-1}\gamma_2 v_{sm}[\![\ddot{u}_n]\!]_{sm}$.

- o) *Two transverse micro-waves,* with propagation velocity $v_n = v_{tm}$. Their amplitude is of the form $[\![\dot{\mathbf{V}}]\!]_{tm} = \chi_e(\mathbf{n}\otimes\mathbf{e} + \mathbf{e}\otimes\mathbf{n}) + \chi_f(\mathbf{n}\otimes\mathbf{f} + \mathbf{f}\otimes\mathbf{n})$, with χ_e and χ_f the components of the amplitude itself. Now, there is a coupled *transverse macro-wave*, obtained by the study of equation (79), of amplitude:

$$[\![\ddot{\mathbf{u}}]\!]_{tm} = \frac{\kappa_* v_{tm}\lambda_6}{v_t^2 - v_{tm}^2}(\chi_e\,\mathbf{e} + \chi_f\,\mathbf{f}), \tag{82}$$

but no discontinuities along Σ in the mass and temperature fields, for equations (78) and (81), as in the classical case, in fact $[\![\dot{\rho}]\!]_{tm} = 0$ and $[\![\ddot{\vartheta}]\!]_{tm} = 0$.

o) *One extensional micro-wave,* whose vector amplitude is

$$[\![\dot{\mathbf{V}}]\!]_{em} = \delta(\zeta\, \mathbf{n}\otimes\mathbf{n} + \mathbf{e}\otimes\mathbf{e} + \mathbf{f}\otimes\mathbf{f});$$

here δ is its scalar amplitude, ζ the constant so defined: $\zeta := \frac{2(v_{tm}^2 - v_{sm}^2 + \lambda_1) - \lambda_2 + \sqrt{D}}{2(\lambda_1+\lambda_2)}$, D the discriminant $D := 12\lambda_1(\lambda_1+\lambda_2) + 9\lambda_8^2 + 4(v_{tm}^2 - v_{sm}^2)(v_{tm}^2 - v_{sm}^2 + 2\lambda_1 - \lambda_2) > 0$ by inspection for inequalities $(30)_{4,5}$. This wave propagates at a constant speed $v_n = v_{em}$, with $v_{em}^2 := v_{tm}^2 + \lambda_1 + \frac{3}{2}\lambda_2 + \frac{1}{2}\sqrt{D}$: $v_{tm}^2 + \lambda_1 + \frac{3}{2}\lambda_2 > 0$ for the same inequalities and so $v_{em} > v_{tm}$; obviously this result holds only if $(\lambda_1+\lambda_2) \neq 0$.
The coupled *macro-wave* is now *longitudinal*, that is,

$$[\![\ddot{\mathbf{u}}]\!]_{em} = \delta\,\varpi\, v_{em}\,\mathbf{n}, \quad \text{with the constant } \varpi := \frac{\kappa_*}{v_l^2 - v_{em}^2}[2\lambda_5 + (\lambda_5+\lambda_6)\zeta], \tag{83}$$

and the discontinuity amplitudes in the mass and temperature variations are, respectively:

$$[\![\dot{\rho}]\!]_{em} = \rho_*\delta\,\varpi \quad \text{and} \quad [\![\ddot{\vartheta}]\!]_{em} = \delta\,\check{\vartheta}, \tag{84}$$

with the constant $\check{\vartheta} := \gamma^{-1} v_{em}^2[\kappa_*\gamma_3(2+\zeta) - \gamma_2\varpi]$.

Remark. The *longitudinal micro-waves* of compaction or distention, usually predicted in the voids theory [1, 38], is here excluded, in general, unless we impose the additional condition of the subsequent point ii) on the constitutive thermoelastic constants.

If $(\lambda_1+\lambda_2) = 0$, we have other two significant subcases:

i) *one purely transverse micro-wave,* of vector amplitude $[\![\dot{\mathbf{V}}]\!]_{pm} = \varrho(\mathbf{I} - \mathbf{n}\otimes\mathbf{n})$ and speed $v_{pm}^2 := v_{sm}^2 - 2\lambda_1$, which corresponds to another *longitudinal macro-, mass, temperature wave* of amplitudes

$$[\![\ddot{\mathbf{u}}]\!]_{pm} = \varrho\,\varsigma\, v_{pm}\,\mathbf{n}, \quad \text{with} \quad \varsigma = \frac{2\kappa_*\lambda_5}{v_l^2 - v_{pm}^2}, \tag{85}$$

$$[\![\dot{\rho}]\!]_{pm} = \rho_*\varrho\,\varsigma, \quad [\![\ddot{\vartheta}]\!]_{pm} = \tilde{\vartheta}\varrho, \quad \text{with} \quad \tilde{\vartheta} := \frac{v_{pm}^2}{\gamma}(2\kappa_*\gamma_3 - \gamma_2\varsigma); \tag{86}$$

ii) *a purely longitudinal micro-wave,* which recovers the quoted prediction of voids theories and which propagates at constant speed $v_{lm}^2 := 2v_{tm}^2 - v_{sm}^2 + \lambda_1$; the amplitude of the discontinuity is $[\![\dot{\mathbf{V}}]\!]_{lm} = \hat{\delta}\,\mathbf{n}\otimes\mathbf{n}$: thus it is a wave of compaction, if the scalar amplitude $\hat{\delta} < 0$, and of distention, if $\hat{\delta} > 0$. In this last case the coupled wave is again *longitudinal* and the connected amplitudes are

$$[\![\ddot{\mathbf{u}}]\!]_{lm} = \hat{\delta}\,\hat{\varpi}\, v_{lm}\,\mathbf{n}, \quad \text{with} \quad \hat{\varpi} = \frac{\kappa_*(\lambda_5+\lambda_6)}{v_l^2 - v_{lm}^2}, \tag{87}$$

$$[\![\dot{\rho}]\!]_{lm} = \rho_*\hat{\delta}\,\hat{\varpi}, \quad [\![\ddot{\vartheta}]\!]_{lm} = \hat{\vartheta}\hat{\delta}, \quad \text{with} \quad \hat{\vartheta} := \frac{v_{lm}^2}{\gamma}(\kappa_*\gamma_3 - \gamma_2\hat{\varpi}). \tag{88}$$

7.2. Homentropic modes

In the particular case when the solid with nano-pores does not conduct heat, *i.e.*, $\mathbf{h} \equiv 0$ whatever $\nabla \vartheta$ we choose, then the energy balance (25) may be written, in the linear theory, in the following form $\theta_* \dot{\eta} = \lambda$ and the jump across a macro-acceleration wave Σ shows that

$$[\![\dot{\eta}]\!] = 0. \tag{89}$$

Thus, we have established the following result: *In a non-conducting thermoelastic body with nano-pores, every macro-acceleration wave is homentropic.*

The jump Eqs. (78) and (80) remain unchanged, while, since $[\![\dot{\vartheta}]\!] \neq 0$, Eqs. (79) and (81) are replaced by the following ones:

$$\left[v_n^2 \mathbf{I} - \mathcal{U}(\mathbf{n} \otimes \mathbf{n}) \right] [\![\ddot{\mathbf{u}}]\!] = -\kappa_* v_n (\lambda_5 \, \mathbf{n} \otimes \mathbf{I} + \lambda_6 \, \mathbf{I} \otimes \mathbf{n}) [\![\dot{\mathbf{V}}]\!] - \gamma_2 v_n [\![\dot{\vartheta}]\!] \mathbf{n}, \tag{90}$$

$$\gamma_1 \, v_n [\![\dot{\vartheta}]\!] = \gamma_2 [\![\ddot{\mathbf{u}}]\!] \cdot \mathbf{n} - \kappa_* \gamma_3 v_n \, \mathbf{I} \cdot [\![\dot{\mathbf{V}}]\!], \tag{91}$$

where the relation $(35)_4$ and the condition (71) were used in the jump of Eq. (39) and in Eq. (89); by substituting Eq. (91) into Eq. (90) we obtain the following jump, similar to (79),

$$\left[v_n^2 \mathbf{I} - \tilde{\mathcal{U}}(\mathbf{n} \otimes \mathbf{n}) \right] [\![\ddot{\mathbf{u}}]\!] = -\kappa_* v_n (\tilde{\lambda}_5 \, \mathbf{n} \otimes \mathbf{I} + \lambda_6 \, \mathbf{I} \otimes \mathbf{n}) [\![\dot{\mathbf{V}}]\!], \tag{92}$$

but where we introduced the instantaneous *homentropic acoustic macro-tensor*

$$\tilde{\mathcal{U}}(\mathbf{n} \otimes \mathbf{n}) := \left[v_t^2 \, \mathbf{I} + \left(\tilde{v}_l^2 - v_t^2 \right) \mathbf{n} \otimes \mathbf{n} \right], \tag{93}$$

with $\tilde{v}_l^2 := v_l^2 - \gamma_2^2 \gamma_1^{-1}$, and the constant $\tilde{\lambda}_5 := \lambda_5 - \gamma_2 \gamma_3 \gamma_1^{-1}$.

The linear algebraic system of Eqs. (78), (92) and (80) for the amplitudes of the macro-acceleration waves Σ in the homentropic case has the same five solutions with the same propagation speeds, as the homothermal one: two shear optical micro-waves, two transverse micro-waves and one extensional micro-wave. The only variation is in the extensional one and consists in the change of constants v_l and λ_5 with $\tilde{v}_l$ and $\tilde{\lambda}_5$, respectively.

Finally, the temperature jump (91) is absent in all shear optical and transverse micro-waves, while in the extensional one we have:

$$[\![\dot{\vartheta}]\!]_{em} = \gamma_1^{-1} [\gamma_2 \varsigma - \kappa_* \gamma_3 (2 + \zeta)] \, \delta, \tag{94}$$

with the constant $\tilde{\omega} := \frac{\kappa_*}{\tilde{v}_l^2 - v_{em}^2} [2\tilde{\lambda}_5 + (\tilde{\lambda}_5 + \lambda_6)\zeta]$.

7.3. Generalized transverse case

This last subsection concerns the behaviour of macro-acceleration waves that are both homothermal and homentropic and which are usually called *generalized transverse waves*. In physical terms these waves are uninfluenced by thermo-mechanical coupling effects in the transmitting matrix material of the porous solid.

Both conditions $[\![\,\dot{\vartheta}]\!] = 0$ and $\gamma = 0$ apply; thus, from Eq. (91) (or from (81)), we obtain the following compatibility condition for the generalized transverse wave:

$$[\![\,\ddot{\mathbf{u}}]\!] \cdot \mathbf{n} = \kappa_* \gamma_3 \gamma_2^{-1} v_n \, \mathbf{I} \cdot [\![\,\dot{\mathbf{V}}]\!]. \tag{95}$$

Also now we have shear optical and transverse micro-waves as in the previous instances; instead, in general, extensional macro-acceleration waves do not occur, unless the condition (95) is satisfied, that is: $\gamma_3(2+\zeta)(v_l^2 - v_{em}^2) = \gamma_2[2\lambda_5 + (\lambda_5 + \lambda_6)\zeta]$.

8. Evolution equations for wave amplitudes

Let us study now the growth or the decay of the macro-acceleration waves Σ which travel through the thermo-elastic material with nano-pores, thus we restrict ourselves to plane waves which are of uniform scalar amplitude with assigned initial value, uniform in the sense that the scalar amplitude does not vary with position on Σ.

For this purpose, we differentiate twice with respect to time each term of Eq. (38) and once those of Eqs. (39) and (41), take into account the balance of micromomentum (40) and form jumps of all equations across the wave Σ to have:

$$\begin{aligned}
&[\![\,\dot{\rho}]\!] = -\rho_* [\![\,\mathrm{Div}\,\ddot{\mathbf{u}}]\!],\\
&\mathbf{d} = v_t^2 [\![\,\Delta\ddot{\mathbf{u}}]\!] + (v_l^2 - v_t^2)[\![\,\nabla\,(\mathrm{Div}\,\ddot{\mathbf{u}})]\!] + \kappa_* \left(\lambda_5 [\![\,\nabla\,(\mathrm{tr}\,\dot{\mathbf{V}})]\!] + \lambda_6 [\![\,\mathrm{Div}\,\dot{\mathbf{V}}]\!]\right) + \gamma_2 [\![\,\nabla\,\dot{\vartheta}]\!],\\
&[\![\,\ddot{\mathbf{V}}]\!] = v_{sm}^2 [\![\,\Delta\mathbf{V}]\!] + 2(v_{tm}^2 - v_{sm}^2)\,\mathrm{sym}\,[\![\,\nabla\,(\mathrm{Div}\,\mathbf{V})]\!] + \lambda_1 [\![\,\nabla^2(\mathrm{tr}\,\mathbf{V})]\!] + \\
&\qquad + \left\{\lambda_1 [\![\,\mathrm{Div}\,(\mathrm{Div}\,\mathbf{V})]\!] + \lambda_2 [\![\,\Delta(\mathrm{tr}\,\mathbf{V})]\!] - \omega\,\mathbf{I}\cdot[\![\,\dot{\mathbf{V}}]\!]\right\}\mathbf{I} - 2\sigma [\![\,\dot{\mathbf{V}}]\!],\\
&\gamma [\![\,\Delta\dot{\vartheta}]\!] + \gamma_1 [\![\,\ddot{\vartheta}]\!] + \gamma_2 [\![\,\mathrm{Div}\,\ddot{\mathbf{u}}]\!] + \gamma_3\,\mathbf{I}\cdot[\![\,\dot{\mathbf{V}}]\!] = 0,
\end{aligned} \tag{96}$$

where $\mathbf{d}\,(\equiv\ d_n\mathbf{n} + d_e\mathbf{e} + d_f\mathbf{f})$ represents the jump in the third time-derivative of the displacement field $\mathbf{u}$.

Algebraic computations, very similar to those carried out in [26, 42], with the use of the Hugoniot-Hadamard compatibility condition (70) and of definitions (76) and (77) of the homothermal acoustic tensors, we obtain the following evolution equations for the propagating wave Σ:

$$[\![\,\dot{\rho}]\!] = \rho_*(v_n^{-1} d_n - \mathrm{Div}\,[\![\,\ddot{\mathbf{u}}]\!]), \tag{97}$$

$$\begin{aligned}
&\left[v_n^2 \mathbf{I} - \mathcal{U}(\mathbf{n}\otimes\mathbf{n})\right]\mathbf{d} = \kappa_* v_n \left\{v_n\left[\lambda_5 \nabla\,(\mathrm{tr}\,[\![\,\dot{\mathbf{V}}]\!]) + \lambda_6 \mathrm{Div}\,[\![\,\dot{\mathbf{V}}]\!]\right] - \left[\lambda_5(\mathrm{tr}\,[\![\,\ddot{\mathbf{V}}]\!])\mathbf{n} + \lambda_6 [\![\,\ddot{\mathbf{V}}]\!]\mathbf{n}\right]\right\} + \\
&\quad + v_n\left\{(v_t^2 - v_l^2)\left[(\mathrm{Div}\,[\![\,\ddot{\mathbf{u}}]\!])\mathbf{n} + \nabla\,([\![\,\ddot{\mathbf{u}}]\!]\cdot\mathbf{n})\right] - 2v_t^2(\nabla\,[\![\,\ddot{\mathbf{u}}]\!])\mathbf{n} - \gamma_2 [\![\,\ddot{\vartheta}]\!]\mathbf{n}\right\} + \gamma_2 v_n^2 \nabla\,[\![\,\dot{\vartheta}]\!],
\end{aligned} \tag{98}$$

$$\begin{aligned}
&\left[v_n^2 \mathcal{I} - \mathcal{C}(\mathbf{n}\otimes\mathbf{n})\right][\![\,\ddot{\mathbf{V}}]\!] = 2v_n(v_{sm}^2 - v_{tm}^2)\,\mathrm{sym}\,\left[(\mathrm{Div}\,[\![\,\dot{\mathbf{V}}]\!])\otimes\mathbf{n} + \nabla\,([\![\,\dot{\mathbf{V}}]\!]\mathbf{n})\right] - \\
&\quad -2\,v_n\left\{v_{sm}^2(\nabla\,[\![\,\dot{\mathbf{V}}]\!])\mathbf{n} + \lambda_1\left\{\mathrm{sym}[\nabla\,(\mathrm{tr}\,[\![\,\dot{\mathbf{V}}]\!])\otimes\mathbf{n}] + (\mathbf{n}\cdot\mathrm{Div}\,[\![\,\dot{\mathbf{V}}]\!])\mathbf{I}\right\}\right\} - \\
&\qquad -2\,v_n\,\lambda_2[\mathbf{n}\cdot\nabla\,(\mathrm{tr}\,[\![\,\dot{\mathbf{V}}]\!])]\mathbf{I} - v_n^2[\omega\,\mathrm{tr}\,[\![\,\dot{\mathbf{V}}]\!]\,\mathbf{I} + 2\sigma\,[\![\,\dot{\mathbf{V}}]\!]],
\end{aligned} \tag{99}$$

$$\begin{aligned}
&\gamma\Theta = v_n\left[2\gamma(\mathbf{n}\cdot\nabla\,[\![\,\dot{\vartheta}]\!]) + \gamma_2 d_n\right] - \\
&\qquad -v_n^2\left(\gamma\Delta[\![\,\vartheta]\!] + \gamma_1 [\![\,\dot{\vartheta}]\!] + \gamma_2 \mathrm{Div}\,[\![\,\ddot{\mathbf{u}}]\!] + \kappa_*\gamma_3 \mathrm{tr}\,[\![\,\dot{\mathbf{V}}]\!]\right),
\end{aligned} \tag{100}$$

with Θ that indicates the jump in the third time-derivative of the temperature change field ϑ.

Therefore, the transport equations in the linearized case will give standard evolution laws of the type $f' = -\mu f$ and hence $f = f_0\, e^{-\mu\phi}$, where μ is a constant, f_0 is the strength of the wave at $\tau = \tau_0$ and ϕ is the increasing distance, measured along the normal to the wave, from the wave front at the same time.

8.1. Evolution of homothermal waves

Now, since for a plane homothermal macro-acceleration wave entering the natural reference placement $\mathcal{B}_*$ the jump of $\dot{\vartheta}$ vanishes for the Fourier condition (72), by developing the analysis of equations (97)-(100) we get the following consequences:

8.1.1. *Shear optical case*

By inserting the micro-wave solution of §7.1 of amplitude $[\![\dot{\mathbf{V}}]\!]_{sm}$ and speed v_{sm}, which have fields $\ddot{u}$, $\dot{\rho}$ and $\ddot{\vartheta}$ continuous through the wave, we obtain that:

$$[\![\ddot{\rho}]\!] = \rho_* d_n v_{sm}^{-1}, \quad \left[v_{sm}^2 \mathbf{I} - \mathcal{U}(\mathbf{n}\otimes\mathbf{n})\right]\mathbf{d} = -\kappa_* v_{sm}\left[\lambda_5(\mathrm{tr}\,[\![\ddot{\mathbf{V}}]\!])\mathbf{n} + \lambda_6[\![\ddot{\mathbf{V}}]\!]\mathbf{n}\right], \tag{101}$$

$$\left[v_{sm}^2\mathcal{I} - \mathcal{C}(\mathbf{n}\otimes\mathbf{n})\right][\![\ddot{\mathbf{V}}]\!] =$$

$$= 2v_{sm}^3\left\{\left[\frac{d\alpha}{dn} + \frac{\sigma\alpha}{v_{sm}}\right](\mathbf{f}\otimes\mathbf{f} - \mathbf{e}\otimes\mathbf{e}) - \left[\frac{d\beta}{dn} + \frac{\sigma\beta}{v_{sm}}\right](\mathbf{e}\otimes\mathbf{f} + \mathbf{f}\otimes\mathbf{e})\right\}, \tag{102}$$

$$\gamma\Theta = \gamma_2 v_{sm} d_n - \kappa_*\gamma_3 v_{sm}^2\,\mathrm{tr}\,[\![\ddot{\mathbf{V}}]\!]; \tag{103}$$

hence, from Eq. (102), it must be $[\![\ddot{V}]\!]_{11} = [\![\ddot{V}]\!]_{12} = [\![\ddot{V}]\!]_{13} = 0$ and $[\![\ddot{V}]\!]_{33} = -[\![\ddot{V}]\!]_{22}$, while $[\![\ddot{V}]\!]_{22}$ and $[\![\ddot{V}]\!]_{23}$ remain undefined,

$$\alpha(\tau) = \alpha_0 \exp\left(-\frac{\sigma}{v_{sm}}\phi\right) \text{ and } \beta(\tau) = \beta_0 \exp\left(-\frac{\sigma}{v_{sm}}\phi\right), \tag{104}$$

with α_0 and β_0 the values at $\tau = \tau_0$. Instead, by analysing Eq. $(101)_2$, we have that $\mathbf{d}_{sm} = \mathbf{0}$ and thus $[\![\ddot{\rho}]\!]_{sm} = 0$ and $\Theta_{sm} = 0$.

This kind of micro-wave does not cause any disturbance in the mechanical and thermal fields and the scalar amplitudes α and β decay to zero as the time interval $(\tau - \tau_0)$ increases indefinitely (because ϕ behaves so).

8.1.2. *Transverse micro-wave*

For the transverse solutions of §7.1, whose amplitude is $[\![\dot{\mathbf{V}}]\!]_{tm}$, speed v_{tm} and jump $[\![\ddot{\mathbf{u}}]\!]_{tm}$ given by equation (82) (while $\dot{\rho}$ and $\ddot{\vartheta}$ are continuous), the algebraic system of evolution Eqs. (97)-(100) reduces to the following one:

$$[\![\ddot{\rho}]\!] = \rho_* v_{tm}^{-1} d_n, \quad \left[v_{tm}^2\mathbf{I} - \mathcal{U}(\mathbf{n}\otimes\mathbf{n})\right]\mathbf{d} = \tag{105}$$

$$= \kappa_* v_{tm}^2 \lambda_6 \frac{v_{tm}^2 + v_t^2}{v_{tm}^2 - v_t^2}\left(\frac{d\chi_e}{dn}\mathbf{e} + \frac{d\chi_f}{dn}\mathbf{f}\right) - \kappa_* v_{tm}\left[\lambda_5(\mathrm{tr}\,[\![\ddot{\mathbf{V}}]\!])\mathbf{n} + \lambda_6[\![\ddot{\mathbf{V}}]\!]\mathbf{n}\right],$$

$$\left[v_{tm}^2\mathcal{I} - \mathcal{C}(\mathbf{n}\otimes\mathbf{n})\right][\![\ddot{\mathbf{V}}]\!] = -4v_{tm}^3\mathrm{sym}\left[\left(\frac{d\chi_e}{dn} + \frac{\sigma\chi_e}{v_{tm}}\right)\mathbf{e} + \left(\frac{d\chi_f}{dn} + \frac{\sigma\chi_f}{v_{tm}}\right)\mathbf{f}\right]\otimes\mathbf{n}, \tag{106}$$

$$\gamma\Theta = \gamma_2 v_{tm} d_n - \kappa_*\gamma_3 v_{tm}^2\mathrm{tr}\,[\![\ddot{\mathbf{V}}]\!]. \tag{107}$$

Therefore, we have from Eq. (106) that all the components of $[\![\ddot{\mathbf{V}}]\!]_{tm}$ are equal to zero except for the undetermined $[\![\ddot{V}]\!]_{12}$ and $[\![\ddot{V}]\!]_{13}$, while the scalar amplitudes are given by

$$\chi_i(\tau) = \chi_{i0} \exp\left(-\frac{\sigma}{v_{tm}}\phi\right), \text{ for } i = e, f, \tag{108}$$

with χ_{i0} the values at $\tau = \tau_0$. Instead, Eqs. (105), (107) and (108) establish that

$$(d_n)_{tm} = [\![\ddot{\rho}]\!]_{tm} = \Theta_{tm} = 0,\ (d_i)_{tm} = \frac{\kappa_* v_{tm} \lambda_6}{v_t^2 - v_{tm}^2}\left([\![\ddot{V}]\!]_{1i} + \frac{\sigma(v_{tm}^2 + v_t^2)}{v_{tm}^2 - v_t^2}\chi_i\right), \text{ for } i = e, f. \tag{109}$$

Also in this case the scalar amplitudes decay to zero as the time interval $(\tau - \tau_0)$ increases indefinitely, but, unlike shear optical waves, we have here a macro-acceleration jump with a third order discontinuity related to the elastic properties of nano-pores and to a part that decays to zero.

8.1.3. Extensional mode

In this case the solutions of §7.1 for the amplitude and the speed are $[\![\ddot{\mathbf{V}}]\!]_{em}$ and v_{em}, respectively, and thus, by applying Eqs (83) and (84), we have:

$$[\![\ddot{\rho}]\!] = \rho_* d_n v_{em}^{-1} - \rho_* v_{em} \varpi \frac{d\delta}{dn}, \qquad \left[v_{em}^2 \mathbf{I} - \mathcal{U}(\mathbf{n}\otimes\mathbf{n})\right]\mathbf{d} = \tag{110}$$

$$= \kappa_* v_{em}\left\{v_{em}\left[2\lambda_5 + (\lambda_5+\lambda_6)\zeta\right]\frac{v_{em}^2 + v_l^2}{v_{em}^2 - v_l^2}\frac{d\delta}{dn}\mathbf{n} - \left[\lambda_5(\mathrm{tr}\,[\![\ddot{\mathbf{V}}]\!])\mathbf{n} + \lambda_6[\![\ddot{\mathbf{V}}]\!]\mathbf{n}\right]\right\},$$

$$\left[v_{em}^2\mathbf{I} - \mathcal{C}(\mathbf{n}\otimes\mathbf{n})\right][\![\ddot{\mathbf{V}}]\!] = -v_{em}\left\{2\frac{d\delta}{dn}\left[\lambda_1\zeta + \lambda_2(2+\zeta)\right] + v_{em}\omega(2+\zeta)\delta\right\}\mathbf{I} +$$

$$+2v_{em}\left\{\frac{d\delta}{dn}\left[(v_{sm}^2 - 2v_{tm}^2)\zeta - \lambda_1(2+\zeta)\right] - v_{em}\sigma\zeta\delta\right\}(\mathbf{n}\otimes\mathbf{n}) \tag{111}$$

$$-2v_{em}\left(v_{sm}^2\frac{d\delta}{dn} + v_{em}\sigma\delta\right)(\mathbf{e}\otimes\mathbf{e} + \mathbf{f}\otimes\mathbf{f}),$$

$$\gamma\Theta = v_{em}\left(2\gamma\check{\vartheta}\frac{d\delta}{dn} + \gamma_2 d_n\right) - v_{em}^2\left(\gamma_1\check{\vartheta}\delta + \gamma_2 v_{em}\varpi\frac{d\delta}{dn} + \kappa_*\gamma_3\,\mathrm{tr}\,[\![\ddot{\mathbf{V}}]\!]\right). \tag{112}$$

The solutions of the system (111) are $[\![\ddot{V}]\!]_{ij} = 0$, if $i \neq j$, and $[\![\ddot{V}]\!]_{33} = [\![\ddot{V}]\!]_{22}$, while, to determine δ, it is necessary to know either of $[\![\ddot{V}]\!]_{11}$ or $[\![\ddot{V}]\!]_{22}$ previously, otherwise it remains undetermined and we cannot say anything about the growth or the decay of this wave; vice versa, if we are able to assign the behaviour of the amplitude δ, we can resolve the remaining two jumps. Moreover, from Eqs. (110) and (112), we obtain that $(d_e)_{em} = (d_f)_{em} = 0$, while also $[\![\ddot{\rho}]\!]_{em}$, $(d_n)_{em}$ and Θ_{em} suffer of the same undeterminacies already spoken about. The micro-wave is then accompanied by second and third order discontinuities in macro-mechanical, mass and thermal fields.

8.1.4. $\lambda_1 + \lambda_2 = 0$ case.

In this peculiar subcase we observed in §7.1:

i) *A purely transverse micro-wave* of amplitude $[\![\dot{\mathbf{V}}]\!]_{pm}$ and speed v_{pm}, the other jumps being given by Eqs. (85) and (86). By performing same developments of previous solutions, we obtain that $[\![\ddot{V}]\!]_{ij} = 0$, if $i \neq j$, and $[\![\ddot{V}]\!]_{33} = [\![\ddot{V}]\!]_{22}$ (which remain undetermined); moreover, $[\![\ddot{V}]\!]_{11} = \frac{2v_{pm}^2\omega}{v_{tm}^2 - 2v_{pm}^2 - \lambda_1}\varrho$, with the scalar amplitudes ϱ given by $\varrho(\tau) = \varrho_0 \exp\left(-\frac{\omega+\sigma}{v_{pm}}\phi\right)$ (ϱ_0 being the value at $\tau = \tau_0$). Thus, it results that $(d_e)_{pm} = (d_f)_{pm} = 0$, while $(d_n)_{pm} = \Gamma\varrho - \frac{2\kappa_*\lambda_5 v_{pm}}{v_{pm}^2 - v_l^2}[\![\ddot{V}]\!]_{22}$, $[\![\ddot{\rho}]\!]_{pm} = \rho_* v_{pm}^{-1} d_n$ and $\gamma\Theta_{pm} = \gamma_2 v_{pm} d_n - 2\kappa_*\gamma_3 v_{pm}^2 [\![\ddot{V}]\!]_{22} - \Pi\,\varrho$, where Γ and Π are constants related to previous defined constitutive constants.
Hence the third order discontinuity of the longitudinal macro-wave, induced by the purely transverse micro-wave, has a first part that decays to zero as the time interval $(\tau - \tau_0)$ go to infinity and a second one related to the elastic properties of nano-pores, as well as discontinuities in the mass and temperature derivatives.

ii) *A purely longitudinal micro-wave* of amplitude $[\![\dot{\mathbf{V}}]\!]_{lm}$ and speed of propagation v_{lm} for which we have that all $[\![\ddot{V}]\!]_{ij} = 0$, if $i \neq j$, $[\![\ddot{V}]\!]_{11}$ remains undefined and $[\![\ddot{V}]\!]_{33} = [\![\ddot{V}]\!]_{22} = \frac{\omega}{v_{lm}^2 - v_{sm}^2 + 2\lambda_1}\hat{\delta}$, with $\hat{\delta}(\tau) = \hat{\delta}_0 \exp\left(-\frac{\omega+\sigma}{2v_{lm}}\phi\right)$ and $\hat{\delta}_0$ its value at $\tau = \tau_0$; in addition, also now $(d_e)_{lm} = (d_f)_{lm} = 0$, while $(d_n)_{lm} = \Xi\,\hat{\delta} - \kappa_* v_{lm}(\lambda_5 + \lambda_6)[\![\ddot{V}]\!]_{11}$, $[\![\ddot{\rho}]\!]_{lm} = \rho_* v_{lm}^{-1} d_n$ and $\gamma\Theta_{lm} = \gamma_2 v_{lm} d_n - \kappa_*\gamma_3 v_{lm}^2 [\![\ddot{V}]\!]_{11} - \Upsilon\,\hat{\delta}$, with Ξ and Υ constants related to constitutive constants.
Therefore, also the pure micro-wave of compaction or distention is accompanied by a third order discontinuities in the mechanical field with a first part that decays to zero with the increasing of the time interval $(\tau - \tau_0)$ and a second one related to nano-pores properties.

8.2. Homentropic and generalized transverse evolution instances

o) *Homentropic macro-acceleration waves:* When the solid with nano-pores does not conduct heat (see §7.2), we have to substitute evolution Eq. (100) with the following one:

$$\gamma_1[\![\ddot{\vartheta}]\!] = \gamma_2 v_n^{-1} d_n - \gamma_2 \mathrm{Div}\,[\![\ddot{\mathbf{u}}]\!] - \kappa_*\gamma_3 \mathrm{tr}\,[\![\ddot{\mathbf{V}}]\!], \tag{113}$$

which is obtained by deriving with respect to the time τ the energy balance $\dot{\eta} = \theta_*^{-1}\lambda$, by using relation $(35)_4$ and by taking its jump.
As we observed in §7.2, discussions about this subcase follow closely those carried out for the homothermal one with respect to shear optical and transverse macro-acceleration waves; instead, for the extensional, purely transverse and purely longitudinal ones the only change consists in the choose of constants $\tilde{v}_l$ and $\tilde{\lambda}_5$ in place of v_l and λ_5: for example, Eq. $(110)_2$ must be substituted by

$$\left[v_{em}^2\mathbf{I} - \mathcal{U}(\mathbf{n}\otimes\mathbf{n})\right]\mathbf{d}_{em} = \tag{114}$$
$$= \kappa_* v_{em}\left\{v_{em}\left[2\tilde{\lambda}_5 + (\tilde{\lambda}_5 + \lambda_6)\zeta\right]\frac{v_{em}^2 + v_l^2}{v_{em}^2 - \tilde{v}_l^2}\frac{d\delta}{dn}\mathbf{n} - \left[\lambda_5(\mathrm{tr}\,[\![\dot{\mathbf{V}}]\!]_{em})\mathbf{n} + \lambda_6[\![\dot{\mathbf{V}}]\!]_{em}\mathbf{n}\right]\right\},$$

while Eq. (113) gives $\gamma_1[\![\ddot{\vartheta}]\!]_{em} = \gamma_2 v_{em}^{-1}(d_n)_{em} - \gamma_2\tilde{\omega} v_{em}\frac{d\delta}{dn} - \kappa_*\gamma_3([\![\ddot{V}]\!]_{11} + 2[\![\ddot{V}]\!]_{22})$.

Hence, with this simple change of constants, the conclusions about the evolution of amplitudes of the macro-acceleration waves remain the same as in the corresponding homothermal case.

o) *Generalized Transverse Case:* This last peculiar wave, both homothermal and homentropic, has the same shear optical and transverse solutions as the homothermal case (see §7.2) and so the solutions of the evolution equations have the same behaviour.

The extensional micro-wave, and so comments about its amplitude evolution, occurrs only if the condition (95) is satisfied, *i.e.*, $\gamma_3(2+\zeta)(v_l^2 - v_{em}^2) = \gamma_2[2\lambda_5 + (\lambda_5 + \lambda_6)\zeta]$.

9. Conclusions

The results showed in this chapter can be outlined as it follows:

a) *Linear thermo-dynamic theory.* We derived the linear theory of a thermoelastic solid with nano-pores which includes inelastic surface effects associated with changes in the deformation of the holes in the vicinity of void boundaries and which generalizes classical voids theories. In order to get the fields equations we used the principles of objectivity and equipresence, besides the compatibility with the Clausius-Duhem inequality.

b) *Micro-vibrations.* The first application to micro-vibrations in absence of dissipation gives origin to three admissible results: a dilatational micro-thermal oscillation and two solutions, both with no thermal vibrations, with the same frequency and with null trace: a shear mode and an extensional mode with constant volume.

c) *Plane waves.* Here we presented the solutions of secular equations governing the propagation of harmonic plane waves in the porous thermoelastic medium: there can exist two shear optical micro-elastic waves, two coupled transverse elastic waves and four coupled longitudinal thermo-elastic waves. The exact or approximate values of the phase speeds, specific losses, attenuation factors and amplitude ratios are discussed for large and small frequencies.

d) *Macro-acceleration waves.* Last investigation regarded the propagation conditions and the growth equations which govern the motion of particular weak singularities, called macro-acceleration waves, for which only jumps of the derivatives of the macro- and micro-displacement of order 2 and of the temperature of order 1 are of interest in the theory.

We observed that, for a linear conducting homogeneous centrosymmetric isotropic material with nano-pores, every macro-acceleration wave is homothermal and only three speeds of propagation are possible: i) one related to two shear-optical micro-modes completely decoupled from the mass and thermoelastic macro-properties of the matrix material and which decay to zero when the time interval increases; ii) the second velocity associated to two transverse micro-modes coupled with a transverse macro-acceleration wave, spreading without perturbing mass and thermal fields: the micro-modes decay still to zero, while the associated macro-ones have a constant part and an added contribution that decays still to zero; iii) the third one linked to one extensional micro-wave coupled with a longitudinal macro-wave and with discontinuities in the second and third order of derivatives of mass and thermal fields.

Instead in the non-conducting case every wave is homentropic, but we obtained the same number of propagation velocities and of macro-acceleration waves as in the previous homothermal instance.

At the end, for generalized transverse macro-acceleration waves, only the extensional micro-mode does not occurr, in general.

Acknowledgements

The support of the Italian "Gruppo Nazionale per la Fisica Matematica" of the "Istituto Nazionale di Alta Matematica" and of the Department of Mechanics and Materials of the "Mediterranean" University of Reggio Calabria (Italy) is gratefully acknowledged.

Author details

Pasquale Giovine
Dipartimento di Meccanica e Materiali - Università degli Studi "Mediterranea", Via Graziella 1, Località Feo di Vito, I-89122 Reggio Calabria, Italy

10. References

[1] Nunziato JW, Cowin SC (1979) A Nonlinear Theory of Elastic Materials with Voids. Arch. Rational Mech. Anal. 72:175-201.
[2] Cowin SC, Nunziato JW (1983) Linear Elastic Materials with Voids. J. of Elasticity. 13:125-147.
[3] Capriz G (1989) Continua with Microstructure. Series: Springer Tracts in Natural Philosophy. New York: Springer-Verlag. 35:92 p.
[4] Giovine P (1996) Porous Solids as Materials with Ellipsoidal Structure. In: Batra RC, Beatty MF, editors. Contemporary Research in the Mechanics and Mathematics of Materials. Barcelona: CIMNE. pp. 335-342.
[5] Capriz G, Podio-Guidugli P (1981) Materials with Spherical Structure. Arch. Rational Mech. Anal. 75:269-279.
[6] Wilmanski K (2002) Mass Exchange, Diffusion and Large Deformations of Poroelastic Materials. In: Capriz C, Ghionna VN, Giovine P, editors. Modeling and Mechanics of Granular and Porous Materials. Series: Modeling and Simulation in Science, Engineering and Technology. Boston: Birkhauser. 12:211-242.
[7] Cowin SC, Weinbaum S, Zeng Y (1995) A Case for Bone Canaliculi as the Anatomical Site of Strain Generated Potentials. J. Biomechanics. 28:1281-1297.
[8] Cowin SC (1998) On Mechanosensation in Bone under Microgravity. Bone. 22:119S-125S Supplement.
[9] Lakes RS (1986) Experimental Microelasticity of Two Porous Solids. Int. J. Solids Structures 22:55-63.
[10] Grioli G (2003) Microstructures as a Refinement of Cauchy Theory. Problems of Physical Concreteness. Continuum Mech. Thermodyn. 15:441-450.
[11] Cieszko M (2004) Extended Description of Pore-Space Structure and Fluid Flow in Porous Materials. Application of Minkowski Space. Proceed. of the XIVth International STAMM'04. Aachen: Shaker Verlag. pp. 93-102.
[12] Giovine P (1999) A Linear Theory of Porous Elastic Solids. Trans. Porous Media. 34:305-318.

[13] Bear J, Bachmat Y (1986) Macroscopic Modelling of Transport Phenomena in Porous Media. Applications to Mass, Momentum and Energy Transport. Trans. Porous Media. 1:241-269.
[14] Ponte-Castañeda P, Zaidman M (1994) Constitutive Models for Porous Materials with Evolving Microstructure. J. Mech. Phys. Solids. 42:1459-1497.
[15] Kailasam M, Ponte-Castañeda P (1998) A General Constitutive Theory for Linear and Nonlinear Particulate Media with Microstructure Evolution. J. Mech. Phys. Solids. 46:427-465.
[16] Kirchner N, Steinmann P (2006) Mechanics of extended continua: modelling and simulation of elastic microstretch materials. Computational Mechanics. 40:651-666.
[17] Mindlin RD (1964) Micro-structure in Linear Elasticity. Arch. Rational Mech. Anal. 16:51-78.
[18] Giovine P, Margheriti L, Speciale MP (2008) On Wave Propagation in Porous Media with Strain Gradient Effects. Computers and Mathematics with Applications. 55/2:307-318.
[19] Giovine P (2003) A Continuum Theory of Soils: Viewed as Peculiar Immiscible Mixtures. Mathematical and Computer Modelling. 37:525-532.
[20] Giovine P (2005) On Adsorption and Diffusion in Microstructured Porous Media. In: Huyghe JM, Raats PAC, Cowin SC, editors. IUTAM Symposium on Physicochemical and Electromechanical Interactions in Porous Media. Series: Solid Mechanics and Its Applications. Dordrecht: Springer. 125:183-191.
[21] Hasselman DPH, Singh JP (1982) Criteria for Thermal Stress Failure of Brittle Structural Ceramics. In: Thermal Stresses I. Northolland, Amsterdam
[22] Truesdell CA, Noll W. (1992) The Non-Linear Field Theories of Mechanics. 2nd edition. Berlin: Springer-Verlag. 591 p.
[23] Capriz G, Giovine P (1997) On Microstructural Inertia. Math. Mod. Meth. Appl. Sciences. 7:211-216.
[24] Capriz G, Giovine P (1997) Remedy to Omissions in a Tract on Continua with Microstructure. Atti XIII° Congresso AIMETA 97, Siena, Meccanica Generale I. pp. 1-6.
[25] Capriz G, Virga EG (1994) On Singular Surfaces in the Dynamics of Continua with Microstructure. Quart. Appl. Math. 52:509-517.
[26] Giovine P (2005) On Acceleration Waves in Continua with Large Pores. Proceed. of the XIVth International STAMM'04. Aachen: Shaker Verlag. pp. 113-124.
[27] Paoletti P (2012) Acceleration waves in complex materials. Discrete and Continuous Dynamical Systems - Series B. 17:637-659.
[28] Capriz G, Podio-Guidugli P, Williams W (1981) On Balance Equations for Materials with Affine Structure. Meccanica. 17:80-84.
[29] Chadwick P, Currie PK (1972) The Propagation of Acceleration Waves in Heat Conducting Elastic Materials. Arch. Rational Mech. Anal. 49:137-158.
[30] Suhubi ES, Eringen AC (1964) Nonlinear Theory of Micro–Elastic Solids - II. Int. J. Engng. Sci. 2:389-404.
[31] Smith GF (1971) On Isotropic Functions of Symmetric Tensors, Skew-Symmetric Tensors and Vectors. Int. J. Engng. Sci. 9:899-916.
[32] Giacobbe V, Giovine P (2009) Plane Waves in Linear Thermoelastic Porous Solids. In: Ganghoffer J-F, Pastrone F, editors. Mechanics of Microstructured Solids: Cellular Materials, Fibre Reinforced Solids and Soft Tissues. Series: Lecture Notes in Applied and Computational Mechanics. Berlin: Springer Verlag. 46:71-80.

[33] Puri P (1972) Plane Waves in Thermoelasticity and Magnetothermoelasticity. Int. J. Eng. Sciences. 10:467-477.

[34] Puri P, Cowin SC (1985) Plane Waves in Linear Elastic Materials with Voids. J. of Elasticity. 15:167-183.

[35] Sharma JN, Kaur D (2006) Plane Harmonic Waves in Generalized Thermoelastic Materials with Voids. Private communication.

[36] Singh J, Tomar SK (2007) Plane Waves in Thermoelastic Material with Voids. Mechanics of Materials. 39:932-940.

[37] Manacorda T (1979) *Introduzione alla Termomeccanica dei Continui.* Quaderni UMI **14** (Pitagora Editrice, Bologna 1979)

[38] Ieşan D (1986) A Theory of Thermoelastic Materials with Voids. Acta Mechanica. 60:67-89.

[39] Mariano PM, Sabatini L (2000) Homothermal Acceleration Waves in Multifield Theories of Continua. Int. J. Non-Linear Mech. 35:963-977.

[40] Wright TW (1973) Acceleration Waves in Simple Elastic Materials. Arch. Rational Mech. Anal. 50:237-277.

[41] Giovine P, Fiorino L (2008) Some Remarks on Acceleration Waves in Porous Solids. In: Manganaro N, Rionero S, editors. Proceed. 14th Int. Conf. on Waves and Stability in Continuous Media, Scicli (Ragusa). Singapore: World Scientific Publishing. pp. 280-286.

[42] Giovine P, Fiorino L (2005) Nano-Pores in Thermoelastic Materials. Atti XVII° Congresso AIMETA 05 di Meccanica Teorica ed Applicata, Mini-symposium on "Nanotechnologies: Building up Structures at the Nano and Meso-Scales". Firenze (2005) p.283. Technical Note n.385 on the CD rom.

Chapter 3

Acoustical Modeling of Laminated Composite Cylindrical Double-Walled Shell Lined with Porous Materials

K. Daneshjou, H. Ramezani and R. Talebitooti

Additional information is available at the end of the chapter

1. Introduction

Although the researchers have done many efforts to perform the numerical model such as FEM (Finite Elements Method) to investigate the wave prorogation through the shells, the analytical vibro-acoustic modeling of the composite shells is unavoidable because of the accuracy of the model in a broadband frequency. Bolton *et. al.* [1] investigated sound transmission through sandwich structures lined with porous materials and following Lee *et. al.* [2] proposed a simplified method to analyze curved sandwich structures. Daneshjou *et. al.* [3-5] studied an exact solution to estimate the transmission loss of orthotropic and laminated composite cylindrical shells with considering all three displacements of the shell. Recently the authors [6] have presented an exact solution of free harmonic wave propagation in a double-walled laminated composite cylindrical shell whose walls sandwich a layer of porous material using an approximate method. This investigation is focused on sound transmission through the sandwich structure, which includes the porous material core between the two laminated composite cylindrical shells to predict the reliable results for all structures used foam as an acoustic treatment.

Wave propagation through a composite cylindrical shell lined with porous materials is investigated, based on classical laminated theory. The porous material is completely modeled using elastic frame. The vibro-acoustic equations of the shell are derived considering both the shell vibration equations and boundary conditions on interfaces. These coupled equations are solved simultaneously to calculate the Transmission Loss (TL). Moreover, the results are verified with a special case where the porosity approaches zero. Finally, the numerical results are illustrated to properly study the geometrical and physical properties of composite and porous material. In addition, the effects of the stacking sequence of composite shells and fiber directions are properly studied.

2. Propagation of sound in porous media

If the porous material is assumed a homogeneous aggregate of the elastic frame and the fluid trapped in pores, its acoustic behavior can be considered by the following two wave equations (See Eq. (22) and Eq. (25) of [1]):

$$\nabla^4 e_s + A_1 \nabla^2 e_s + A_2 e_s = 0 \tag{1}$$

$$\nabla^2 \varpi + \xi_t^2 \varpi = 0 \tag{2}$$

Eqs. (1) and (2) determine 2 elastic longitudinal waves and 1 rotational wave, respectively. In Eqs. (1) and (2) ∇ is the vector differential operator, $e_s = \nabla.\bar{\mathbf{u}}$ is the solid volumetric strain, $\bar{\mathbf{u}}$ is the displacement vector of the solid, $\varpi = \nabla \times \bar{\mathbf{u}}$ is the rotational strain in the solid phase, $A_1 = \omega^2(\varphi\hat{\rho}_{22} + \mu\hat{\rho}_{11} - 2\chi\hat{\rho}_{12})/(\varphi\mu - \chi^2)$, $A_2 = \omega^4(\hat{\rho}_{22}\hat{\rho}_{11} - (\hat{\rho}_{12})^2)/(\varphi\mu - \chi^2)$, and ξ_t is the wave number of the shear wave (See Eq. 10). $\hat{\rho}_{11}$, $\hat{\rho}_{12}$ and $\hat{\rho}_{22}$ are equivalent masses given by:

$$\hat{\rho}_{11} = \rho_1 + \rho_a - j\sigma_r\phi^2\left(\frac{1}{\omega} + \frac{4j\alpha_\infty^2\kappa_v\rho_0}{\sigma_r^2\Lambda^2\phi^2}\right) \tag{3}$$

$$\hat{\rho}_{12} = -\rho_a + j\sigma_r\phi^2\left(\frac{1}{\omega} + \frac{4j\alpha_\infty^2\kappa_v\rho_0}{\sigma_r^2\Lambda^2\phi^2}\right) \tag{4}$$

$$\hat{\rho}_{22} = \phi\rho_0 + \rho_a - j\sigma_r\phi^2\left(\frac{1}{\omega} + \frac{4j\alpha_\infty^2\kappa_v\rho_0}{\sigma_r^2\Lambda^2\phi^2}\right) \tag{5}$$

where j is the imaginary unit $j^2 = -1$. ρ_1 and ρ_0 are the densities of the solid and fluid parts of the porous material. Moreover, parameters α_∞, κ_v, Λ, σ_r, ϕ and ω are tortuosity, air viscosity, viscous characteristic length, flow resistivity, porosity, and angular frequency, respectively. χ, φ and μ represent material properties: $\chi = (1-\phi)G$, $\mu = \phi G$, $\varphi = A + 2\delta$, $\delta = E/2(1+\nu)$ and $A = \nu E/(1+\nu)(1-2\nu)$. E and v are the in vacuo Young's modulus and Poisson's ratio of the bulk solid phase, respectively. Assuming that pores are shaped in cylindrical form, an expression for G is:

$$G = \rho_0 c_2^2 \left\{ 1 + \left[(\varsigma - 1)\sqrt{\phi\sigma_r} / N_{\mathrm{Pr}}^{0.5}\sqrt{-2j\omega\rho_0\alpha_\infty} \right] \left[\frac{J_1\left(2N_{\mathrm{Pr}}^{0.5}\sqrt{-2\omega\rho_0\alpha_\infty j/\phi\sigma_r}\right)}{J_0\left(2N_{\mathrm{Pr}}^{0.5}\sqrt{-2\omega\rho_0\alpha_\infty j/\phi\sigma_r}\right)} \right] \right\}^{-1} \tag{6}$$

where ς is the ratio of specific heats, c_2 is the speed of sound in the fluid phase of porous materials, N_{Pr} represents the Prandtl number, and J_0 and J_1 are Bessel functions of the first kind, zero and first order, respectively. ρ_a is the inertial coupling term:

$$\rho_a = \phi\rho_0(\alpha_\infty - 1) \tag{7}$$

The complex wave numbers of the two compression (longitudinal) waves, ξ_α and ξ_β are:

$$\xi_\alpha^2 = \frac{\omega^2}{2(\varphi\mu - \chi^2)}(\varphi\hat{\rho}_{22} + \mu\hat{\rho}_{11} - 2\chi\hat{\rho}_{12} + \sqrt{\wp}) \tag{8}$$

$$\xi_\beta^2 = \frac{\omega^2}{2(\varphi\mu - \chi^2)}(\varphi\hat{\rho}_{22} + \mu\hat{\rho}_{11} - 2\chi\hat{\rho}_{12} - \sqrt{\wp}) \tag{9}$$

where

$$\wp = (\varphi\hat{\rho}_{22} + \mu\hat{\rho}_{11} - 2\chi\hat{\rho}_{12})^2 - 4(\varphi\mu - \chi^2)(\hat{\rho}_{22}\hat{\rho}_{11} - \hat{\rho}_{12}^2) \tag{10}$$

and the wave number of the shear (rotational) wave is:

$$\xi_t^2 = \frac{\omega^2}{\delta}(\frac{\hat{\rho}_{22}\hat{\rho}_{11} - \hat{\rho}_{12}^2}{\hat{\rho}_{22}}) \tag{11}$$

δ is the shear modulus of the porous material.

3. Simplified method

As the full method is too complicated to model the porous layer in the curved sandwich structures, thus a simplified method is expanded for this category of structures [2]. The foundation of this approximate method considers the strongest wave between those ones. It includes two steps. At the first step a flat double laminated composite with infinite extents with the same cross sectional construction is considered using the full method. Then, only the strongest wave number is chosen from the results and the material is modeled using the wave number and its corresponding equivalent density. Thus, the material is modeled as an equivalent fluid.

The strain energy which is related to the displacement in the solid and fluid phases can be defined for each wave component. The energy terms can be represented as follows; E_{1s} and E_{1f} for the airborne wave, E_{3s} and E_{3f} for the frame wave and E_{5s} for the shear wave, which the subscripts s and f represent the solid and fluid phase, respectively. For each new problem, comparing the ratios of the energy carried by the frame wave and the shear wave to the airborne wave in the fluid and solid phases: i.e., $\frac{E_{1f}}{E_{1f}}, \frac{E_{1s}}{E_{1f}}, \frac{E_{3f}}{E_{1f}}, \frac{E_{3s}}{E_{1f}}, \frac{E_{5s}}{E_{1f}}$ show the strongest wave component in the entire frequency range.

4. Model specification

Figure 1 shows a schematic of the cylindrical double shell of infinite length subjected to a plane wave with an incidence angle γ. The radii and the thicknesses of the shells are $R_{i,e}$ and $h_{i,e}$ in which the subscripts i and e represent the inner and outer shells. A concentric layer of porous material is installed between the shells. The acoustic media in the outside and the inside of the shell are represented by density and speed of sound: (s_1, c_1) outside and (s_3, c_3) inside.

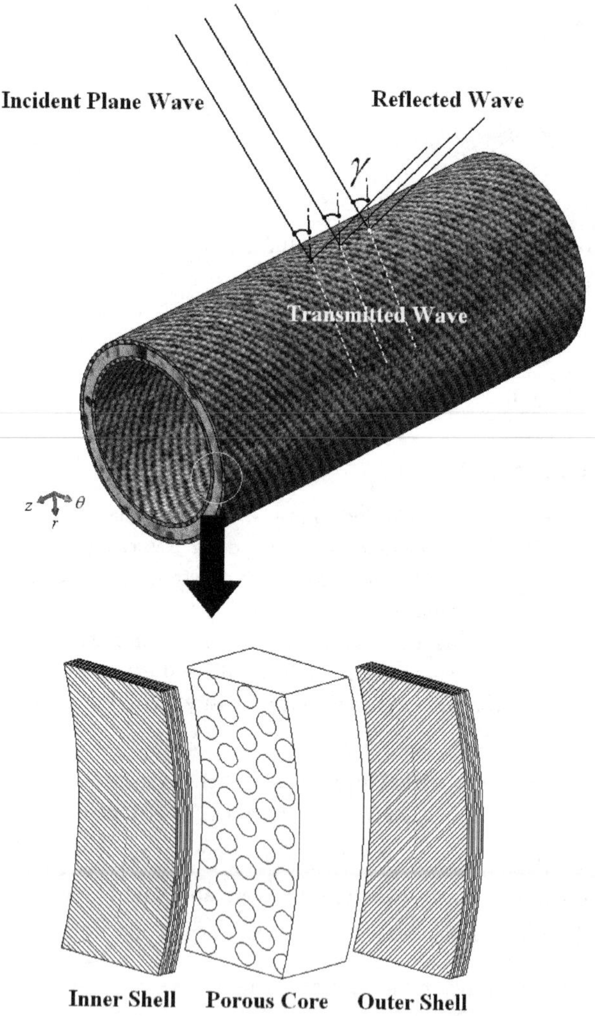

Figure 1. Schematic diagram of the double-walled cylindrical composite shell lined with porous materials

5. Applying full method to two-dimensional problem

For a two-dimensional problem as shown in the $x-y$ plane of Fig. 2, the potential of the incident wave can be expressed as [1]:

$$\Im_i = e^{-j(\xi_x x+\xi_{1y} y)} \tag{12}$$

where $\xi_x = \xi_1 \sin\gamma$, $\xi_{1y} = \xi_1 \cos\gamma$, $\xi_1 = \omega / c_1$, c_1 is the speed of sound in incident Medium, γ is the angle of incidence.

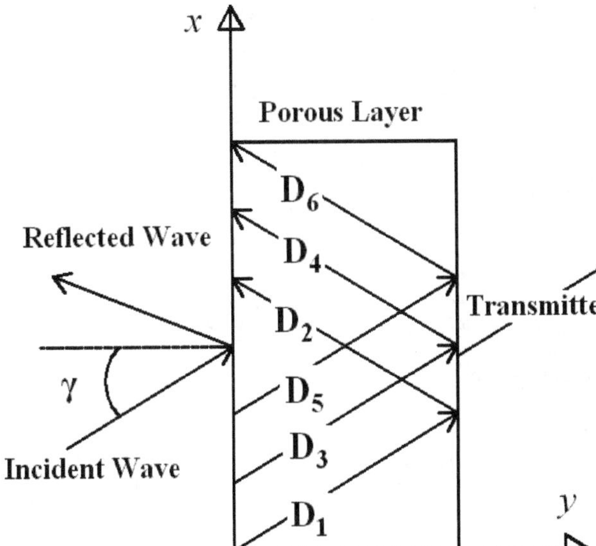

Figure 2. Illustration of wave propagation in the porous layer

Three kinds of the waves propagate in porous material, therefore six traveling waves, which have the same trace wave numbers, are induced by an oblique incident wave in a finite depth layer of porous material, as shown in Fig. 2. The $x-$ and $y-$ direction components of the displacements and stresses of the solid and fluid phases were derived by Bolton et. al. [1]. The displacements in the solid phase are:

$$\hat{u}_x = j\xi_x e^{-j\xi_x x}\left[\frac{D_1}{\xi_\alpha^2}e^{-j\xi_{\alpha y}y} + \frac{D_2}{\xi_\alpha^2}e^{j\xi_{\alpha y}y} + \frac{D_3}{\xi_\beta^2}e^{-j\xi_{\beta y}y} + \frac{D_4}{\xi_\beta^2}e^{j\xi_{\beta y}y}\right]$$

$$-j\frac{\xi_{ty}}{\xi_t^2}e^{-j\xi_x x}\left[D_5 e^{-j\xi_{ty}y} - D_6 e^{j\xi_{ty}y}\right] \tag{13}$$

$$\hat{u}_y = je^{-j\xi_x x}\left[\frac{\xi_{\alpha y}}{\xi_\alpha^2}D_1 e^{-j\xi_{\alpha y}y} - \frac{\xi_{\alpha y}}{\xi_\alpha^2}D_2 e^{j\xi_{\alpha y}y} + \frac{\xi_{\beta y}}{\xi_\beta^2}D_3 e^{-j\xi_{\beta y}y} - \frac{\xi_{\beta y}}{\xi_\beta^2}D_4 e^{j\xi_{\beta y}y}\right]$$

$$+j\frac{\xi_x}{\xi_t^2}e^{-j\xi_x x}\left[D_5 e^{-j\xi_{ty}y} + D_6 e^{j\xi_{ty}y}\right] \tag{14}$$

The displacements in the fluid phase are:

$$\hat{U}_x = j\xi_x e^{-j\xi_x x}\left[b_1\frac{D_1}{\xi_\alpha^2}e^{-j\xi_{\alpha y}y} + b_1\frac{D_2}{\xi_\alpha^2}e^{j\xi_{\alpha y}y} + b_2\frac{D_3}{\xi_\beta^2}e^{-j\xi_{\beta y}y} + b_2\frac{D_4}{\xi_\beta^2}e^{j\xi_{\beta y}y}\right]$$

$$-jg\frac{\xi_{ty}}{\xi_t^2}e^{-j\xi_x x}\left[D_5 e^{-j\xi_{ty}y} - D_6 e^{j\xi_{ty}y}\right] \tag{15}$$

$$\hat{U}_y = je^{-j\xi_x x}\left[b_1\frac{\xi_{\alpha y}}{\xi_\alpha^2}D_1 e^{-j\xi_{\alpha y}y} - b_1\frac{\xi_{\alpha y}}{\xi_\alpha^2}D_2 e^{j\xi_{\alpha y}y} + b_1\frac{\xi_{\beta y}}{\xi_\beta^2}D_3 e^{-j\xi_{\beta y}y} - b_1\frac{\xi_{\beta y}}{\xi_\beta^2}D_4 e^{j\xi_{\beta y}y}\right]$$

$$+jg\frac{\xi_x}{\xi_t^2}e^{-j\xi_x x}\left[D_5 e^{-j\xi_{ty}y} + D_6 e^{j\xi_{ty}y}\right] \tag{16}$$

The stresses in the solid phase are:

$$\hat{\sigma}_y^s = e^{-j\xi_x x}\begin{bmatrix}\left(2\delta\frac{\xi_{\alpha y}^2}{\xi_\alpha^2} + A + b_1\chi\right)D_1 e^{-j\xi_{\alpha y}y} + \left(2\delta\frac{\xi_{\alpha y}^2}{\xi_\alpha^2} + A + b_1\chi\right)D_2 e^{j\xi_{\alpha y}y} \\ +\left(2\delta\frac{\xi_{\beta y}^2}{\xi_\beta^2} + A + b_2\chi\right)D_3 e^{-j\xi_{\beta y}y} + \left(2\delta\frac{\xi_{\beta y}^2}{\xi_\beta^2} + A + b_2\chi\right)D_4 e^{j\xi_{\beta y}y} \\ +2\delta\frac{\xi_x\xi_{ty}}{\xi_t^2}\left(D_5 e^{-j\xi_{ty}y} - D_6 e^{j\xi_{ty}y}\right)\end{bmatrix} \tag{17}$$

$$\hat{\tau}_{xy} = e^{-j\xi_x x}\delta\begin{bmatrix}\frac{2\xi_x\xi_{\alpha y}}{\xi_\alpha^2}\left(D_1 e^{-j\xi_{\alpha y}y} - D_2 e^{j\xi_{\alpha y}y}\right) + \frac{2\xi_x\xi_{\beta y}}{\xi_\beta^2}\left(D_3 e^{-j\xi_{\beta y}y} - D_4 e^{j\xi_{\beta y}y}\right) \\ +\frac{\left(\xi_x^2 - \xi_{ty}^2\right)}{\xi_t^2}\left(D_5 e^{-j\xi_{ty}y} + D_6 e^{j\xi_{ty}y}\right)\end{bmatrix} \tag{18}$$

The stresses in the fluid phase are:

$$\hat{\sigma}^f = e^{-j\xi_x x}\left[(\chi + b_1\mu)D_1 e^{-j\xi_{\alpha y}y} + (\chi + b_1\mu)D_2 e^{j\xi_{\alpha y}y} + (\chi + b_2\mu)D_3 e^{-j\xi_{\beta y}y} + (\chi + b_2\mu)D_4 e^{j\xi_{\beta y}y}\right] \tag{19}$$

where $\xi_{\alpha y}=\sqrt{\xi_\alpha^2-\xi_x^2}$, $\xi_{\beta y}=\sqrt{\xi_\beta^2-\xi_x^2}$, $\xi_{ty}=\sqrt{\xi_t^2-\xi_x^2}$, $j=\sqrt{-1}$,
$b_1=\frac{(\hat{\rho}_{11}\mu-\hat{\rho}_{12}\chi)}{(\hat{\rho}_{22}\chi-\hat{\rho}_{12}\mu)}-\frac{(\phi\mu-\chi^2)}{\left[\omega^2(\hat{\rho}_{22}\chi-\hat{\rho}_{12}\mu)\right]}\xi_\alpha^2$, $b_2=\frac{(\hat{\rho}_{11}\mu-\hat{\rho}_{12}\chi)}{(\hat{\rho}_{22}\chi-\hat{\rho}_{12}\mu)}-\frac{(\phi\mu-\chi^2)}{\left[\omega^2(\hat{\rho}_{22}\chi-\hat{\rho}_{12}\mu)\right]}\xi_\beta^2$, and
$g=-\frac{\hat{\rho}_{12}}{\hat{\rho}_{22}}$.

These complex relations and six constants D_1-D_6 have to be determined by applying boundary conditions (BCs). When the elastic porous material is bonded directly to a panel, there exist six BCs. Also, the transverse displacement and in-plane displacement at the neutral axis can be followed as [1]:

$$w_t(x,t)=W_t(x)e^{j\omega t} \text{ and } w_p(x,t)=W_p(x)e^{j\omega t} \tag{20}$$

$W_t(x)$ and $W_p(x)$ are transverse and in-plane displacements. Four BCs are obtained from the interface compatibility:

$$V_y=j\omega W_t(x),\ \hat{u}_y=W_t(x),\ \hat{U}_y=W_t(x) \text{ and } \hat{u}_x=W_p(x)(-/+)\frac{h_p}{2}\frac{dW_t(x)}{dx} \tag{21}$$

where h_p is the panel thickness and V_y is the normal acoustic particle velocity. Two BCs are obtained from the equations of motion, followed as:

$$(+/-)\hat{p}_{amb}(-/+)q_p-j\xi_x\frac{h_p}{2}\hat{\tau}_{xy}=(\hat{D}\xi_x^4-\omega^2 I)W_t-\hat{B}\xi_x^3 jW_p \tag{22}$$

$$(+/-)\hat{\tau}_{yx}=(\hat{A}\xi_x^2-\omega^2 I)W_p-\hat{B}\xi_x^3 jW_t \tag{23}$$

$\hat{p}_{amb}$ is the acoustic pressures applied on the panel, q_p is the normal force per unit panel area exerted on the panel by the elastic porous material $q_p=-\hat{\sigma}_y^s-\hat{\sigma}^f$. The Inertia term, I, and the extensional, coupling and bending stiffness, $\hat{A}$, $\hat{B}$ and $\hat{D}$ are [8]:

$$I=\sum_{l=1}^{L}\rho^{(l)}(y_l-y_{l-1}) \tag{24}$$

$$\hat{A}=\sum_{l=1}^{L}\hat{Q}^{(l)}(y_l-y_{l-1}) \tag{25}$$

$$\hat{B}=\frac{1}{2}\sum_{l=1}^{L}\hat{Q}^{(l)}(y_l^2-y_{l-1}^2) \tag{26}$$

$$\hat{D} = \frac{1}{3}\sum_{l=1}^{L}\hat{Q}^{(l)}(y_l^3 - y_{l-1}^3) \tag{27}$$

where $\rho^{(l)}$ is the mass density of the lth layer of the shell per unit midsurface area and y_l is the distance from the midsurface to the surface of the lth layer having the farthest y coordinate. The material constant $\hat{Q}^{(l)}$ is defined as $\hat{Q}^{(l)} = \frac{E_1^{(l)}}{1 - v_{12}v_{21}}$ where $E_1^{(l)}$ is module of elasticity in the direction 1, and $v_{12}^{(l)}$ and $v_{21}^{(l)}$ are Poisson's ratios in the directions 1 and 2 of the lth ply, respectively. The fiber coordinates of ply is described, as 1 and 2, where direction 1 is parallel to the fibers and 2 is perpendicular to them. In BCs, the first signs are appropriate when the porous material is attached to the positive $y-$ facing surface of the panel, and the second signs when the porous material is attached to the negative $y-$ facing surface.

6. Prediction of ratios of the energy

The energy related to the waves in the fluid phase and solid phase are descript as follows [2].

The airborne wave:

$$E_{1f} = \frac{1}{2}\left[\phi.\left|(\chi + b_1\mu)\cdot b_1 \frac{\xi_{\alpha y}^2}{\xi_\alpha^2} D_1^2\right|\right] \tag{28}$$

$$E_{1s} = \frac{1}{2}\left[(1-\phi).\left|\left(2\delta\frac{\xi_{\alpha y}^2}{\xi_\alpha^2} + A + b_1\chi\right)\cdot\frac{\xi_{\alpha y}^2}{\xi_\alpha^2} D_1^2\right|\right] \tag{29}$$

And the frame wave:

$$E_{3f} = \frac{1}{2}\left[\phi.\left|(\chi + b_2\mu)\cdot b_2 \frac{\xi_{\beta y}^2}{\xi_\beta^2} D_3^2\right|\right] \tag{30}$$

$$E_{3s} = \frac{1}{2}\left[(1-\phi).\left|\left(2\delta\frac{\xi_{\beta y}^2}{\xi_\beta^2} + A + b_2\chi\right)\cdot\frac{\xi_{\beta y}^2}{\xi_\beta^2} D_3^2\right|\right] \tag{31}$$

$$E_{5s} = \frac{1}{2}\left[(1-\phi).\left|2\delta\left(\frac{\xi_{ty}^2}{\xi_t^2}\right)^2 D_5^2\right|\right] \tag{32}$$

where the subscripts f and s represents the fluid and solid phases, respectively.

7. Formulation of the problem

In the external space, the wave equation becomes [3]:

$$c_1\nabla^2(p^I+p_1^R)+\frac{\partial^2(p^I+p_1^R)}{\partial t^2}=0 \tag{33}$$

where p^I and p_1^R are the acoustic pressures of the incident and reflected waves and ∇^2 is the Laplacian operator in the cylindrical coordinate system, and c_1 is the speed of sound of outside medium. The wave equations in the fluid phase of the porous layer and internal space are the same as Eq. (33) with different variable names.

The shell motions are described by classic theory, fully considering the displacements in all three directions. Let the axial coordinate be z, the circumferential direction be θ and the normal direction to the middle surface of the shell be r. Equations of motion in the axial, circumferential and radial directions of a laminated composite thin cylindrical shell in cylindrical coordinate can be written as below [7]:

$$\frac{\partial N_\alpha}{\partial z}+\frac{1}{R_{i,e}}\frac{\partial N_{\alpha\beta}}{\partial\theta}+q_{\alpha_{i,e}}=-\hat{I}_{i,e}\frac{\partial^2 u^0_{i,e}}{\partial t^2} \tag{34}$$

$$\frac{\partial N_{\alpha\beta}}{\partial z}+\frac{1}{R_{i,e}}\frac{\partial N_\beta}{\partial\theta}+\frac{1}{R_{i,e}}\left[\frac{\partial M_{\alpha\beta}}{\partial z}+\frac{1}{R_{i,e}}\frac{\partial M_\beta}{\partial\beta}\right]+q_{\beta i,e}=-\hat{I}_{i,e}\frac{\partial^2 v^0_{i,e}}{\partial t^2} \tag{35}$$

$$\frac{\partial^2 M_\alpha}{\partial z^2}-\frac{\partial N_\beta}{R_{i,e}}+2\frac{1}{R_{i,e}}\frac{\partial^2 M_{\alpha\beta}}{\partial\theta\partial z}+\frac{1}{R_{i,e}^2}\frac{\partial^2 M_\beta}{\partial\theta^2}+q_{z_{i,e}}=-\hat{I}_{i,e}\frac{\partial^2 w^0_{i,e}}{\partial t^2} \tag{36}$$

In which the subscripts i and e represent the inner and outer shells. q_α, q_β, and q_z are external pressure components. u^0, v^0, and w^0 are the displacements of the shell at the neutral surface in the axial, circumferential, and radial directions respectively. The inertia terms are followed as:

$$\hat{I}=\hat{I}_1+\hat{I}_2\left(\frac{1}{R}\right)\text{where } \left[\hat{I}_1,\hat{I}_2\right]=\sum_{l=1}^{L}\int_{y_{l-1}}^{y_l}\rho^{(l)}\left[1,z\right]dz \tag{37}$$

R is cylindrical radius. Mid-surface strain and curvature can be expressed as:

$$\varepsilon_{0\alpha} = \frac{\partial u^0}{\partial z}, \quad \varepsilon_{0\beta} = \frac{1}{R}\left\{\frac{\partial v^0}{\partial \theta} + w^0\right\} \text{ and } \gamma_{0\alpha\beta} = \frac{\partial v^0}{\partial z} + \frac{1}{R}\frac{\partial u^0}{\partial \theta} \tag{38}$$

$$\kappa_\alpha = -\frac{\partial^2 w^0}{\partial z^2}, \quad \kappa_\beta = \frac{1}{R^2}\left\{\frac{\partial v^0}{\partial \theta} - \frac{\partial^2 w^0}{\partial \theta^2}\right\} \text{and } \kappa = \frac{1}{R}\left\{\frac{\partial v^0}{\partial z} - 2\frac{\partial^2 w^0}{\partial z \partial \theta}\right\} \tag{39}$$

The forces and moments are:

$$\begin{bmatrix} N_\alpha \\ N_\beta \\ N_{\alpha\beta} \\ M_\alpha \\ M_\beta \\ M_{\alpha\beta} \end{bmatrix} = \begin{bmatrix} A_{11} & A_{12} & A_{16} & B_{11} & B_{12} & B_{16} \\ A_{21} & A_{22} & A_{26} & B_{21} & B_{22} & B_{26} \\ A_{61} & A_{62} & A_{66} & B_{61} & B_{62} & B_{66} \\ B_{11} & B_{12} & B_{16} & D_{11} & D_{12} & D_{16} \\ B_{21} & B_{22} & B_{26} & D_{21} & D_{22} & D_{26} \\ B_{61} & B_{62} & B_{66} & D_{61} & D_{62} & D_{66} \end{bmatrix} \begin{bmatrix} \varepsilon_{0\alpha} \\ \varepsilon_{0\beta} \\ \gamma_{0\alpha\beta} \\ \kappa_\alpha \\ \kappa_\beta \\ \kappa \end{bmatrix} \tag{40}$$

where the extensional, coupling and bending stiffness, $A_{\tilde{p}\tilde{q}}$, $B_{\tilde{p}\tilde{q}}$ and $D_{\tilde{p}\tilde{q}}$ are:

$$A_{\tilde{p}\tilde{q}} = \sum_{l=1}^{L} Q_{\tilde{p}\tilde{q}}^{(l)}(y_l - y_{l-1}) \quad \tilde{p},\tilde{q} = 1,2,3 \tag{41}$$

$$B_{\tilde{p}\tilde{q}} = \frac{1}{2}\sum_{l=1}^{L} Q_{\tilde{p}\tilde{q}}^{(l)}(y_l^2 - y_{l-1}^2) \quad \tilde{p},\tilde{q} = 1,2,3 \tag{42}$$

$$D_{\tilde{p}\tilde{q}} = \frac{1}{3}\sum_{l=1}^{L} Q_{\tilde{p}\tilde{q}}^{(l)}(y_l^3 - y_{l-1}^3) \quad \tilde{p},\tilde{q} = 1,2,3 \tag{43}$$

$Q_{\tilde{p}\tilde{q}}^{(l)}$, material constant, is the function of physical properties of each ply. The displacement components of the inner and outer shell at an arbitrary distance r from the midsurface along the axial, the circumferential and the radial directions are [8]:

$$u_e^0 = \sum_{n=0}^{\infty} u_{ne}^0 \cos(n\theta)\exp[j(\omega t - \xi_{1z} z)] \tag{44}$$

$$v_e^0 = \sum_{n=0}^{\infty} v_{ne}^0 \sin(n\theta)\exp[j(\omega t - \xi_{1z} z)] \tag{45}$$

$$w_e^0 = \sum_{n=0}^{\infty} w_{ne}^0 \cos(n\theta)\exp[j(\omega t - \xi_{1z} z)] \tag{46}$$

$$u_i^0 = \sum_{n=0}^{\infty} u_{ni}^0 \cos(n\theta)\exp[j(\omega t - \xi_{3z}z)] \tag{47}$$

$$v_i^0 = \sum_{n=0}^{\infty} v_{ni}^0 \sin(n\theta)\exp[j(\omega t - \xi_{3z}z)] \tag{48}$$

$$w_i^0 = \sum_{n=0}^{\infty} w_{ni}^0 \cos(n\theta)\exp[j(\omega t - \xi_{3z}z)] \tag{49}$$

In Eqs. (44 - 46) ξ_1, wave number in the external medium, is defined as $\xi_1 = \frac{\omega}{c_1}$, $\xi_{1z} = \xi_1 \sin\gamma$, $\xi_{1r} = \xi_1 \cos\gamma$ and in Eqs. (47 - 49) ξ_3, wave number in the internal cavity, is expressed as $\xi_3 = \frac{\omega}{c_3}$, $\xi_{3z} = \xi_{1z}$, $\xi_{3r} = \sqrt{\xi_3^2 - \xi_{3z}^2}$.

The boundary conditions at the two interfaces between the shells and fluid are [2]:

$$\frac{\partial}{\partial r}(p^I + p_1^R) = -s_1 \frac{\partial^2 w_e^0}{\partial t^2} \quad @ r = R_e \tag{50}$$

$$\frac{\partial}{\partial r}(p_2^T + p_2^R) = -s_2 \frac{\partial^2 w_e^0}{\partial t^2} \quad @ r = R_e \tag{51}$$

$$\frac{\partial}{\partial r}(p_2^T + p_2^R) = -s_2 \frac{\partial^2 w_i^0}{\partial t^2} \quad @ r = R_i \tag{52}$$

$$\frac{\partial}{\partial r}(p_3^T) = -s_3 \frac{\partial^2 w_i^0}{\partial t^2} \quad @ r = R_i \tag{53}$$

s_1 and s_3 are densities of outside and inside acoustic media, respectively. s_2 is the equivalent density of the porous material and can be obtained from the Simplified method as suggested in [2]. The harmonic plane incident wave p^I can be expressed in cylindrical coordinates as [2]:

$$p^I(r,z,\theta,t) = p_0 e^{j(\omega t - \xi_{1z}z)} \sum_{n=0}^{\infty} \varepsilon_n (-j)^n J_n(\xi_{1r} r)\cos(n\theta) \tag{54}$$

where p_0 is the amplitude of the incident wave, $n = 0,1,2,3,...$ indicates the circumferential mode number, $\varepsilon_n = 1$ for $n = 0$ and 2 for $n = 1,2,3,...$, and J_n is the Bessel function of the first kind of order n.

Considering the circular cylindrical geometry, the pressures are expanded as:

$$p_1^R(r,z,\theta,t) = e^{j(\omega t - \xi_{1z} z)} \sum_{n=0}^{\infty} p_{n1}^R H_n^2(\xi_{1r} r)\cos(n\theta) \tag{55}$$

$$p_2^T(r,z,\theta,t) = e^{j(\omega t - \xi_{2z} z)} \sum_{n=0}^{\infty} p_{n2}^T H_n^1(\xi_{2r} r)\cos(n\theta) \tag{56}$$

$$p_2^R(r,z,\theta,t) = e^{j(\omega t - \xi_{2z} z)} \sum_{n=0}^{\infty} p_{n2}^R H_n^2(\xi_{2r} r)\cos(n\theta) \tag{57}$$

$$p_3^T(r,z,\theta,t) = e^{j(\omega t - \xi_{3z} z)} \sum_{n=0}^{\infty} p_{n3}^T H_n^1(\xi_{3r} r)\cos(n\theta) \tag{58}$$

where H_n^1 and H_n^2 are the Hankel functions of the first and second kind of order n, $\xi_{2z} = \xi_{3z} = \xi_{1z}$, and $\xi_{2r} = \sqrt{\xi_2^2 - \xi_z^2}$. The wave number, ξ_2, in the porous core can be obtained from the simplified method as suggested in [2].

Substitution of the expressions in the displacements of the shell (Eq. (44)-Eq.(49)) and the acoustic pressures (Eq. (54)-Eq.(58)) equations into six shell equations (Eq. (34)-Eq.(36)) and four boundary conditions (Eq. (50)-Eq.(53)) yields ten equations, which can be decoupled for each mode if the orthogonality between the trigonometric functions is utilized. These ten equations can be sorted into a form of a matrix equation as follow:

$$\left[\hbar\right]\left\{u_{ni}^0, v_{ni}^0, w_{ni}^0, u_{ne}^0, v_{ne}^0, w_{ne}^0, p_{n1}^R, p_{n2}^T, p_{n2}^R, p_{n3}^T\right\}^T = \left\{\bar{\lambda}\right\} \tag{59}$$

where $[\hbar]$ is a 10×10 matrix with components given in Appendix and $\{\bar{\lambda}\}$ is:

$$\left\{\bar{\lambda}\right\} = \left\{0,0,0,0,0,-p_0\,\varepsilon_n(-j)^n J_n(\xi_{1r} R_e)\xi_{1r}, -p_0\varepsilon_n(-j)^n \frac{dJ_n(\xi_{1r} R_e)}{dr}\xi_{1r}, 0,0,0\right\}^T \tag{60}$$

The ten unknown coefficients p_{n1}^R, p_{n2}^T, p_{n2}^R, p_{n3}^T, u_{ne}^0, v_{ne}^0, w_{ne}^0, u_{ni}^0, v_{ni}^0 and w_{ni}^0 are obtained in terms of p_0 with solving Eq. (59) for each mode n, which then can be substituted back into Eqs. (44) to (49) and (54) to (58) to find the displacements of the shell and the acoustic pressures in series forms.

8. Calculation of transmission losses (TLs)

The transmission coefficient, $\tau(\gamma)$, is the ratio of the amplitudes of the incident and transmitted waves. $\tau(\gamma)$ is obviously a function of the incidence angle γ defined by [2]:

$$\tau(\gamma) = \sum_{n=0}^{\infty} \frac{\mathrm{Re}\left\{p_{n3}^{T} H_n^1(\xi_{3r} R_i)(j\omega w_{1n}^0)^*\right\} s_1 c_1 \pi R_i}{\varepsilon_n R_e \cos(\gamma) p_0^2} \tag{61}$$

where $\varepsilon_n = 1$ for $n = 0$ and $\varepsilon_n = 2$ for $n = 1,2,3,\ldots$. $\mathrm{Re}\{\ \}$ and the superscript $*$ represents the real part and the complex conjugate of the argument.

To consider the random incidences, $\tau(\gamma)$ can be averaged according to the Paris formula [9]:

$$\overline{\tau} = 2\int_0^{\gamma_m} \tau(\gamma)\sin\gamma\cos\gamma d\gamma \tag{62}$$

where γ_m is the maximum incident angle. Integration of Eq. (62) is conducted numerically by Simpson's rule. Finally, the average TLs is obtained as:

$$TL_{avg} = 10\log\frac{1}{\overline{\tau}} \tag{63}$$

In the following, the averaged TLs of the structure is calculated in terms of the 1/3 octave band for random incidences.

9. Convergence algorithm

Eqs. (44) to (49) and (54) to (58) are obtained in series form. Therefore, enough numbers of modes should be included in the analysis to make the solution converge. Therefore, an iterative procedure is constructed in each frequency, considering the maximum iteration number. Unless the convergence condition is met, it iterates again. When the TLs calculated at two successive calculations are within a pre-set error bound, the solution is considered to have converged. An algorithm for the calculation of TLs at each frequency is followed as:

REPEAT

$TL_n = 10\log\frac{1}{\overline{\tau}}$

Set $n = n+1$

UNTIL $|TL_{n+1}| - |TL_n| < 10^{-6}$

10. Numerical results

Parametric numerical studies of transmission loss (TLs) are conducted for a double-walled composite laminated cylindrical shell lined with porous material specified as follows, considering 1/3 octave band frequency. Table 1 presents the geometrical and environmental properties of a sandwich cylindrical structure. Each layer of the laminated composite shells are made of graphite/epoxy, see Table 2. The plies were arranged in a $\left[0^{\circ},45^{\circ},90^{\circ},-45^{\circ},0^{\circ}\right]_{s}$ pattern.

The results are verified by those investigated by the authors' previous work for an especial case in which the porous material properties go into fluid phase (In other world, the porosity is close to 1). The comparison of these results shown in Fig. 3, indicates a good agreement.

The calculated transmission loss for the laminated composite shell is compared with those of other authors for a special case of isotropic materials. In other word, in this model the mechanical properties of the lamina in all directions are chosen the same as an isotropic material such as Aluminum, and then the fiber angles are approached into zero. Fig. 4 compares the TL values of the special case of laminated composite walls obtained from present model and those of aluminum walls from Lee's study [2]. The results show an excellent agreement.

We are going to verify the model in behavior comparing the results of cylindrical shell in a case where the radius of the cylindrical shell becomes large or the curvature becomes negligible with the results of the flat plate done by Bolton [1] (See Fig. 5). It should be also noted that both structures sandwich a porous layer and have the same thickness. Although it is not expected to achieve the same results as the derivation of the shell equations is quite different comparing with derivation of plate equations, however, the comparison between the two curves indicate that they behave in a same trend in the broadband frequencies.

	Ambient	Outer Shell	Porous Core	Inner Shell	Cavity
Material	Air	Composite	Porous Material	Composite	Air
Density (kg/m³)	1.21	-	-	-	0.94
Speed of Sound (m/s)	343	-	-	-	389
Radius (mm)	-	172.5	-	150	-
Thickness (mm)	-	2	20	3	-
Bulk Density of Solid Phase* (kg/m³)	-	-	30	-	-
Bulk Young's Modulus* (kPa)	-	-	800	-	-
Bulk Poisson's Ratio (-)	-	-	0.4	-	-
Flow Resistivity (MKs)	-	-	25000	-	-
Tortuosity (-)	-	-	7.8	-	-
Porosity (-)	-	-	0.9	-	-
Loss Factor (-)	-	-	0.265	-	-

* *in vacuo.*

Table 1. Geometrical and environmental properties [2]

	Graphite/epoxy	Glass/epoxy
Axial Modulus (GPa)	137.9	38.6
Circumferential Modulus (GPa)	8.96	8.2
Shear Modulus (GPa)	7.1	4.2
Density (kg/m³)	1600	1900
Major Poisson's Ratio (-)	0.3	0.26

Table 2. Orthotropic properties [3]

Figure 6 indicates that whenever the radius of the shell descends, the TL of the shell is ascending in low frequency region. It is due to the fact that the flexural rigidity of the cylinder will be increased with reduction of shells radii. In addition, decreasing the radius of the shell leads to weight reduction and then in high frequency especially in Mass-controlled region the power transmission into the structure increases.

Figure 7 shows the effect of the composite material on TL. Materials chosen for the comparison are graphite/epoxy and glass/epoxy (Table 2). The figure shows that material must be chosen properly to enhance TL at Stiffness-control zone. The results represent a desirable level of TL at Stiffness-control zone (Lower frequencies) for graphite/epoxy. It is readily seen that, in higher frequency, as a result of density of materials, the TL curves are ascending. Therefore, the TL of glass/epoxy is of the highest condition in the Mass-controlled region.

It is well anticipated that increase of porous layer thickness leads to increase of TL. As illustrated in Fig. 8, a considerable increase due to thickening the porous layer is obtained. As it is well obvious from this figure, the weight increase of about 12% ($h_c = 60$ mm) and 25% ($h_c = 100$ mm), the averaged TL values are properly increased about 35% and 60%, respectively in broadband frequency. It is a very interesting result that can encourage engineers to use these structures in industries.

Figure 9 shows a comparison between the transmission loss for a ten-layered composite shell and an aluminum shell with the same radius and thickness. Since, the composite shell is stiffer than the aluminum one, its TL is upper than that of aluminum shell in the Stiffness-controlled region. However, as a result of lower density of composite shell, it does not appear to be effective as an aluminum shell in Mass-controlled region.

The effect of stacking sequence is shown in Fig. 10. Two patterns $\left[0^\circ,90^\circ,0^\circ,90^\circ,0^\circ\right]_s$ and $\left[90^\circ,0^\circ,90^\circ,0^\circ,90^\circ\right]_s$ are defined to designate stacking sequence of plies. The arrangements of layers are so effective on TLs curve, especially on Stiffness-controlled region. However, no clear discrepancy is depicted in Mass-controlled region.

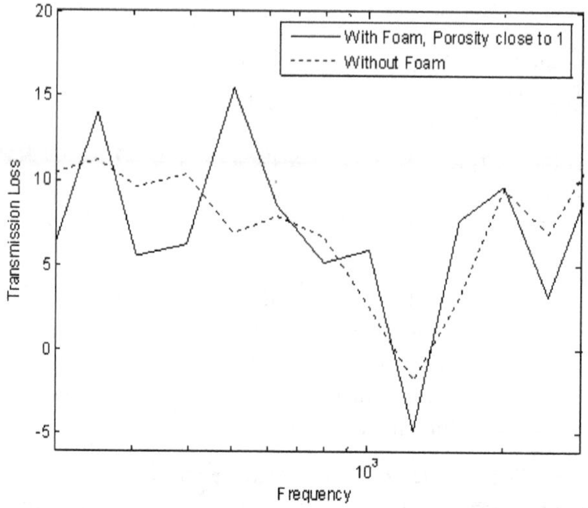

Figure 3. Comparison of cylindrical double-walled shell with and without foam where porosity close to 1

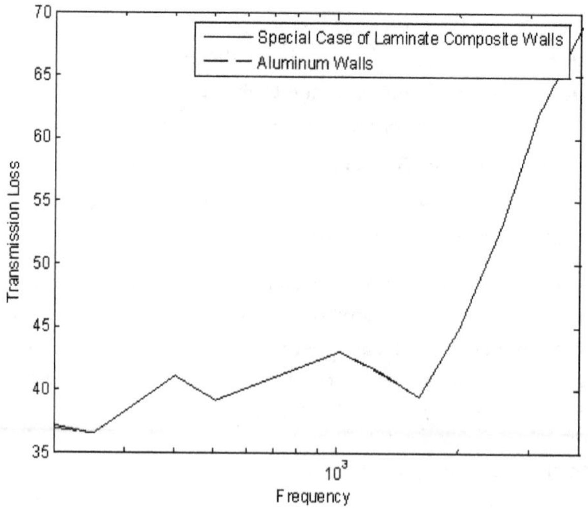

Figure 4. Comparison of cylindrical isotropic double-walled shell with special case of laminated composite double-walled shell ($R_i = 150$ mm, $R_e = 200$ mm, $h_i = 3$ mm, $h_e = 2$ mm, and $h_c = 47.5$ mm)

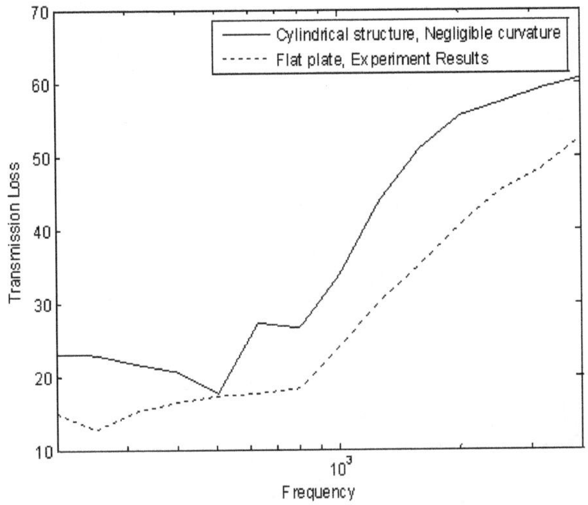

Figure 5. Comparison of an isotropic cylindrical double-walled shell with a negligible curvature and a double-panel structure

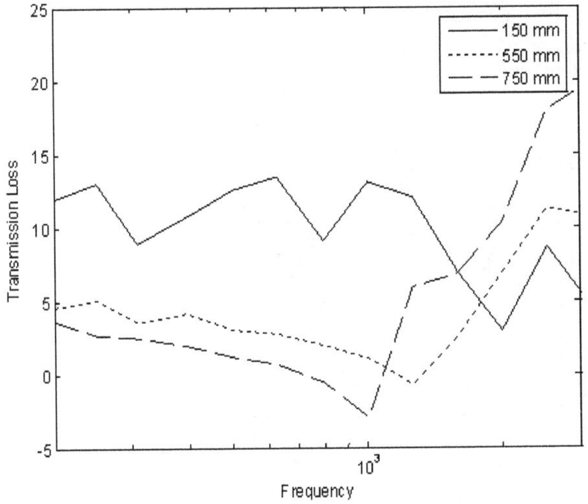

Figure 6. Comparison of a cylindrical double-walled shell lined with porous material with different inner shell's radius

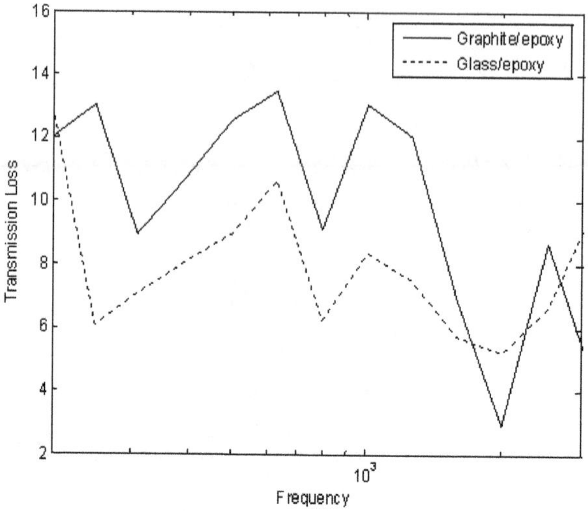

Figure 7. Comparison of a TL curves for ten-layer laminated composite shell with different material

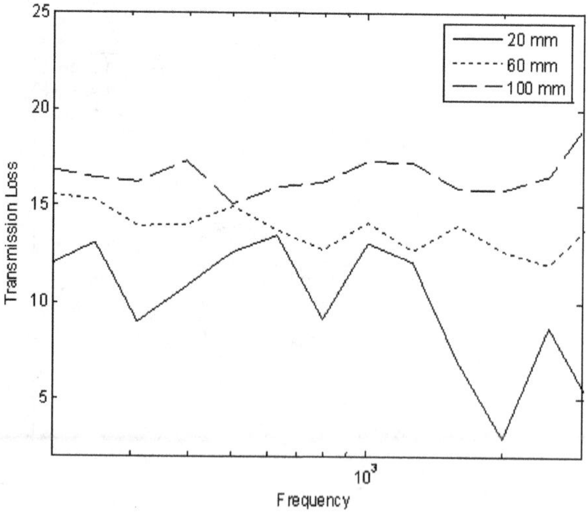

Figure 8. Comparison of a cylindrical double-walled shell lined with porous material with different core thickness

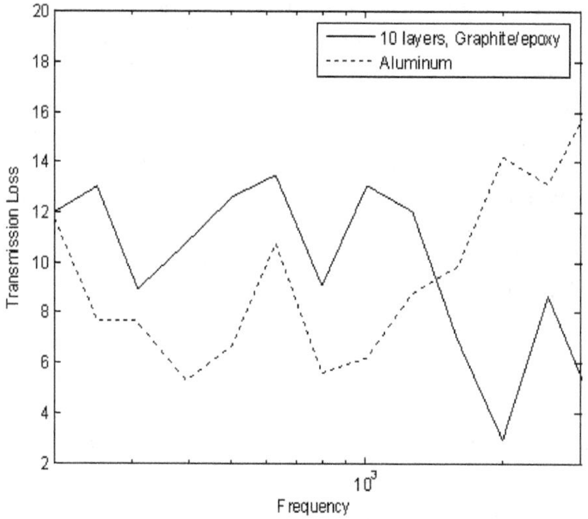

Figure 9. Comparison between Aluminum and ten-layer laminated composite shell

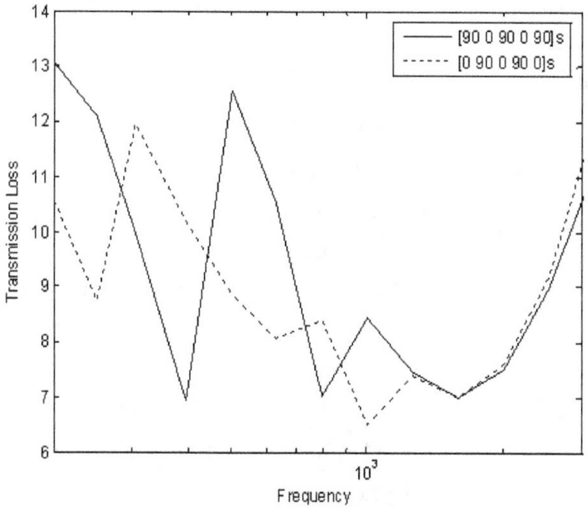

Figure 10. TL curves for the ten-layered composite shell with respect to stacking sequence

11. Conclusions

Transmission losses (TLs) of double-walled composite laminated shells sandwiching a layer of porous material were calculated. It is also considered the acoustic-structural coupling effect as well as the effect of the multi-waves in the porous layer. In order to make the problem solvable, one dominant wave was used to model the porous layer. In general the comparisons indicated the benefits of porous materials. Also, a considerable increase due to thickening the porous layer was obtained. For example, the weight increase of about 12% and 25% may respectively lead to an increase of 35% or 60% in amount of averaged TL values in broadband frequency. In addition, it was shown that increasing the axial modulus of plies made the TL be increased in low frequency range. Moreover, the comparison of double-walled shell with a gap and the one sandwiched with porous materials (where the porosity is close to 1) indicated a good agreement. Eventually the arrangement of layers in laminated composite can be so effective in Stiffness-controlled region. Therefore, optimizing the arrangement of layers can be useful in future study.

Author details

K. Daneshjou
Department of Mechanical Engineering, Iran University of Science and Technology, Tehran, Iran

H. Ramezani
Young Researchers Club, Islamic Azad University - Shahrerey branch, Tehran, Iran

R. Talebitooti
Department of Automotive Engineering, Iran University of Science and Technology, Tehran, Iran

Nomenclature

A_1, A_2	Physical and geometrical factors
$A_{\tilde{p}\tilde{q}}$, $B_{\tilde{p}\tilde{q}}$, $D_{\tilde{p}\tilde{q}}$, $\hat{A}$, $\hat{B}$, $\hat{D}$	Stiffness coefficients
$D_1 - D_6$	Wave equations constants
E	Bulk Young's modulus
$E_1^{(l)}$	Module of elasticity in the direction 1 of the l th ply
E_{1f}, E_{1s}, E_{3s}, E_{3f}, E_{5s}	The strain energy associated with the displacement in the solid and fluid phases
J_n	Bessel function of the first kind of order n
H_n^1, H_n^2	Hankel functions of the first and second kind of order n
I, $\hat{I}$, $\hat{I}_1$, $\hat{I}_2$	Inertia terms
N_{Pr}	Prandtl number
N_α, N_β, $N_{\alpha\beta}$, M_α, M_β, $M_{\alpha\beta}$	Forces and Moments
$\hat{Q}^{(l)}$, $Q_{\tilde{p}\tilde{q}}^{(l)}$	Material constant
R_i, R_e	Radii of inner and outer shell
V_y	Normal acoustic particle velocity

W^I, W^T	The incident and transmitted power flow per unit length of the shell
$W_t(x)$, $W_p(x)$	Transverse and in-plane displacements
c_1, c_3	Speed of sound in external and cavity media
c_2	Speed of sound in the fluid phase of porous materials
e_s	Solid volumetric strain
h_i, h_e	Shell wall thickness of inner and outer shell
h_p	Thickness of Panel
n	Circumferential mode number
q_α, q_β, q_z	External pressure components
$\hat{p}_{amb}$	Acoustic pressures applied on the panel
p_0	Pressure amplitude of the incident wave
$p_1^R, p_2^T, p_2^R, p_3^T$	Incident, Reflected and transmitted pressures in external, porous layer and cavity media
s_1, s_2, s_3	Density of external medium, equivalent density of the porous material and internal medium
u^0, v^0, w^0	Displacements of the shell in the axial, circumferential, and radial directions
$\bar{\mathbf{u}}$	Displacement vector of the solid
y_l	Distance from the midsurface to the surface of the lth layer having the farthest y coordinate
z, θ, r	Cylindrical coordinate
α_∞	Tortuosity
ε_n	Neumann factor
κ_v	Air viscosity
δ	Shear modulus of the porous material
ρ_0	Densities of the fluid parts of the porous material
$\rho^{(l)}$	Mass density of the lth layer of the shell per unit midsurface area
ρ_1	Bulk density of the solid phase
ρ_a	Inertial coupling term
$\hat{\rho}_{11}, \hat{\rho}_{12}, \hat{\rho}_{22}$	Equivalent masses
Λ	Viscous characteristic length
σ_r	Flow resistivity
ϕ	Porosity
∇^2	Laplacian operator in the cylindrical coordinate system

γ	Angle of incidence
γ_m	Maximum incident angle
χ, φ, μ	Material properties
ω	Angular frequency
ϖ	Rotational strain in the solid phase
ξ_1, ξ_3	Wave number in external and cavity media
ξ_α, ξ_β, ξ_t	Complex wave numbers of the two compression and one shear waves
$\Im_i$	Potential of the incident wave, 2D dimension
$\hat{u}_x$, $\hat{u}_y$, $\hat{U}_x$, $\hat{U}_y$	Displacement components in the solid and fluid phases, 2D dimension
$\hat{\sigma}_y^s$, $\hat{\tau}_{xy}$, $\hat{\sigma}^f$	Stresses in the solid and fluid phases, 2D dimension
$v_{12}^{(l)}$, $v_{21}^{(l)}$	Poisson's ratios in the directions 1 and 2 of the lth ply, 2D dimension
$\hat{\nu}$	Bulk Poisson's ratio
$\hat{\eta}$	Loss factor
ς	Ratio of specific heats
$\mathrm{Re}\{\ \}$,*	Real part and the complex conjugate

Appendix

The non-zero components of the matrix $[\hbar]$ appearing in equation (59) are followed as:

$$\hbar_{1,1} = A_{11}\left(-j\xi_z\right)^2 + 2A_{16}\frac{1}{R_i}\left(-j\xi_z\right)(-n) + A_{66}\frac{1}{R_i^2}(-n^2) - I_i\omega^2$$

$$\hbar_{1,2} = A_{12}\frac{1}{R_i}\left(-j\xi_z\right)(n) + A_{16}\left(-j\xi_z\right)^2 + B_{12}\frac{1}{R_i^2}\left(-j\xi_z\right)(n) + B_{16}\frac{1}{R_i}\left(-j\xi_z\right)^2 + A_{62}\frac{1}{R_i^2}(-n^2) + A_{66}\frac{1}{R_i}\left(-j\xi_z\right)(n) + B_{62}\frac{1}{R_i^3}(-n^2) + B_{66}\frac{1}{R_i^2}\left(-j\xi_z\right)(n)$$

$$\hbar_{1,3} = A_{12}\frac{1}{R_i}\left(-j\xi_z\right) - B_{11}\left(-j\xi_z\right)^3 - B_{12}\frac{1}{R_i^2}\left(-j\xi_z\right)(-n^2) - 3B_{16}\frac{1}{R_i}\left(-j\xi_z\right)^2(-n) + A_{62}\frac{1}{R_i^2}(-n) - B_{62}\frac{1}{R_i^3}(n^3) - 2B_{66}\frac{1}{R_i^2}(-n^2)\left(-j\xi_z\right)$$

$$\hbar_{2,1} = A_{12}\frac{1}{R_i}\left(-j\xi_z\right)(n) + A_{16}\left(-j\xi_z\right)^2 + B_{12}\frac{1}{R_i^2}\left(-j\xi_z\right)(n) + B_{16}\frac{1}{R_i}\left(-j\xi_z\right)^2 + A_{62}\frac{1}{R_i^2}(-n^2) +$$
$$A_{66}\frac{1}{R_i}\left(-j\xi_z\right)(n) + B_{62}\frac{1}{R_i^3}(-n^2) + B_{66}\frac{1}{R_i^2}\left(-j\xi_z\right)(n)$$

$$\hbar_{2,2} = A_{22}\frac{1}{R_i^2}(-n^2) + 2A_{26}\frac{1}{R_i}\left(-j\xi_z\right)(n) + 2B_{22}\frac{1}{R_i^3}(-n^2) + 4B_{26}\frac{1}{R_i^2}\left(-j\xi_z\right)(n) + A_{66}\left(-j\xi_z\right)^2 +$$
$$2B_{66}\frac{1}{R_i}\left(-j\xi_z\right)^2 + D_{22}\frac{1}{R_i^4}(-n^2) + 2D_{23}\frac{1}{R_i^3}\left(-j\xi_z\right)(n) + D_{33}\frac{1}{R_i^2}\left(-j\xi_z\right)^2 - I_i\omega^2$$

$$\hbar_{2,3} = A_{22}\frac{1}{R_i^2}(-n) - B_{12}\frac{1}{R_i}(-n)\left(-j\xi_z\right)^2 - B_{22}\frac{1}{R_i^3}(n^3) - 3B_{23}\frac{1}{R_i^2}\left(-j\xi_z\right)(-n^2) + A_{62}\frac{1}{R_i}\left(-j\xi_z\right) +$$
$$-B_{61}\left(-j\xi_z\right)^3 - 2B_{66}\frac{1}{R_i}(-n)\left(-j\xi_z\right)^2 + B_{22}\frac{1}{R_i^3}(-n) - D_{21}\frac{1}{R_i^2}\left(-j\xi_z\right)^2(-n) - D_{22}\frac{1}{R_i^4}(n^3) -$$
$$3D_{23}\frac{1}{R_i^3}\left(-j\xi_z\right)(-n^2) + B_{32}\frac{1}{R_i^2}\left(-j\xi_z\right) - D_{61}\frac{1}{R_i}\left(-j\xi_z\right)^3 - 2D_{33}\frac{1}{R_i^2}\left(-j\xi_z\right)^2(-n)$$

$$\hbar_{3,1} = A_{12}\frac{1}{R_i}\left(-j\xi_z\right) - B_{11}\left(-j\xi_z\right)^3 - B_{12}\frac{1}{R_i^2}\left(-j\xi_z\right)(-n^2) - 3B_{16}\frac{1}{R_i}\left(-j\xi_z\right)^2(-n) +$$
$$A_{62}\frac{1}{R_i^2}(-n) + -B_{62}\frac{1}{R_i^3}(n^3) - 2B_{66}\frac{1}{R_i^2}(-n^2)\left(-j\xi_z\right)$$

$$\hbar_{3,2} = A_{22}\frac{1}{R_i^2}(-n) - B_{12}\frac{1}{R_i}(-n)\left(-j\xi_z\right)^2 - B_{22}\frac{1}{R_i^3}(n^3) - 3B_{23}\frac{1}{R_i^2}\left(-j\xi_z\right)(-n^2) +$$
$$A_{62}\frac{1}{R_i}\left(-j\xi_z\right) + -B_{61}\left(-j\xi_z\right)^3 - 2B_{66}\frac{1}{R_i}(-n)\left(-j\xi_z\right)^2 + B_{22}\frac{1}{R_i^3}(-n) -$$
$$D_{21}\frac{1}{R_i^2}\left(-j\xi_z\right)^2(-n) - D_{22}\frac{1}{R_i^4}(n^3) - 3D_{23}\frac{1}{R_i^3}\left(-j\xi_z\right)(-n^2) +$$
$$B_{32}\frac{1}{R_i^2}\left(-j\xi_z\right) - D_{61}\frac{1}{R_i}\left(-j\xi_z\right)^3 - 2D_{33}\frac{1}{R_i^2}\left(-j\xi_z\right)^2(-n)$$

$$\hbar_{3,3} = A_{22}\frac{1}{R_i^2} - 2B_{21}\frac{1}{R_i}\left(-j\xi_z\right)^2 - 2B_{22}\frac{1}{R_i^3}(-n^2) - 4B_{23}\frac{1}{R_i^2}\left(-j\xi_z\right)(-n) + -D_{11}\left(-j\xi_z\right)^4 -$$
$$2D_{12}\frac{1}{R_i^2}(-n^2)\left(-j\xi_z\right)^2 + 4D_{16}\frac{1}{R_i}\left(-j\xi_z\right)^3(-n) + 4D_{62}\frac{1}{R_i^3}\left(-j\xi_z\right)(n^3) +$$
$$4D_{33}\frac{1}{R_i^2}\left(-j\xi_z\right)^2(-n^2) + D_{22}\frac{1}{R_i^4}\left(n^4\right) - I_i\omega^2$$

$$\hbar_{3,8} = H_n^1(\xi_{2r}R_i), \;, \; \hbar_{3,9} = H_n^2(\xi_{2r}R_i), \;\; \hbar_{3,10} = H_n^1(\xi_{3r}R_i)$$

$$\hbar_{4,4} = A_{11}\left(-j\xi_z\right)^2 + 2A_{16}\frac{1}{R_e}\left(-j\xi_z\right)(-n) + A_{66}\frac{1}{R_e^2}(-n^2) - I_e\omega^2$$

$$\hbar_{4,5} = A_{12}\frac{1}{R_e}\left(-j\xi_z\right)(n) + A_{16}\left(-j\xi_z\right)^2 + B_{12}\frac{1}{R_e^2}\left(-j\xi_z\right)(n) + B_{16}\frac{1}{R_e}\left(-j\xi_z\right)^2 + A_{62}\frac{1}{R_e^2}(-n^2) +$$
$$A_{66}\frac{1}{R_e}\left(-j\xi_z\right)(n) + B_{62}\frac{1}{R_e^3}(-n^2) + B_{66}\frac{1}{R_e^2}\left(-j\xi_z\right)(n)$$

$$\hbar_{4,6} = A_{12}\frac{1}{R_e}\left(-j\xi_z\right) - B_{11}\left(-j\xi_z\right)^3 - B_{12}\frac{1}{R_e^2}\left(-j\xi_z\right)(-n^2) - 3B_{16}\frac{1}{R_e}\left(-j\xi_z\right)^2(-n) +$$
$$A_{62}\frac{1}{R_e^2}(-n) + -B_{62}\frac{1}{R_e^3}(n^3) - 2B_{66}\frac{1}{R_e^2}(-n^2)\left(-j\xi_z\right)$$

$$\hbar_{5,4} = A_{12}\frac{1}{R_e}\left(-j\xi_z\right)(n) + A_{16}\left(-j\xi_z\right)^2 + B_{12}\frac{1}{R_e^2}\left(-j\xi_z\right)(n) + B_{16}\frac{1}{R_e}\left(-j\xi_z\right)^2 + A_{62}\frac{1}{R_e^2}(-n^2) +$$
$$A_{66}\frac{1}{R_e}\left(-j\xi_z\right)(n) + B_{62}\frac{1}{R_e^3}(-n^2) + B_{66}\frac{1}{R_e^2}\left(-j\xi_z\right)(n)$$

$$\hbar_{5,5} = A_{22}\frac{1}{R_e^2}(-n^2) + 2A_{26}\frac{1}{R_e}\left(-j\xi_z\right)(n) + 2B_{22}\frac{1}{R_e^3}(-n^2) + 4B_{26}\frac{1}{R_e^2}\left(-j\xi_z\right)(n) + A_{66}\left(-j\xi_z\right)^2 +$$
$$2B_{66}\frac{1}{R_e}\left(-j\xi_z\right)^2 + D_{22}\frac{1}{R_e^4}(-n^2) + 2D_{23}\frac{1}{R_e^3}\left(-j\xi_z\right)(n) + D_{33}\frac{1}{R_e^2}\left(-j\xi_z\right)^2 - I_e\omega^2$$

$$\hbar_{5,6} = A_{22}\frac{1}{R_e^2}(-n) - B_{12}\frac{1}{R_e}(-n)\left(-j\xi_z\right)^2 - B_{22}\frac{1}{R_e^3}(n^3) - 3B_{23}\frac{1}{R_e^2}\left(-j\xi_z\right)(-n^2) +$$
$$A_{62}\frac{1}{R_e}\left(-j\xi_z\right) + -B_{61}\left(-j\xi_z\right)^3 - 2B_{66}\frac{1}{R_e}(-n)\left(-j\xi_z\right)^2 + B_{22}\frac{1}{R_e^3}(-n) -$$
$$D_{21}\frac{1}{R_e^2}\left(-j\xi_z\right)^2(-n) - D_{22}\frac{1}{R_e^4}(n^3) - 3D_{23}\frac{1}{R_e^3}\left(-j\xi_z\right)(-n^2) + B_{32}\frac{1}{R_e^2}\left(-j\xi_z\right) -$$
$$D_{61}\frac{1}{R_e}\left(-j\xi_z\right)^3 - 2D_{33}\frac{1}{R_e^2}\left(-j\xi_z\right)^2(-n)$$

$$\hbar_{6,4} = A_{12}\frac{1}{R_e}\left(-j\xi_z\right) - B_{11}\left(-j\xi_z\right)^3 - B_{12}\frac{1}{R_e^2}\left(-j\xi_z\right)(-n^2) - 3B_{16}\frac{1}{R_e}\left(-j\xi_z\right)^2(-n) +$$
$$A_{62}\frac{1}{R_e^2}(-n) + -B_{62}\frac{1}{R_e^3}(n^3) - 2B_{66}\frac{1}{R_e^2}(-n^2)\left(-j\xi_z\right)$$

$$\hbar_{6,5} = A_{22}\frac{1}{R_e^2}(-n) - B_{12}\frac{1}{R_e}(-n)\left(-j\xi_z\right)^2 - B_{22}\frac{1}{R_e^3}(n^3) - 3B_{23}\frac{1}{R_e^2}\left(-j\xi_z\right)(-n^2) +$$
$$A_{62}\frac{1}{R_e}\left(-j\xi_z\right) + -B_{61}\left(-j\xi_z\right)^3 - 2B_{66}\frac{1}{R_e}(-n)\left(-j\xi_z\right)^2 + B_{22}\frac{1}{R_e^3}(-n) -$$
$$D_{21}\frac{1}{R_e^2}\left(-j\xi_z\right)^2(-n) - D_{22}\frac{1}{R_e^4}(n^3) - 3D_{23}\frac{1}{R_e^3}\left(-j\xi_z\right)(-n^2) +$$
$$B_{32}\frac{1}{R_e^2}\left(-j\xi_z\right) - D_{61}\frac{1}{R_e}\left(-j\xi_z\right)^3 - 2D_{33}\frac{1}{R_e^2}\left(-j\xi_z\right)^2(-n)$$

$$\hbar_{6,6} = A_{22}\frac{1}{R_e^2} - 2B_{21}\frac{1}{R_e}\left(-j\xi_z\right)^2 - 2B_{22}\frac{1}{R_e^3}(-n^2) - 4B_{23}\frac{1}{R_e^2}\left(-j\xi_z\right)(-n) +$$
$$-D_{11}\left(-j\xi_z\right)^4 - 2D_{12}\frac{1}{R_e^2}(-n^2)\left(-j\xi_z\right)^2 + 4D_{16}\frac{1}{R_e}\left(-j\xi_z\right)^3(-n) +$$
$$4D_{62}\frac{1}{R_e^3}\left(-j\xi_z\right)(n^3) + 4D_{33}\frac{1}{R_e^2}\left(-j\xi_z\right)^2(-n^2) + D_{22}\frac{1}{R_e^4}\left(n^4\right) - I_e\omega^2$$

$$\hbar_{6,7} = H_n^2(\xi_{1r}R_e),\ \hbar_{6,8} = H_n^1(\xi_{2r}R_e),\ \hbar_{6,9} = -H_n^2(\xi_{2r}R_e),\ \hbar_{7,6} = -s_1\,\omega^2$$

$$\hbar_{7,7} = H_n^{2'}(\xi_{1r}R_e)\xi_{1r},\ \hbar_{8,6} = -s_2\,\omega^2,\ \hbar_{8,8} = H_n^{1'}(\xi_{2r}R_e)\xi_{2r},\ \hbar_{8,9} = p_{n2}^R H_n^{2'}(\xi_{2r}R_e)\xi_{2r}$$

$$\hbar_{9,3} = -s_2\,\omega^2,\ \hbar_{9,8} = H_n^{1'}(\xi_{2r}R_i)\xi_{2r},\ \hbar_{9,9} = H_n^{2'}(\xi_{2r}R_i)\xi_{2r},\ \hbar_{10,3} = -s_3\,\omega^2,$$
$$\hbar_{10,10} = H_n^{1'}(\xi_{3r}R_i)\xi_{3r}$$

Here:

$$(\ )' = \frac{d}{dr},\ \xi_z = \xi_{1z} = \xi_{2z} = \xi_{3z},\ \xi_{2r} = \sqrt{\xi_2^2 - \xi_z^2}..$$

12. References

[1] Bolton J S, Shiau N M, Kang Y J (1996) Sound Transmission through Multi-Panel Structures Lined with Elastic Porous Materials. Journal of Sound and Vibration. 191: 317-347.

[2] Lee J H, Kim J (2001) Simplified method to solve sound transmission through structures lined with elastic porous material. J. of the Acoust. Soc of Am. 110(5): 2282-2294.

[3] Daneshjou K, Nouri A, Talebitooti R (2006) Sound transmission through laminated composite cylindrical shells using analytical model. Arch. Appl. Mech. 77: 363-379.

[4] Daneshjou K, Talebitooti R, Nouri A (2007) Analytical Model of Sound Transmission through Orthotropic Double Walled Cylindrical Shells. CSME Transaction. 32(1).

[5] Daneshjou K, Talebitooti R, Nouri A (2009) Investigation model of sound transmission through orthotropic cylindrical shells with subsonic external flow. Aerospace Science and Technology. 13: 18-26.
[6] Daneshjou K, Ramezani H, Talebitooti R (2011) Wave Transmission through Laminated Composite Double-Walled Cylindrical Shell Lined with Porous Materials. Applied Mathematics and Mechanics - English Edition, 32(6): 1-16.
[7] Qatu M S (2004)Vibration of Laminated Shells and Plates. Amsterdam: Elsevier Academic.
[8] Leissa A (1973) Vibration of Shells. Scientific and technical Information center, NASA, Washington D.C.
[9] Pierce A D (1981) Acoustics. New York: McGraw-Hill.

Chapter 4

Dispersion Relations and Modal Patterns of Wave in a Cylindrical Shell

Yu-Cheng Liu, Yun-Fan Hwang and Jin-Huang Huang

Additional information is available at the end of the chapter

1. Introduction

The dispersion relation of a circular cylindrical shell had been a subject of great interest for several decades [1-3]. In some earlier works, numerical procedures were used exclusively in the computation of dispersion relations. Recently, Karczub [4] obtained an analytical expression for the dispersion relations from Flügge shell theory by using a symbolic algebra package, Mathematica. Karczub's main concern was to check the agreement between his analytical results and results obtained previously from numerical methods, and limited to the harmonic orders $n \geqq 1$. The agreement was excellent. The axisymmetric waves (n=0) are important in the transmission of longitudinal waves, and are particularly important in acoustics due to their high radiation efficiency. In order to have complete analytical solutions for the shell dispersion relations, the solutions for the n=0 case are included and discussed in this paper.

The solutions of the modal patterns for each of the propagating and non-propagating modes are also important and are not discussed in Karczub's paper. These solutions are crucial to determine the vibration of a finite shell under various admissible boundary conditions and arbitrary external forces. The dispersion relations and the associated eigenvectors are also the means by which to construct transfer matrices for vibroacoustic transmission in cylindrical shell structures or pipe-hose systems [5]. The traditional and standard method of finding eigenvectors was given in detail by Leissa [6]; the eigenvector was normalized in such a way that its radial component of the displacement was unity, while the longitudinal and circumferential components were expressed as displacement ratios relative to the radial displacement. In some cases, these ratios could become exceedingly large and the eigenvector thus could deviate significantly from the eigenvector commonly used in mathematical physics.

In order to improve this problem for finding the normalized eigenvectors with norms equal to unity, an alternative method by using the built-in eigenvalue problem in any numerical package is proposed. A continuous variation of mode patterns for all kind of wave types in the desired frequency range will be rapidly obtained, and the physical meanings of wave propagation can also be shown more clearly and straightforward [8].

2. Methods to determine the propagating mode wavenumbers

2.1. Expression of equation of motion of cylindrical shell

The coordinate system which $\{u,v,w\}$ are presented as the axial, circumferential, and radial direction, respectively, is shown in Fig. 1.

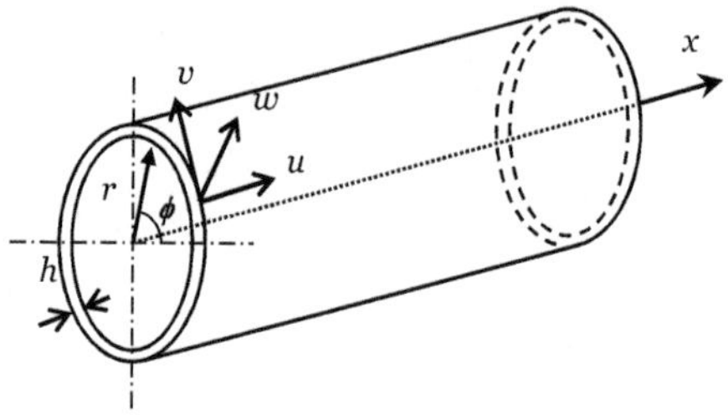

Figure 1. Coordinates and displacement orientation.

The equations of motion for the free vibration of a cylindrical shell can be written and shown as below

$$\left[\begin{bmatrix}\tilde{L}_{11} & \tilde{L}_{12} & \tilde{L}_{13}\\ \tilde{L}_{21} & \tilde{L}_{22} & \tilde{L}_{23}\\ \tilde{L}_{31} & \tilde{L}_{32} & \tilde{L}_{33}\end{bmatrix}+\frac{\rho(1-\nu^2)}{E}\frac{\partial^2}{\partial t^2}\begin{bmatrix}1&0&0\\0&1&0\\0&0&1\end{bmatrix}\right]\begin{Bmatrix}u(x,\phi,t)\\ v(x,\phi,t)\\ w(x,\phi,t)\end{Bmatrix}=\begin{Bmatrix}0\\0\\0\end{Bmatrix} \quad (1)$$

where $\tilde{L}_{ij}$, (i,j)=1~3, are the spatial operators by using the Flügge theory. These parameters can be listed as below

$$\begin{aligned}
&\tilde{L}_{11}=-\frac{\partial^2}{\partial x^2}-\frac{(1+\beta^2)(1-\nu)}{2a^2}\frac{\partial^2}{\partial \phi^2}\ ,\ \tilde{L}_{12}=-\frac{1+\nu}{2a}\frac{\partial^2}{\partial x\partial \phi}\\
&\tilde{L}_{13}=-\frac{\nu}{a}\frac{\partial}{\partial x}+\beta^2 a\frac{\partial^3}{\partial x^3}-\beta^2\frac{1-\nu}{2a}\frac{\partial^3}{\partial x\partial \phi^2}\\
&\tilde{L}_{21}=\tilde{L}_{12}\ ,\ \tilde{L}_{22}=-\frac{(1-\nu)(1+3\beta^2)}{2}\frac{\partial^2}{\partial x^2}-\frac{1}{a^2}\frac{\partial^2}{\partial \phi^2}\ ,\\
&\tilde{L}_{23}=-\frac{1}{a^2}\frac{\partial}{\partial \phi}+\beta^2\frac{(3-\nu)}{2}\frac{\partial^3}{\partial x^2\partial \phi}\\
&\tilde{L}_{31}=-\tilde{L}_{13}\ ,\ \tilde{L}_{32}=-\tilde{L}_{23}\ ,\\
&\tilde{L}_{33}=\frac{1}{a^2}+\beta^2\left(a^2\frac{\partial^4}{\partial x^4}+2\frac{\partial^4}{\partial x^2\partial \phi^2}+\frac{1}{a^2}\frac{\partial^4}{\partial \phi^4}\right)+\left(\frac{\beta}{a}\right)^2\left(1+2\frac{\partial^2}{\partial \phi^2}\right)
\end{aligned} \quad (2)$$

where a is the mean shell radius, ν is Poisson ratio, E is Young's modulus of elasticity, ρ is the mass density, h is the shell thickness, and $\beta=h/(a(12)^{0.5})$ is a non-dimensional thickness parameter which is proportional to the ratio between the thickness and radius.

In order to transfer the expression from time and displacement domain into frequency and wavenumber domain, the following spectral representation can be further to use [7]

$$f(x,\phi,t)=\sum_{n=-\infty}^{\infty} e^{in\phi}\int_{-\infty}^{\infty}\int_{-\infty}^{\infty} F(\alpha,n,\omega)e^{i(\alpha x-\omega t)}d\alpha d\omega \tag{3}$$

The dimensionless spectral listed in Eq.(1) can becomes as following

$$\left\{\begin{bmatrix} L_{11}(s,n) & L_{12}(s,n) & L_{13}(s,n) \\ L_{12}(s,n) & L_{22}(s,n) & L_{23}(s,n) \\ -L_{13}(s,n) & -L_{23}(s,n) & L_{33}(s,n) \end{bmatrix} - \Omega^2 \begin{bmatrix} 1 & 0 & 0 \\ 0 & 1 & 0 \\ 0 & 0 & 1 \end{bmatrix}\right\} \cdot \begin{Bmatrix} U(s,n,\Omega) \\ V(s,n,\Omega) \\ W(s,n,\Omega) \end{Bmatrix} \cdot e^{i(\alpha x-\omega t)} \cdot e^{in\phi} = 0 \tag{4}$$

where $\Omega=(\omega a/c_L)$ is the dimensionless frequency, $c_L=[E/(\varrho(1-\nu^2))]^{0.5}$ is the longitudinal wave speed, $s=\alpha a$ is the dimensionless wavenumber, and n is the circumferential wavenumber, respectively. L_{ij}, (i,j)=1~3, are the spectral spatial operators and listed as follow

$$L_{11}=s^2+[(1-\nu)/2](1+\beta^2)n^2 \quad , \quad L_{12}=n[(1+\nu)/2]s$$
$$L_{13}=-i\left\{\left(\nu-[(1-\nu)/2]\beta^2n^2\right)s+\beta^2s^3\right\} \quad , \quad L_{22}=[(1-\nu)/2](1+3\beta^2)s^2+n^2$$
$$L_{23}=-i\left\{n+n[(3-\nu)/2]\beta^2s^2\right\} \quad , \quad L_{33}=[1+\beta^2(1-2n^2)]+\beta^2\left(s^4+2n^2s^2+n^4\right)$$

2.2. Analytical expressions for dispersion relations

The relationship between s and Ω is so called dispersion relations of cylindrical shell. When a value of dimensionless angular frequency, Ω, is given, non-trivial solutions for the matrix form in the Eq.(4) exist at certain values of s. Finally, the specified roots can be integrated to the polynomial of s^2. All the calculating processes and expressions can be rapidly built up by using any kind of symbolic algebra package, such as Matlab or Mathematica. According the physical characteristics of acoustic propagation, two conditions can further be separated and discussed as following.

2.2.1. *n=0*

When n=0, two independent sets of equations and corresponding characteristic equation can be calculated and shown as below:

1. For axial and radial displacements:

$$\left\{\begin{bmatrix} L_{11}(s,n) & L_{13}(s,n) \\ -L_{13}(s,n) & L_{33}(s,n) \end{bmatrix} - \Omega^2 \begin{bmatrix} 1 & 0 \\ 0 & 1 \end{bmatrix}\right\} \cdot \begin{bmatrix} U(s,0,\Omega) \\ W(s,0,\Omega) \end{bmatrix} = 0 \tag{5}$$

For the non-trivial solution for {$U(s,0,\Omega)$, $W(s,0,\ \Omega)$}, the determinant of matrix in Eq. (5) should be zero ad obtained as below:

$$g_6(s^2)^3 + g_4(s^2)^2 + g_2(s^2) + g_0 = 0 \tag{6}$$

where the coefficients, g_i, are the function of n and Ω, and shown as

$$g_6 = \beta^2\left(1+\beta^2\right) \ , \ g_4 = \beta^2\left(2\nu - \Omega^2\right),$$
$$g_2 = 1+\beta^2+\nu^2-\Omega^2 \ , \ g_0 = \Omega^2\left(\Omega^2-\beta^2-1\right)$$

The solutions of Eq.(5) represent a combination of longitudinally and radially expanding and contracting motions; these are the so-called breathing-modes. The expressions of polynomial s^2 is obtained in Eq.(6).

Six characteristic roots are given

$$\begin{aligned}&\text{Branch pair 1}: s = \pm\sqrt{\Gamma_1 + 2\Gamma_2}\\&\text{Branch pair 2}: s = \pm\sqrt{\Gamma_1 - \Gamma_2 + \left(\sqrt{3}\Gamma_3\right)i}\\&\text{Branch pair 3}: s = \pm\sqrt{\Gamma_1 - \Gamma_2 - \left(\sqrt{3}\Gamma_3\right)i}\end{aligned} \tag{7}$$

where

$$\Gamma_1 = -g_4/\left(3g_6\right) \ , \ \Gamma_2 = \Lambda_2 - \Lambda_3 \ , \ \Gamma_3 = \Lambda_2 + \Lambda_3$$
$$\Lambda_1 = \left(\lambda_2 + \sqrt{4\lambda_1^3 + \lambda_2^2}\right)^{1/3} \ , \ \Lambda_2 = \Lambda_1/\left(6g_6 2^{(1/3)}\right) \ , \ \Lambda_3 = \lambda_1/\left(3g_6\Lambda_1 2^{(2/3)}\right)$$
$$\lambda_1 = -g_4^2 + 3g_2g_6 \ , \ \lambda_2 = -2g_4^3 + 9g_2g_4g_6 - 27g_0g_6^2$$

2. For the circumferential displacement:

The 2nd equation in Eq. (4) can be separated and shown as:

$$\left\{L_{22}(s,n) - \Omega^2\right\}\cdot\left[V\left(s,0,\Omega\right)\right] = 0 \tag{8}$$

In the same way, under the non-trivial solution for $V(s,0,\Omega)$, the new relation can be rewritten as

$$\left[(1-\nu)(1+3\beta^2)/2\right]s^2 - \Omega^2 = 0 \tag{9}$$

Obviously, the solution of Eq.(8) represents a uniformly circumferential motion, which is therefore a torsional-mode. The corresponding characteristic roots may be considered the 4th branch pair and show as following

$$\text{Branch pair 4}: s = \pm\Omega/\sqrt{(1-\nu)(1+3\beta^2)/2} \tag{10}$$

This torsional waves are known for being non-dispersive, since the wave speed is

$$c_T = \omega / \alpha = c_L \Omega / s = c_S \sqrt{1 + 3\beta^2}$$

where c_S is the shear wave speed. When the thickness is approximated to zero, torsional wave and shear wave speed will almost be the same. Therefore, in Flügge's theory, the shear wave speed depends slightly on the thickness and radius ratio.

2.2.2. $n \geqq 1$

Under the non-trivial solution of Eq. (4), a 4th polynomial of s^2 can be obtained as

$$g_8(s^2)^4 + g_6(s^2)^3 + g_4(s^2)^2 + g_2(s^2) + g_0 = 0 \tag{11}$$

These roots yield eight branches of dispersion curves, which are grouped into four branch pairs:

$$\begin{aligned} &\text{Branch pair 1}: s = \pm\left(A_1 + A_2 + A_4\right)^{1/2} \\ &\text{Branch pair 2}: s = \pm\left(A_1 + A_2 - A_4\right)^{1/2} \\ &\text{Branch pair 3}: s = \pm\left(-A_1 + A_2 + A_4\right)^{1/2} \\ &\text{Branch pair 4}: s = \pm\left(-A_1 + A_2 - A_4\right)^{1/2} \end{aligned} \tag{12}$$

where g_i and A_i are the coefficients and listed in the Appendix. Since A_i's are functions of n and Ω, these four branch pairs can be analytically determined once the given values of n and Ω are specified.

2.3. Methods to determine the propagating mode pattern

As mentioned earlier, the solutions regarding the modal patterns (eigenvectors) for all four root branch pairs are important. The conventional method has been frequently used for a long time. In this method, the mode shapes are expressed as the ratios among U, V and W. By returning to Eq.(4) after each of the roots ($\pm s_i$, i=1 to 4) have been found for a given Ω, these ratios can be found by solving any two of the three simultaneous equations. Finally, the third can be discarded. Detail processes can be obtained from Leissa. Because the value of the ratio U/W or V/W can become quite large when either U or V is the predominant component of the mode, this method is not the most convenient and natural way to express the eigenvectors, particularly when eigenfunction expansion is used to solve vibration problems.

A new alternative method, which solves for the eigenvectors by using a commercially available linear algebra package, will be introduced. This method is extremely simple and straightforward. The kernel of processing is by using the built-in formulation of eigenvalue and eigenvector problem in the numerical package.

According to the eigenvalue problem, the eigenvectors can be found by solving the following homogeneous equations, and the simple flow can be introduce by following:

Step 1. For a specific frequency, dimensionless frequency, Ω, can be obtained and rewritten as Ω_s. In this way, eight dimensionless wavenumber, s^2, can be also found easily from Eqs. (7), (10), or (12). Then the homogeneous equation by substituting any one of s^2 into Eq. (4) can be obtained as below:

$$\begin{bmatrix} \overset{*}{L}_{11}(s,n)-\Omega_s^2 & \overset{*}{L}_{12}(s,n) & \overset{*}{L}_{13}(s,n) \\ \overset{*}{L}_{12}(s,n) & \overset{*}{L}_{22}(s,n)-\Omega_s^2 & \overset{*}{L}_{23}(s,n) \\ \overset{*}{L}_{13}(s,n) & \overset{*}{L}_{23}(s,n) & \overset{*}{L}_{33}(s,n)-\Omega_s^2 \end{bmatrix} \begin{Bmatrix} U(s,n,\Omega) \\ V(s,n,\Omega) \\ W(s,n,\Omega) \end{Bmatrix} = \begin{Bmatrix} 0 \\ 0 \\ 0 \end{Bmatrix} \tag{13}$$

where the asterisk in superscript of L_{ij} represents that all the coefficients are known.

Step 2. After slightly rearranging, Eq.(13) can be further modified as

$$\begin{bmatrix} \overset{*}{L}_{11}(s,n) & \overset{*}{L}_{12}(s,n) & \overset{*}{L}_{13}(s,n) \\ \overset{*}{L}_{12}(s,n) & \overset{*}{L}_{22}(s,n) & \overset{*}{L}_{23}(s,n) \\ \overset{*}{L}_{13}(s,n) & \overset{*}{L}_{23}(s,n) & \overset{*}{L}_{33}(s,n) \end{bmatrix} \begin{Bmatrix} U(s,n,\Omega) \\ V(s,n,\Omega) \\ W(s,n,\Omega) \end{Bmatrix} = \Omega_s{}^2 \begin{Bmatrix} U(s,n,\Omega) \\ V(s,n,\Omega) \\ W(s,n,\Omega) \end{Bmatrix} \tag{14}$$

Step 3. If Ω_s^2 is assumed to be unknown for the time being, Eq.(14) can be rewritten to be a typical linear eigenvalue problem and shown as:

$$\begin{bmatrix} \overset{*}{L}_{11}(s,n) & \overset{*}{L}_{12}(s,n) & \overset{*}{L}_{13}(s,n) \\ \overset{*}{L}_{12}(s,n) & \overset{*}{L}_{22}(s,n) & \overset{*}{L}_{23}(s,n) \\ \overset{*}{L}_{13}(s,n) & \overset{*}{L}_{23}(s,n) & \overset{*}{L}_{33}(s,n) \end{bmatrix} \begin{Bmatrix} U(s,n,\Omega) \\ V(s,n,\Omega) \\ W(s,n,\Omega) \end{Bmatrix} = 0 \tag{15}$$

In most cases in Eq. (15), for any given value of s (the analytical root), there are three distinct eigenvalues in Ω^2 by using the built-in eigenvector function in the numerical package, but only one of them will be equal to Ω_s^2. Consequently, the eigenvector solution of Eq.(13) is one of the three eigenvectors of Eq.(15), which corresponding eigenvalue is equal to Ω_s^2. This is because, among the three eigenvalues, only the one that is equal to Ω_s will yield the specific s given here, the other two will yield different values of s and are not what we want. When s is specified, the corresponding eigenvector can further calculated by using the numerical solution package, such as MatLab.

2.4. Comparsion between alternative and conventional method for modal pattern

The dispersion relations were calculated using the thickness-radius factor β=0.0058 and the Poisson ratio υ=0.3. For n=2, the dispersion relation and corresponding modal patterns of Branch pairs 1 are obtained in Fig. 2.(a) and 2(b). The consistency of modal patterns between the alternative and conventional method can also be demonstrated in Fig. 2.(c). The physical characteristics and variations of cylindrical shell at specific circumferential number can be obtained clearly and straightforward.

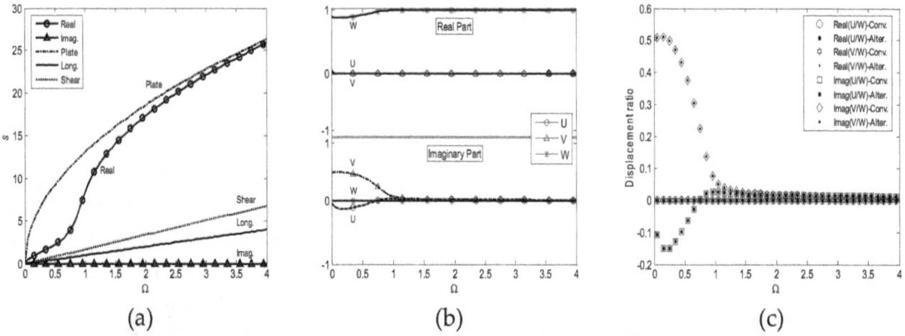

Figure 2. (a) The dispersion curve for Branch 1 roots; (b) The eigenvectors obtained from the current method expressed as [*U V W*]′; (c) Comparison between the eigenvectors (of Branch 1 roots) obtained from the conventional and the current methods (*n*=2).

3. Result and discussion

In order to discuss the characteristics of wave propagation in dispersion relations and modal pattern, the numerical results are calculated by assuming the dimensionless thickness parameter, β, to be equal to 0.0058, and the Poisson ratio equals to 0.3. According to Karczub and Leissa, for the usual range of parameters and $n \geqq 1$, the roots of Eq.(11) were found to have the four following types:

$$\begin{aligned} &TYPE-1: \quad s = \pm(b_1) \\ &TYPE-2: \quad s = \pm(ib_2) \\ &TYPE-3: \quad s = \pm(c+id) \\ &TYPE-4: \quad s = \pm(c-id) \end{aligned} \qquad (16)$$

where b_1, b_2, c and d are positive real numbers. The propagating wave of cylindrical shell is dedicated only by Type 1. Type 2 root constitutes no spatially oscillating term, which represents the nearfield distortion close to the boundary. Type 3 and Type 4 roots represent various types of evanescence waves. However, these four types of roots, do not have a one-to-one correspondence with the four branch pairs shown in Eq.(12). For example, in the case of n=2, the curve of Branch pair 1 begins as a Type 3 (evanescence waves) and then gradually turns into Type 1 (propagating waves) after crossing its cut-off frequency. When n=0 or n=1, the curves of branch pair 1 are Type 1 roots (do not have a cut-off frequency), and they remain as Type 1 roots for all frequencies.

The dispersion curves for all the roots and the corresponding eigenvectors plotted in the figures shown hereafter are those with positive real parts only. In the case of Type 2 roots which have no real part, only the positive imaginary part will be plotted. The roots with negative real parts are not plotted; their eigenvectors are basically the same as those of positive real roots, except there will be sign changes in the longitudinal components if the signs of the radial and circumferential components are chosen to be the same as those plotted.

3.1. The case when n=0

The dispersion relations and the associated eigenvectors are shown in Fig. 3 and 4, respectively. For simplicity, the eigenvectors $[U'\ V'\ W']$ are plotted by using the absolute values of the displacement components, i.e., $[|U|\ |V|\ |W|]'$. The dispersion relations for the longitudinal and bending waves of a plate, and those for shear waves are also shown for reference. It shows that at frequencies much lower than the ring frequency (Ω<<1), Branch 1 represents a non-dispersive longitudinal wave, from which it gradually deviates as Ω approaches the neighborhood of unity; Branch 1's wave eventually approaches that of a plate bending wave when Ω>>1. These are illustrated by the mode shapes shown in Fig. 4(a). Below ring frequency, see the mode shape in Fig. 4(b), Branch 2 (Branch pair 2 root with positive real part) represents the Type 4 root, but it becomes a Type 1 root after crossing the ring frequency and represents the longitudinal wave above the ring frequency, say Ω>1. Therefore, when Ω>1, the Branch 2 curve takes over the role of representing the longitudinal wave which was previously the role of Branch 1's curve. Below the ring frequency, Branch 3 (Branch pair 3 root with positive real part) represents the Type 3 root, but above the ring frequency, it represents the Type 2 root. The modal components of the Branch 3 roots are predominantly radial (see Fig. 4(c)). Branch 4 (Branch pair 4 root with positive real part) represents the characteristic roots of Eq. (13), which therefore represents the non-dispersive torsional wave and has nearly the same dispersion relation as shear waves. The mode shape of torsional mode is not shown; it is simply $[U\ V\ W]'=[0\ 1\ 0]'$.

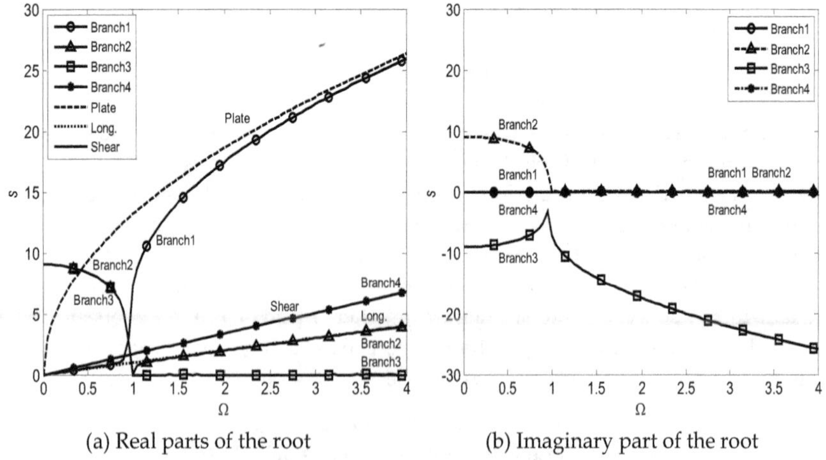

Figure 3. The dispersion relations for n=0 circular harmonic

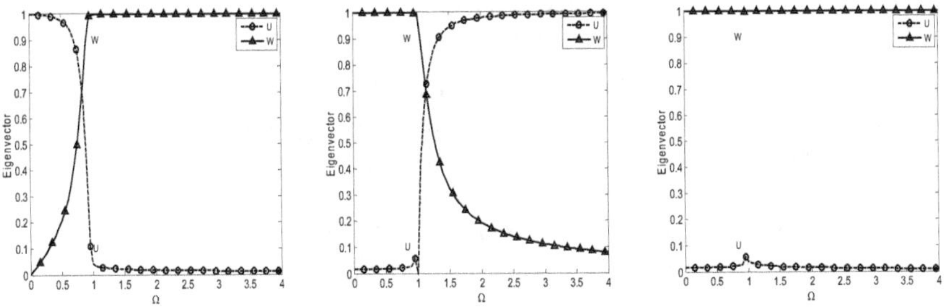

(a) Branch 1 roots (b) Branch 2 roots (c) Branch 3 roots

Figure 4. The normalized eigenvectors of n=0 circular harmonic (no V component)

3.2. The case when n=1

Fig. 5 and 6 show the dispersion relations and the associated eigenvectors, respectively, for the n=1 circular harmonics. The equivalent curves for Bernoulli beam (labeled as "Berbeam"), Timoshenko beam (labeled as "Timbeam"), and the longitudinal and shear wave curves of a plate are also shown for reference. Branch 1 has only real roots (Type 1 roots); the dispersion curve representing beam bending (when Ω<0.5) agrees well with that of the Timoshenko beam. The Bernoulli beam, on the other hand, agrees with the shell and Timoshenko beam curves only up to Ω=0.03; its applicable frequency range is therefore limited. At frequencies above Ω=0.5, the originally beam-like bending wave gradually transitions to a plate-like bending wave with a predominantly radial modal displacement.

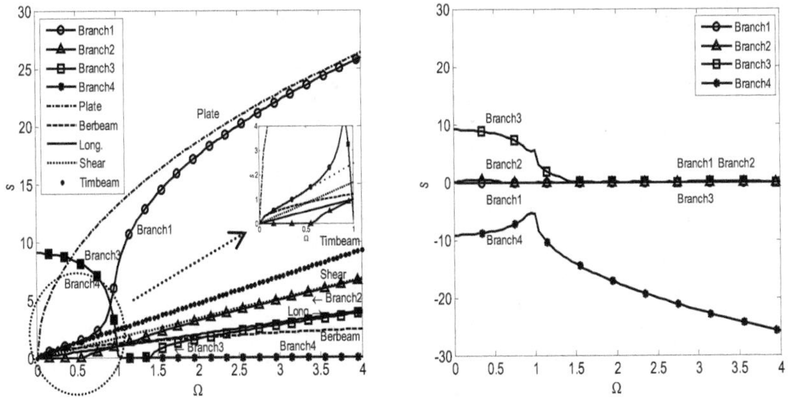

(a) Real part of the root (b) Imaginary part of the root

Figure 5. The dispersion relations for n=1 circular harmonic

Below Ω=0.5, the Branch 2 curve is purely imaginary, and represents the Type 2 root. When Ω>0.5, the roots become positive real (Type 1) and gradually approach the shear wave curve, with a predominantly circumferential displacement component when $\Omega>1.5$ (see Fig. 6(b)). Obviously, this branch has a cut off frequency which is about $\Omega=0.5$.

Below the ring frequency, the Branch 3 curve indicates a Type 3 root with a slow and exponentially diminishing wave. When $\Omega>1.4$ it becomes a Type 1 root, and the Branch 3 curve approaches that of longitudinal wave (see Fig. 5 and 6(c)) at further higher frequencies. The propagating waves in Branch 3 thus has a cutoff frequency of about $\Omega=1.4$. The Branch 4 curve indicates a Type 4 root below the ring frequency, but above ring frequency it has only the imaginary part (Type 2) and represents the transversal near field distortion (see Fig. 6(d)).

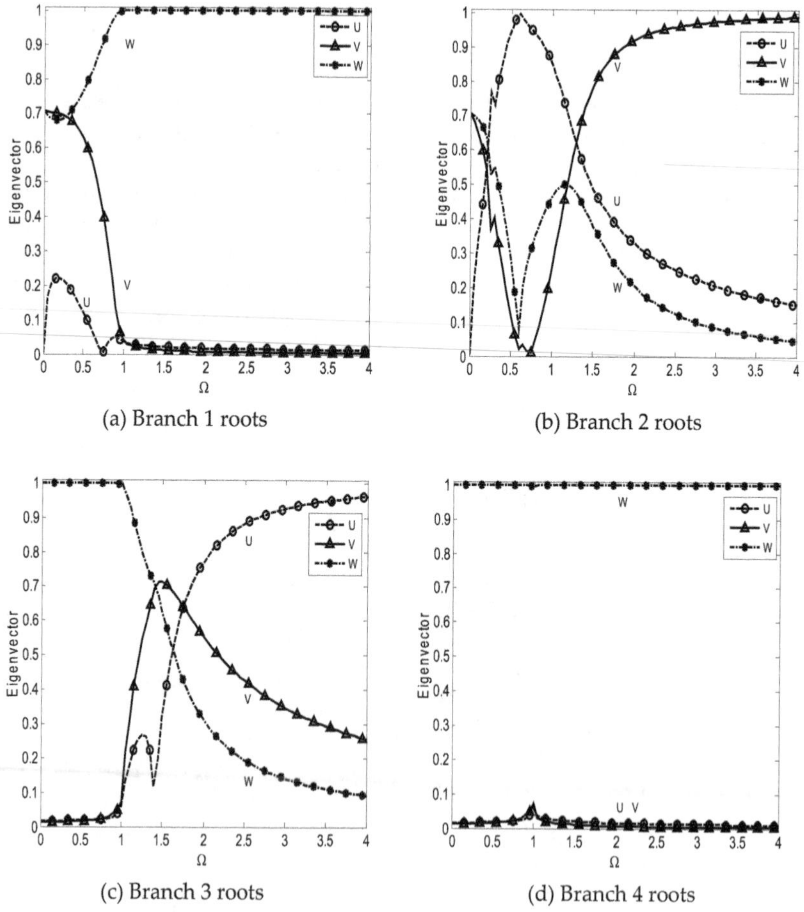

(a) Branch 1 roots (b) Branch 2 roots

(c) Branch 3 roots (d) Branch 4 roots

Figure 6. The normalized eigenvectors of $n=1$ circular harmonic.

3.3. The case when n=2

The dispersion relations for the n=2 circular harmonic case are shown in Fig. 7. At very low frequencies, say Ω<0.016, the roots in Branch 1 are the Type 3 roots. At higher frequencies, the roots become Type 1 and represent propagating waves (which thus have a cut off frequency at Ω≒0.016). Fig. 8(a) shows that below ring frequency, Ω<1, the modal patterns of Branch 1 roots have three well-coupled displacement components, although the predominant component is radial. When Ω>1, the modal displacement becomes predominantly radial and the propagating speed asymptotically approaches that of a plate.

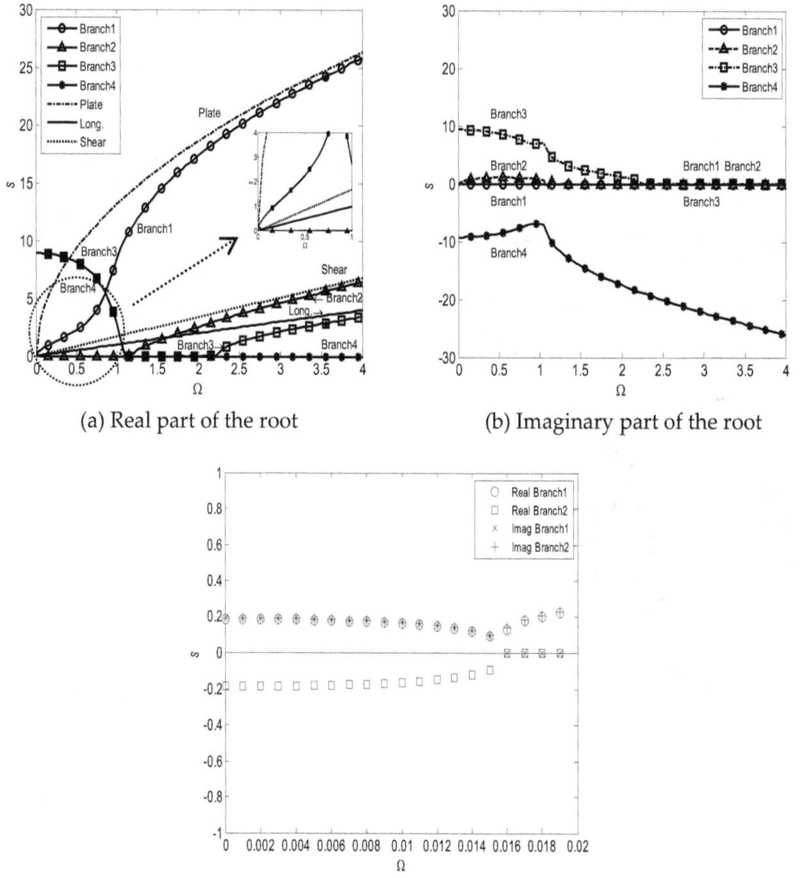

(a) Real part of the root

(b) Imaginary part of the root

(c) The dispersion curves of Branch 1 and 2 below Ω=0.02

Figure 7. The dispersion relations for n=2 circular harmonic

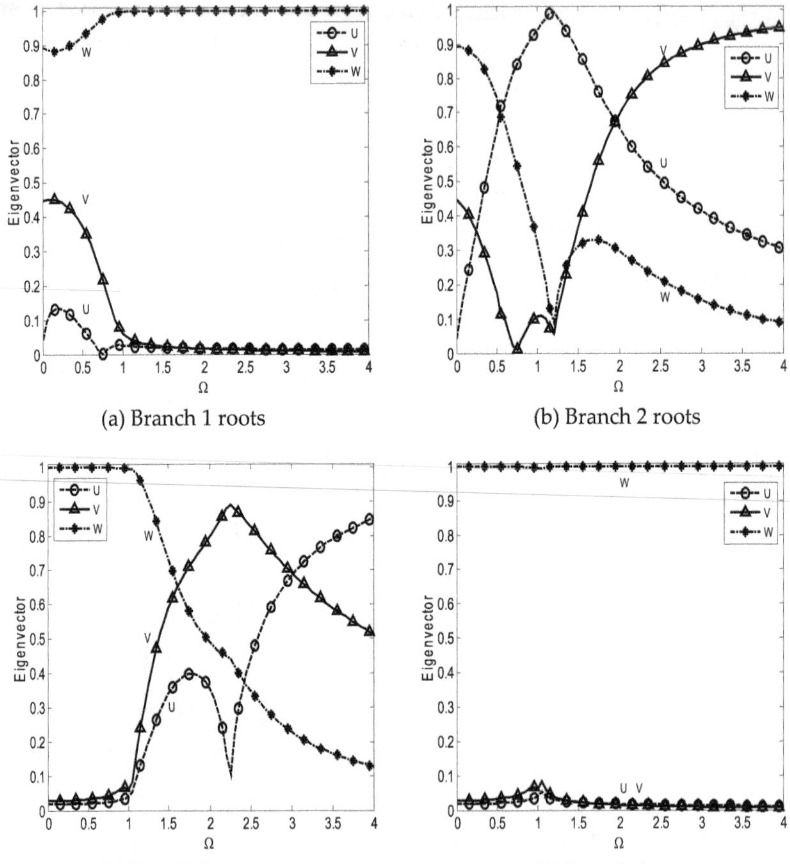

(a) Branch 1 roots (b) Branch 2 roots

(c) Branch 3 roots (d) Branch 4 roots

Figure 8. The normalized eigenvectors of n=2 circular harmonic

At very low frequencies, say $\Omega<0.016$, the roots in Branch 2 are the Type 4 roots. Between say $\Omega=0.016$ and the ring frequency, the roots of this branch are Type 2; they represent the near field distortion of the three coupled components. When $\Omega>1.15$, they become Type 1 propagating waves with the wave speeds that asymptotically approach that of the shear wave, and their modal patterns reflect predominantly circumferential shear motions. The curve of Branch 3 indicates a Type 3 root when $0<\Omega<1.15$, a Type 2 root when $1.15<\Omega<2.2$, and a Type 1 root when $\Omega>2.2$. This is a propagating wave with a high cutoff frequency (at $\Omega\fallingdotseq2.2$), and at higher frequencies its wave speed approaches asymptotically to that of the longitudinal wave. Branch 4 represents the Type 4 root when $0<\Omega\leqq1.15$, which, however, turns into Type 2 and represents the near field distortion when $\Omega>1.15$.

4. Conclusion

The classic problem of the dispersion of waves in a cylindrical shell had been re-examined with analytical solutions obtained by using a symbolic algebra package. The previous work by Karczub did not include the analytical solutions for the dispersion relations of axisymmetric waves (of circular harmonic order $n=0$). An axisymmetric wave contains both longitudinal and transverse components; the former are important in the transmission of longitudinal vibration, and the latter are particularly important in acoustics due to their higher acoustic radiation efficiency than the corresponding waves of higher circular harmonic orders ($n>0$). In this way, a complete set of analytical solutions which based on Flügge shell theory is now available for all orders of circular harmonics, $n=0, 1, 2, \ldots, \infty$.

A considerable effort has been expended on the solutions for the modal patterns of all propagating and non-propagating modes, because a complete set of properly normalized eigenvectors is crucial to the solution of the vibration problem of a finite shell under various admissible boundary conditions. The eigenvectors obtained by the conventional method are not conveniently normalized as are those commonly used in mathematical physics. A new alternative method to find normalized eigenvectors with norms equal to unity has been proposed and discussed.

Use of analytical solutions has demonstrated the capability of a straightforward continuous tracking on the frequency-dependent changes of the root types and of the corresponding modal patterns for each branch of the dispersion curves associated with a particular analytical root. Therefore, a parallel display of the dispersion curves and of the associated modal patterns used in this paper has provided more insight about the wave phenomena in a cylindrical shell.

Appendix

The coefficients in Eq. (11) are listed as follows:

$$\begin{aligned} g_0 = {} & n^2\beta^2 v_1\left(1+\beta^2\right)\left\{-n^2+2n^3-n^4\right\} \\ & +\Omega^2 n^2\left\{\left[\beta^4 v_1+2\beta^2 v_2+v_1\right]+n^2\left[-2\beta^4 v1-\beta^2\left(4v_2-3v_1\right)+v_1\right]\right. \\ & \left.+n^3\beta^2\left[\beta^2 v_1+\left(2v_2-v_1\right)\right]\right\} \\ & +\Omega^4\left\{\left[1+\beta^2\right]+n^2\left[\beta^2\left(v_3-2v_4\right)+\left(v_1-2v_2\right)\right]+n^4\beta^2\right\}-\Omega^6 \end{aligned} \tag{A.1}$$

$$\begin{aligned} g_2 = {} & n^2\beta^2 v_1\left\{\left[3\beta^4 v_1+7\beta^2 v_1+2\left(v-2\right)\right]-2n^2\left[3\beta^4 v_1+\beta^2\left(3v_1+2v_2\right)+\left(v-4\right)\right]\right. \\ & \left.+n^3\left[3\beta^4 v_1-\beta^2\left(v_1-4v_2\right)-4\right]\right\} \\ & +\Omega^2\left\{\left[3\beta^4 v_1+\beta^2\left(2v-3\right)+\left(v^2-v_1+2v_2\right)\right]\right. \\ & +n^2\beta^2\left[3\beta^2 v_1\left(v_3-2v_4\right)+2\left(3v_1-4v_2\right)+2v_1\right] \\ & \left.+n^4\left[\beta^2 v_1\left(-7v_3+8v_4\right)+3\left(2v_2-v_1\right)\right]\right\} \\ & +\Omega^4\left\{\left[-3\beta^2 v_1+\left(v_1-2v_2\right)\right]+2n^2\beta^2\right\} \end{aligned} \tag{A.2}$$

$$\begin{aligned} g_4 = {} & \left\{v_1\left[-3\beta^4+\beta^2\left(-4+3v^2\right)+2v_1v_3\right]+3n^2\beta^2\left[-\beta^2\left(2v_1\left(-2+v^2\right)+v_2\right)+2v_1\right]\right. \\ & \left.+n^4\beta^2 v_1\left[\beta^4 v^2+6\beta^2 v_2-6\right]\right\} \\ & +\Omega^2\beta^2\left\{\left[\left(11v_1-4v_2\right)+v_1\right]+n^2\left[\beta^2\left(1+7v_1+v_1^2\right)+3\left(2v_2-v_1\right)\right]\right\} \\ & +\Omega^4\beta^2 \end{aligned} \tag{A.3}$$

$$\begin{aligned} g_6 = {} & \beta^2 v_1\left\{\left[2v\left(1+3\beta^2\right)\right]+n^2\left[9\beta^4 v_1+\beta^2\left(8v_2-5v_1\right)-4\right]\right\} \\ & +\Omega\beta^{22}\left\{v_1\left(-1+5\beta^2\right)+2v_2\left(-1+\beta^2\right)\right\} \end{aligned} \tag{A.4}$$

$$g_8 = \left\{\beta^2 v_1\left[3\beta^4-2\beta^2-1\right]\right\} \tag{A.5}$$

where

$$\begin{aligned} v_1 &= \frac{-1+v}{2}, & v_2 &= \frac{-2+v}{2} \\ v_3 &= \frac{1+v}{2}, & v_4 &= \frac{2+v}{2} \end{aligned} \tag{A.6}$$

The coefficients in Eq. (12) are given in the following.

$$A_1 = \left(\sqrt{\Lambda_1 + \Lambda_2}\right)/2, \qquad A_2 = -\Lambda_1/4$$
$$A_3 = \left(\sqrt{\Lambda_5 - \Lambda_6}\right)/2, \qquad A_4 = \left(\sqrt{\Lambda_5 + \Lambda_6}\right)/2 \tag{A.7}$$

where

$$\Lambda_1 = 2^{1/3}\lambda_4/(3g_8\lambda_7^{1/3}) + \lambda_7^{1/3}/(3g_8 2^{1/3}), \qquad \Lambda_2 = (\lambda_1^2/4) - (2\lambda_2/3),$$
$$\Lambda_3 = (\lambda_1^2/2) - (4\lambda_2/3), \qquad \Lambda_4 = -\lambda_1^3 + 4\lambda_6 - 8\lambda_3, \tag{A.8}$$
$$\Lambda_5 = -\Lambda_1 + \Lambda_3, \qquad \Lambda_6 = \Lambda_4/(4\sqrt{\Lambda_1 + \Lambda_2})$$

and

$$\lambda_1 = g_6/g_8, \qquad \lambda_2 = g_4/g_8, \qquad \lambda_3 = g_2/g_8, \qquad \lambda_4 = g_4^2 - 3g_2g_6 + 12g_0g_8,$$
$$\lambda_5 = 2g_4(g_4 - 4g_2g_6 - 36g_0g_8) + 27(g_0g_6^2 + g_8g_2^2), \tag{A.9}$$
$$\lambda_6 = g_4g_6/g_8^2, \qquad \lambda_7 = \lambda_5 + \sqrt{-4\lambda_4^3 + \lambda_5^2}$$

Author details

Yu-Cheng Liu, Yun-Fan Hwang and Jin-Huang Huang
Feng Chia University, Electroacoustic Graduate Program, Taiwan, R.O.C.

5. References

[1] Forsberg K. (1963). Influence of boundary conditions on the modal characteristics of thin cylindrical shells. *Journal of the Acoustical Society of America,* Vol.119, No.6 , pp. 1898-1899, ISSN 1729-8806.

[2] Fuller C. R. (1981). The effects of wall discontinuities on the propagation of flexural waves in cylindrical shells. Journal of Sound and Vibration, Vol.75, pp. 207-228, ISSN 0022-460X.

[3] Hwang Y. F. (2002). Structural-acoustic wave transmission and reflection in a hose- pipe system, Proceedings of 2002 ASME International Mechanical Engineering Congress and Exposition, IMECE2002-32683, pp. 91-96, , ISBN 0791836436.

[4] Karczub D. G. (2006). Expressions for direct evaluation of wave number in cylindrical shell vibration studies using the Flügge equations of motion. Journal of the Acoustical Society of America, Vol.119, No.6, pp. 3553-3557, ISSN 0001-4966.

[5] Lesmez M. W., Wiggert D. C., and Hatfield F. J. (1990). Modal analysis of vibrations in liquid-filled piping systems. Journal of Fluids Engineering ASME, Vol. 112, No. 3, pp. 311-318, ISSN 0098-2202.

[6] Leissa A. W. (1997). Vibration of Shells, Acoustical Society of America, ISBN 1563962934.

[7] Skelton E. A. & James J. H. (1997). Theoretical Acoustics of Underwater Structures, Imperial College Press, ISBN 1860940854.

[8] Liu Y. C., Hwang Y. F., and Huang J. H. (2009). Modes of wave propagation and dispersion relations in a cylindrical shell. ASME Journal of Vibration and Acoustic, Vol. 131, No. 4, 041011-1~9, ISSN 1048-9002.

Section 2

Wave Features in Classical Media and Structures

Chapter 5

Spatio-Temporal Feature in Two-Wave and Multi-Wave Propagations

Zhengde Dai, Jun Liu, Gui Mu, Murong Jiang

Additional information is available at the end of the chapter

1. Introduction

It is well known that the wave propagation depends mainly on its velocity and frequency in one direction for a single wave[1-3]. There are many literatures devoted researches of single wave propagation such as solitary wave, periodic wave, chirped wave, rational wave etc [4-6]. However, what can be happen when two and even more waves with different features propagate together along different directions? In the past decades, many methods have been proposed for seeking two waves and multi-wave solutions to nonlinear models in modern physics. Recently, some effective and straight methods have been proposed such as homoclinic test approach(HTA)[7-8], extended homoclinic test approach(EHTA)[9-10] and three wave method [11-12]. These methods were applied to many nonlinear models. Several exact waves with different properties have been found out, such as periodic solitary wave, breather solitary wave, breather homoclinic wave, breather heteroclinic wave, cross kink wave, kinky kink wave, periodic kink wave, two-solitary wave, doubly periodic wave, doubly breather solitary wave, and so on. Because of interaction between waves with different features in propagation process of two-wave or multi-wave, some new phenomena have been discovered and numerically simulated, for example, resonance and non resonance, fission and fusion, bifurcation and deflexion etc. Furthermore, similar to the bifurcation theory of differential dynamical system, constant equilibrium solution of nonlinear evolution equation and propagation velocity of a wave as parameters are introduced to original equation, and then by using the small perturbation of parameter at a special value, two-wave or multi-wave propagation occurs new spatiotemporal change such as bifurcation of breather multi-soliton, periodic bifurcation and soliton degeneracy and so on.

This chapter mainly focus on explanation of different test methods and comprehensive applications to two-wave or multi-wave propagation. New methods will be described such as HTA, EHTA, Three-wave method and parameter small perturbation method. The spatiotemporal variety in exact two-wave and multi-wave propagation will be investigated and numerically simulated. In this chapter, some important models such as shallow water wave propagation models under the transverse long-wave disturbance Potential

Kadomtsev-Petviashvili equation [13-17], Kadomtsev-Petviashvili [18-27] equation, and specially, Kadomtsev-Petviashvili with positive dispersion equation are considered. Applied new methods to these equations, some new two-wave and multi-wave are obtained and spatiotemporal variety in multi-wave propagation is investigated and numerically simulated.

This chapter consists of four sections. In section 1, we introduce some methods including homoclinic test approach (HTA), extended homoclinic test approach(EHTA), three-wave method and parameter small perturbation method. In section 2, we consider the potential Kadomtsev-Petviashvili equation and investigate the exact periodic kink-wave and degeneracy of soliton. In section 3, we consider the Kadomtsev-Petviashvili equation and investigate periodic bifurcation, deflexion of two solitary waves. In section 4, we consider the Kadomtsev-Petviashvili equation with positive dispersion. By using three-wave method, we obtain some breather kinds of multi-solitary wave solutions, and investigate the fission and fusion of multi-wave.

2. Some methods for seeking two-wave and multi-wave

2.1. Homoclinic test approach

Consider a (2+1) dimensional nonlinear evolution equation of the general form

$$F(u, u_t, u_x, u_y, \cdots) = 0 \tag{1}$$

where F is a polynomial of $u(x,y,t)$ and its derivatives, t represents time variable and x, y represent spatial variables. Assume that there exists a transformation of unknown function such that Eq.(1) becomes a bilinear equation in the following form

$$G(D_t, D_x, D_y, \cdots) f \cdot f = 0 \tag{2}$$

where G is a general polynomial in $D_t, D_x, D_y, \cdots$, where the Hirota's bilinear operator D-operator is defined by[2]

$$D_x^m D_t^n f(x,t) \cdot g(x,t) = \left(\frac{\partial}{\partial x} - \frac{\partial}{\partial x'}\right)^m \left(\frac{\partial}{\partial t} - \frac{\partial}{\partial t'}\right)^n [f(x,t) g(x',t')]|_{x'=x, t'=t} \tag{3}$$

Traditionally, one obtains two solition solutions with the assumption

$$f = 1 + e^{\eta_1} + e^{\eta_2} + d e^{\eta_1 + \eta_2} \tag{4}$$

where d is a constant, $\eta_1 = k_1 x + l_1 y + c_1 t$, $\eta_2 = k_2 x + l_2 y + c_2 t$, and $k_j, l_j, c_j, j = 1,2$ are real numbers. If we treat k_1 and k_2 as complex numbers by taking $k_1 = -k_2 = ik, i^2 = -1$ and assuming $l_1 = l_2 = l$ and $c_1 = c_2 = c$, Eq.(4) can be convert into the following form

$$f = 1 + \cos(kx) e^{ly+ct} + d e^{2ly+2ct} \tag{5}$$

In order to get more forms solution of Eq.(1), we use a more general Ansätz which contains complete variables of Eq.(2) replacing Eq.(5):

$$f = 1 + \cos(\eta_1) e^{\eta_2} + d e^{2\eta_2} \tag{6}$$

where

$$\begin{aligned} \eta_1 &= k_1 x + l_1 y + c_1 t \\ \eta_2 &= k_2 x + l_2 y + c_2 t \end{aligned} \tag{7}$$

where all of $d, k_j, l_j, c_j, j = 1, 2$ may be real numbers or complex numbers. This process is similar to the procedure that one can obtain the homoclinic orbit of the defocusing nonlinear Schrödinger equation from its dark-hole soliton solutions[1]. As a result, we call this skill as "Homoclinic test approach".

To derive analytic expression, we can take the following procedure in detail: inserting Eq.(6) into Eq.(2), then equating the coefficients of the same kind terms to zero, subsequently, solving the resulting algebraic equations to determine the relationship between variables $k_j, l_j, j = 1, 2, \cdots$ with the help of symbolic computation software such as Maple. In Eq.(7), noting the cos functions is meaningful because we often take into account periodic effect in real physical background. Indeed, we can observe this solution is periodic breathing from their profile.

2.2. Extended homoclinic test approach

After substituting Eq.(6) into Eq.(2), we can get that whether the Eq.(2) has the nonzero solution. Furthermore, we do some mathematical simplicity. Rewrite Eq.(6) as follows:

$$f = e^{\eta_2}(e^{-\eta_2} + \cos(\eta_1) + de^{\eta_2}) \tag{8}$$

or

$$f = e^{\eta_2}[\sqrt{d}\cosh(\eta_2 + ln(\sqrt{d})) + \cos(\eta_1)] \tag{9}$$

We replace Eq.(9) with

$$f = \sqrt{d}\cosh(\eta_2 + ln(\sqrt{d})) + b_1\cos(\eta_1) \tag{10}$$

where b_1, d are constants. Now, factoring out the exponentials produces: Exploiting this Ans*ä*tz to obtain the new solutions of nonlinear evolution equation is called "Extended Homoclinic test approach". To derive analytic expression, we can take the following procedure in detail: inserting Eq.(10) into Eq.(2), then equating the coefficients of the same kind terms to zero, subsequently, solving the resulting algebraic equations to determine the relationship between variables $k_j, l_j, j = 1, 2, \cdots$ with the help of symbolic computation software such as Maple.

2.3. Three-wave method

Multi-wave solutions are important because they reveal the interactions between the inner-waves and the various frequency and velocity components. The whole multi-wave solution, for instance, may sometimes be converted into a single soliton of very high energy that propagates over large regions of space without dispersing and an extremely destructive wave is therefore produced of which the tsunami is a good example. Since all double-wave solutions can be found by using the exp-function method proposed by Fu and Dai [18], we propose an extension of the three-soliton method [6] in this section (called the three-wave method) for finding coupled wave solutions. Consider a high dimensional nonlinear evolution equation of the general form

$$F(u, u_t, u_x, u_y, u_z, u_{xx}, \cdots) = 0 \tag{11}$$

where $u = u(x, y, z, t)$ and F is a polynomial u of and its derivatives, t represents time variable and x, y, z represent spatial variables. The three-wave method operates as follows.
Step 1: By Painleve analysis, a transformation

$$u = T(f) \tag{12}$$

is made for some new and unknown function f.
Step 2: Convert Eq. (11) into Hirota bilinear form:

$$H(D_t, D_x, D_y, D_z, \cdots) f \cdot f = 0 \tag{13}$$

where D is identical to the Eq.(3).
Step 3: Traditionally, we taking the following Ansätz to obtain the three soliton solution

$$f = 1 + e^{\xi_1} + e^{\xi_2} + e^{\xi_3} + a_{12} e^{\xi_1+\xi_2} + a_{13} e^{\xi_1+\xi_3} + a_{23} e^{\xi_2+\xi_3} + a_{123} e^{\xi_1+\xi_2+\xi_3} \tag{14}$$

where

$$\xi_j = a_j x + b_j y + c_j z + d_j t, \qquad j = 1,2,3 \tag{15}$$

a_{13}, a_{23}, a_{123} are the constants. Eq.(14) can be rewritten as

$$f = e^{\frac{\eta_1}{2}} (e^{-\eta_1} + e^{\eta_2} + e^{\eta_3} + e^{\eta_4} + a_{12} e^{\eta_4} + a_{13} e^{\eta_3} + a_{23} e^{-\eta_2} + a_{123} e^{\eta_1}) \tag{16}$$

where

$$\eta_1 = \frac{\xi_1+\xi_2+\xi_3}{2}, \qquad \eta_2 = \frac{\xi_1-\xi_2-\xi_3}{2}, \qquad \eta_3 = \frac{-\xi_1+\xi_2-\xi_3}{2}, \qquad \eta_4 = \frac{-\xi_1-\xi_2+\xi_3}{2} \tag{17}$$

Thus, this three soliton Ansätz contains four variables η_1, η_2, η_3 and η_4. Here, we treat it in a different way. We factor out the $e^{\frac{\eta_1}{2}}$ in Eq.(16) and decrease the numbers of variables to three terms. On the other hand, we set some parameters in a complex way. At last, the above analysis allows us to construct the following assumptions:

$$f = e^{-\xi_1} + \delta_1 \cos(\xi_2) + \delta_2 \cosh(\xi_3) + \delta_3 e^{\xi_1} \tag{18}$$

or

$$f = e^{-\xi_1} + \delta_1 \cos(\xi_2) + \delta_2 \sinh(\xi_3) + \delta_3 e^{\xi_1} \tag{19}$$

where $\delta_j, j = 1,2,3$ are constants. In fact, from Eq.(18) or Eq.(19), it is easily seen that it only contains three wave variables. As a result, we call this method "three wave method". It is obvious that three-wave method is the extension and improvement of the traditional three-soliton method.

Step 4: Substitute Eq.(18) (or Eq.(19)) into Eq.(13), and collect the coefficients of $e^{-\xi_1}$, e^{ξ_1}, $\sin(\xi_2)$, $\cos(\xi_2)$, $\cosh(\xi_3)$ and $\sinh(\xi_3)$. Then equate the coefficients of these terms to zero and obtain a set of over-determined algebraic equations in a_j, b_j, c_j and $d_j, j = 1,2,3$.
Step 5: Solve the set of algebraic equations in Step 4 using Maple and solve for a_j, b_j, c_j, d_j and $\delta_j, j = 1,2,3$.

Step 6: Substituting the identified values into Eq.(12) and Eq.(13). Thus, we can deduce the exact multi-wave solutions of Eq.(11).

2.4. Introducing parameters and small perturbation method

In this section, we still consider a high dimensional nonlinear evolution equation of the general form

$$F(u, u_t, u_x, u_y, u_z, u_{xx}, \cdots) = 0 \tag{20}$$

where $u = u(x,y,z,t)$ and F is a polynomial u of and its derivatives, t represents time variable and x, y, z represent spatial variables. Here, we introduce a parameter to Eq.(20) in two ways.

(i). Initial solution as a parameter is introduced to Eq.(20). We assume that the u_0 is an initial solution of Eq.(20), then we use the transformation

$$u = u_0 + T(f) \tag{21}$$

to convert nonlinear evolution equation (20) into Hirota bilinear form which contains the small perturbation parameters u_0 :

$$H(u_0, D_t, D_x, D_y, D_z, \cdots) f \cdot f = 0 \tag{22}$$

where D is also the Hirota D-operator. The next step is to exploit the existent method to solving the Eq.(22). The perturbation parameter u_0 plays an important role to the resulting solution, where the spatiotemporal feature in multi-wave propagation including velocity and direction even the shape will change as u_0 makes small perturbation.

(ii). The velocity of a wave variable as a parameter is introduced to Eq.(20). We can assume that $\xi = z - \alpha t$ in Eq.(20), where ξ is a new wave variable and α is its velocity. Then Eq.(20) can be became to

$$F_1(u, \alpha, u_\xi, u_x, u_y, u_{xx}, \cdots) = 0 \tag{23}$$

Then solving Eq.(23). In this case, the spatiotemporal feature in multi-wave propagation for Eq.(20) may happen outstanding change as α makes a small perturbation.

3. Potential Kadomtsev-Petviashvili equation

Potential Kadomtsev-Petviashvili (PKP) equation studied in this section is described as

$$u_{xt} + 6u_x u_{xx} + u_{xxxx} + u_{yy} = 0 \tag{24}$$

where $u : R_x \times R_y \times R_t^+ \to R$. It is well known that the PKP equation arose in number of remarkable non-linear problems both in physics and mathematics, the solutions of PKP equation have been studied extensively in various aspects. By applying various methods and techniques to PKP equation, exact traveling wave solutions, linearly solitary wave solutions, soliton-like solutions and some numerical solutions have been obtained[1-3,5-7]. Recently, two soliton and periodic soliton solutions were presented, resonance and non-resonance interactions between periodic soliton and different line solitons were investigated[6].

In this section, we use homoclinic test method and extended homoclinic test technique to seek periodic soliton solution, exact periodic kink-wave solution, periodic soliton solution and doubly periodic solution of PKP equation. Furthermore, it is explicitly exhibited that the feature of solution is different varying with direction of wave propagation on the x -axis, periodic soliton is degenerated into doubly periodic wave when the direction of a wave propagation changes from progressing to the left into right.

3.1. Degenerative periodic solitary wave

In this section, the periodic solition solution is constructed by homoclinic test technique and bilinear form method and the degeneracy of solitary wave is investigated.

Setting $\xi = x - \alpha t$ in Eq.(24) yields

$$u_{yy} - \alpha u_{\xi\xi} + 6u_\xi u_{\xi\xi} + u_{\xi\xi\xi\xi} = 0 \tag{25}$$

where α is a wave velocity. Let $u = (\ln F)_\xi$, then Eq.(25) can be reduced into the following bilinear form

$$[D_y^2 - \alpha D_\xi^2 + D_\xi^4 - A]F \cdot F = 0 \tag{26}$$

where A is an integration constant, $D_x^m D_t^k$ is defined in Eq.(3). With regard to Eq.(26), using the homoclinic test technique, we can seek the solution in the form

$$F = 1 + b_1(e^{ip\xi} + e^{-ip\xi})e^{\Omega y+\gamma} + b_2 e^{2\Omega y+2\gamma} \tag{27}$$

where $A, p, \Omega, \gamma, b_1$ and b_2 are all real to be determined below.

Substituting Eq.(27) into Eq.(26) yields the exact solution of Eq.(24) in the form

$$u = \frac{-2b_1 p e^{\Omega y+\gamma}\sin(p\xi)}{1 + 2b_1\cos(p\xi)e^{\Omega y+\gamma} + b_2 e^{2\Omega y+2\gamma}} \tag{28}$$

where parameters A, b_1, b_2, Ω, p and γ satisfy dispersive relations

$$A = 0 \qquad \Omega^2 = -p^4 - \alpha p^2 \qquad b_1^2 = \frac{\Omega^2 b_2}{\Omega^2 - 3p^4} \tag{29}$$

Obviously, $\alpha < 0$ is required so that the conditions $\Omega^2 > 0, b_1^2 > 0$ and $0 < p^2 < -\alpha$ can be satisfied. Taking $\xi = x - \alpha t, b_2 = 1, \gamma = 0$ in Eq.(28), then $b_1 = \sqrt{\frac{\Omega^2}{\Omega^2 - 3p^4}} > 1$, the exact solution to PKP equation can be expressed as

$$u = \frac{-b_1 p\sin(p(x-\alpha t))}{b_1\cos(p(x-\alpha t)) + \cosh(\Omega y)} \tag{30}$$

where

$$\begin{cases} \alpha < 0 \\ 0 < p^2 < -\frac{\alpha}{4} \\ \Omega^2 = -p^4 - \alpha p^2 \\ b_1^2 = \frac{\Omega^2}{\Omega^2 - 3p^4} \end{cases} \tag{31}$$

Here, we choice $0 < p^2 < -\frac{\alpha}{4}$ such that $b_1^2 > 0$, then Eq.(30) is the periodic soliton solution of PKP equation which is a periodic traveling wave progressing to the left with velocity $|\alpha|$ on the x-axis, and meanwhile is a soliton on the y-axis (see Fig.1).

Making a variable transformation $\xi = x - \alpha t, \eta = iy$ in Eq.(24), then it can be transformed into the following form

$$u_{\eta\eta} - (-\alpha)u_{\xi\xi} - 6u_\xi u_{\xi\xi} - u_{\xi\xi\xi\xi} = 0 \tag{32}$$

Letting $u = (\ln F)_\xi$, being similar to the way of dealing with PKP equation in above, we take

$$F = 1 + b_3(e^{ip_1\xi} + e^{-ip_1\xi})e^{\Omega_1\eta+\gamma_1} + b_4 e^{2\Omega_1\eta+2\gamma_1}$$

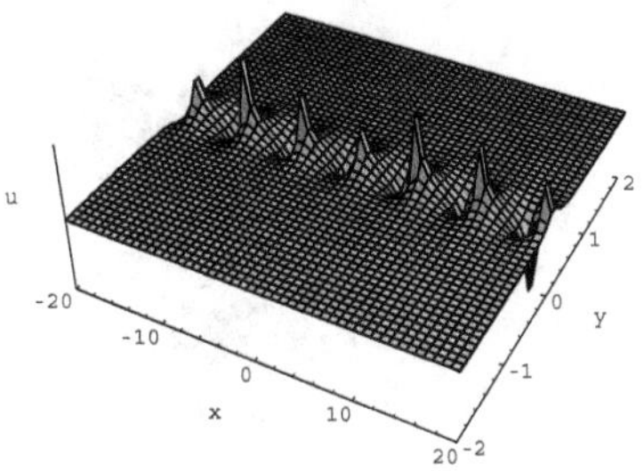

Figure 1. The periodic soliton solution which is a periodic wave progressing to the left with velocity $|\alpha| > 0$ on the x-axis ,and meanwhile is a soliton on the y-axis

By computing, the exact solution of Eq.(32) is given by

$$u = -\frac{2b_3 p_1 \sin(p_1\xi)e^{\Omega_1\eta+\gamma_1}}{1 + 2b_3\cos(p_1\xi)e^{\Omega_1\eta+\gamma_1} + b_4 e^{2\Omega_1\eta+2\gamma_1}} \tag{33}$$

where parameters b_3, b_4, Ω_1, p_1 satisfy following dispersive relations

$$\Omega_1^2 = p_1^4 + \alpha p_1^2 \qquad b_3^2 = \frac{\Omega_1^2 b_4}{\Omega_1^2 + 3p_1^4} \tag{34}$$

We can see that $\Omega_1^2 \geq 0$ always holds for every real p_1 as long as $\alpha \geq 0$.

Taking $\xi = x - \alpha t, \eta = iy$ in Eq.(33), the exact solution to PKP equation $(\alpha \geq 0)$ is expressed as

$$u = -\frac{2b_3 p \sin(p_1(x - \alpha t))}{2b_3\cos(p_1(x-\alpha t)) + (e^{-i\Omega_1 y - \gamma_1} + b_4 e^{i\Omega_1 y+\gamma_1})} \tag{35}$$

where

$$\begin{cases} \alpha \geq 0, \quad p_1^2 \geq 0 \\ \Omega_1^2 = p_1^4 + \alpha p_1^2 \\ b_3^2 = \frac{\Omega_1^2 b_4}{\Omega_1^2 + 3p_1^4} \end{cases} \tag{36}$$

In particularly, taking $\gamma_1 = 0, b_4 = 1$ in Eq.(35) yields

$$u = -\frac{b_3 p_1 \sin(p_1(x-\alpha t))}{b_3\cos(p_1(x-\alpha t)) + \cos(\Omega_1 y)} \tag{37}$$

Obviously, both $\cos p_1(x - \alpha t)$ and $\cos(\Omega_1 y)$ are periodic, so the solution given by Eq.(37) is an doubly periodic solution which is a periodic traveling wave progressing to the right with velocity α on the x-axis, and meanwhile is a periodic standing wave on the y-axis (see Fig.2).

It is important that we can take the same p and p_1 such that $p^2 = p_1^2 = |\frac{\alpha}{k}| = p_0^2$ in Eq.(30) and Eq.(37) from the constraint condition Eq.(31) and Eq.(36) respectively, where $k > 4$ is an

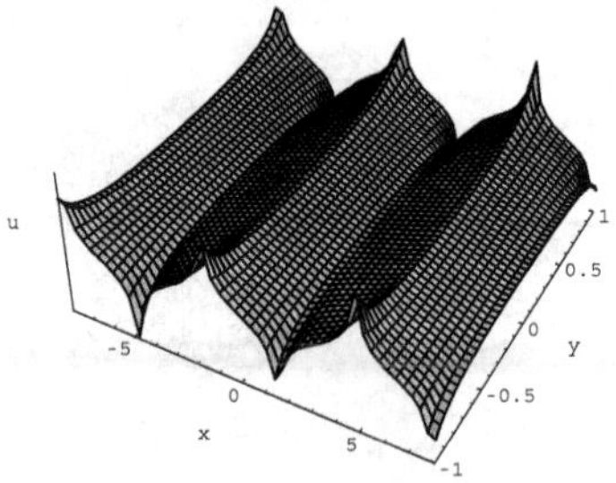

Figure 2. The doubly periodic solution which is a periodic wave progressing to the right with velocity α on the x-axis, and meanwhile is a periodic standing wave on the y-axis.

arbitrary real number. Therefore, we obtain a periodic soliton solution and an doubly periodic solution which have the same period with x-direction as follows

$$\begin{cases} u = \dfrac{-b_1 p_0 \sin(p_0(x-\alpha t))}{b_1 \cos(p_0(x-\alpha t)) + \cosh(\Omega y)}, & \alpha < 0 \\ u = -\dfrac{b_3 p_0 \sin(p_0(x-\alpha t))}{b_3 \cos(p_0(x-\alpha t)) + \cos(\Omega_1 y)}, & \alpha \geq 0 \end{cases} \tag{38}$$

It is easy to find that the feature of solution of Eq.(24) is different from Eq.(38) on the arbitrary small neighborhood of velocity $\alpha = 0$. There exists a periodic soliton solution which is a periodic with x-direction, and a soliton with y-direction as well as $\alpha < 0$. When $\alpha \geq 0$, this periodic soliton degenerates into a doubly periodic wave which are periodic with both x-direction and y-direction. However, period with x-direction is preserved identically.

3.2. Exact periodic kink-wave solution

In this section, a new type of periodic kink-wave solution for PKP equation is obtained using extended homoclinic test technique.

Taking a transformation

$$u = (\ln F)_x \tag{39}$$

Then Eq.(24) can be transformed as

$$[D_y^2 + D_x D_t + D_x^4] F \cdot F = 0 \tag{40}$$

In this case, we let

$$F = e^{-\Omega(x+\alpha t)-\beta y} + b_1 \cos(p(x-\alpha t)+\beta y) + b_2 e^{\Omega(x+\alpha t)} \tag{41}$$

Following the procedure of extended homoclinic test technique, we derive the periodic kink-wave solution of PKP equation(see Fig.3)

$$u = \frac{p[-e^{-p(x+11p^2t)} - b_1 \sin(p(x-11p^2t \pm 3py)) + \frac{b_1^2}{20} e^{p(x+11p^2t)}]}{e^{-p(x+11p^2t)} + b_1 \cos(p(x-11p^2t \pm 3py)) + \frac{b_1^2}{20} e^{p(x+11p^2t)}} \tag{42}$$

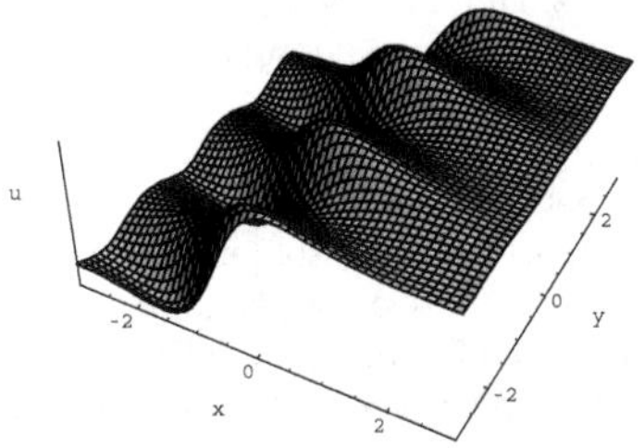

Figure 3. The periodic kink-wave solution

It can be rewritten as

$$u = \frac{p[\frac{b_1}{2\sqrt{5}}e^{p(x+11p^2t)} - 2\sqrt{5}\sin(p(x-11p^2t \pm 3py)) - \frac{2\sqrt{5}}{b_1}e^{-p(x+11p^2t)}]}{\frac{b_1}{2\sqrt{5}}e^{p(x+11p^2t)} + 2\sqrt{5}\cos(p(x-11p^2t \pm 3py)) + \frac{2\sqrt{5}}{b_1}e^{-p(x+11p^2t)}} \tag{43}$$

Specially, taking $b_1 = 2\sqrt{5}k$ and $\gamma = \ln k, k > 1$, yields a periodic kink solution as follows

$$u(x,y,t) = \frac{p[2\sinh(p(x+11p^2t)+\gamma) - \sin(p(x \pm 3py - 11p^2t))]}{2\cosh(p(x+11p^2t)+\gamma) + \cos(p(x \pm 3py - 11p^2t))} \tag{44}$$

It shows a periodic kink-wave whose speed is $11p^2$ and period is $2\pi/3p$ of space variable y. It exhibits elastic interaction between a solitary wave and periodic wave with the same speed in opposed direction each other. It is an interesting phenomenon in fluid mechanics (see Fig.3).

4. Kadomtsev-Petviashvili equation

The (2+1)-dimensional (two spatial and one temporal) generalization of Korteweg-de Vries equation (KdV) was given by Kadomtsev and Petviashvili to discuss the stability of (1+1)-dimensional soliton to the transverse long-wave disturbances, which is known as KP equation or the (2+1)-dimensional KdV equation. There are two distinct versions of KP equation, which can be written in normalized form as follows[4]:

$$u_t - 3(u^2)_x - u_{xxx} + s^2\partial_x^{-1}u_{yy} = 0 \tag{$*$}$$

with the operator ∂_x^{-1} defined by

$$\partial_x^{-1}f(x) = \int_{-\infty}^{x} f(\xi)d\xi$$

The propagation property of solitons depends essentially on the sign of s^2 in equation. The coefficient is defined as follows: $s = \pm i, i^2 = -1$ for negative dispersion and $s = \pm 1$ for positive dispersion. Here $u = u(x,y,t)$ is a scalar function, x and y are respectively the longitudinal and transverse spatial coordinates, the subscripts x, y, t denote partial derivatives. When $s = \pm i$, it is usually called KPI, while for $s = \pm 1$, it is called KPII. KP equation is the natural generalization of the well known KdV equation ($u_t - 6uu_x - u_{xxx} = 0$)

from one to two spatial dimensions. It arises naturally in many other applications, particularly in plasma physics, gas dynamics, and elsewhere. Both KPI and KPII are exactly integrable via the Inverse Scattering Transformation. Kadomtsev and Petviashvili have shown that the line soliton of the KP equation is unstable in the case of positive dispersion and is stable for the negative dispersion[3,22]. The solutions of KP equation have been studied extensively in various aspects. The inclined periodic soliton solution and the lattice soliton solution were expressed as exact imbricate series of rational soliton solutions to KP equation with positive dispersion[22]. Resonant interaction of two-soliton among three obliquely oriented solitons in higher dimension was first studied by Miles[26]. Y.Kodama and Dai have proved that the KP equation provides line solitons in shallow water and these solitons can be of resonance[7, 27].

In this section, spatial-temporal bifurcation for KP equation is considered, several types of exact solutions to KP equation were constructed by bilinear form and homoclinic test approach. It is explicitly analyzed that the feature of the solitary wave is different on the both sides of equilibrium solution $u_0 = -\frac{1}{6}$, which is a unique periodic bifurcation point for KPI and deflexion point of soliton for KPII. As for KPI, when the equilibrium u_0 varies from one side of $-\frac{1}{6}$ to another side, two-solitary wave changes into doubly periodic wave. Whereas, the y-periodic solitary wave changes into $x-t$-periodic solitary wave for KPII, the propagation direction of periodic solitary wave occurs outstanding deflexion. In addition, some new type of multi-wave solutions are obtained using three-wave method.

4.1. Spatiotemporal bifurcation and deflexion of the soliton

Kadomtsev-Petviashvili (KP) equation in normalized variable $u(x, y, t)$ reads

$$u_{xt} - u_{xxxx} - 3(u^2)_{xx} - s^2 u_{yy} = 0 \tag{45}$$

where $u : R_x \times R_y \times R_t \to R$. It is easily to note that $u = u_0$ is an equilibrium solution of KP equation, where u_0 is an arbitrary constant.

Now, we consider KPI equation

$$u_{xt} - u_{xxxx} - 3(u^2)_{xx} - u_{yy} = 0 \tag{46}$$

Setting $\xi = i(x - t)$ gives

$$u_{yy} - u_{\xi\xi} - 3(u^2)_{\xi\xi} + u_{\xi\xi\xi\xi} = 0 \tag{47}$$

Let $u = u_0 - 2(\ln F)_{\xi\xi}$, Eq.(46) can be transformed into the following bilinear form

$$[D_y^2 - (1 + 6u_0)D_\xi^2 + D_\xi^4 - A]F \cdot F = 0 \tag{48}$$

Using "homoclinic test technique", we are going to seek the solution of the form

$$F = 1 + b_1(e^{ip\xi} + e^{-ip\xi})e^{\Omega y+\gamma} + b_2 e^{2\Omega y+2\gamma} \tag{49}$$

where $A, p, \Omega, \gamma, b_1$ and b_2 are all real.

Substituting Eq.(49) into Eq.(48) with Eq.(47) yields the exact solution of Eq.(47) of the form

$$u = u_0 + \frac{2p^2[4b_1^2 + b_1(e^{ip\xi} + e^{-ip\xi})(b_2 e^{\Omega y+\gamma} + e^{-\Omega y-\gamma})]}{[b_1(e^{ip\xi} + e^{-ip\xi}) + (b_2 e^{\Omega y+\gamma} + e^{-\Omega y-\gamma})]^2} \tag{50}$$

where parameters satisfy

$$A = 0 \qquad \Omega^2 = -p^4 - (1+6u_0)p^2 \qquad b_1^2 = \frac{\Omega^2 b_2}{\Omega^2 - 3p^4} \tag{51}$$

It is obviously that $u_0 < -\frac{1}{6}$ is required so that the conditions $\Omega^2 > 0, b_1^2 > 0$ and $p^2 < -\frac{1+6u_0}{4}$ can be satisfied and u_0 is a free parameter.

In Eq.(50), taking $\xi = i(x-t)$, we obtain the two-soliton solution of KPI equation (see Fig.4)

$$u_1(x,y,t) = u_0 + \frac{2p^2[4b_1^2 + b_1(e^{p(x-t)} + e^{-p(x-t)})(b_2 e^{\Omega y+\gamma} + e^{-\Omega y-\gamma})]}{[b_1(e^{p(x-t)} + e^{-p(x-t)}) + (b_2 e^{\Omega y+\gamma} + e^{-\Omega y-\gamma})]^2} \tag{52}$$

where

$$\begin{cases} u_0 < -\frac{1}{6} \\ p^2 < -\frac{1+6u_0}{4} \\ \Omega^2 = -p^4 - (1+6u_0)p^2 \\ b_1^2 = \frac{\Omega^2 b_2}{\Omega^2 - 3p^4} \end{cases} \tag{53}$$

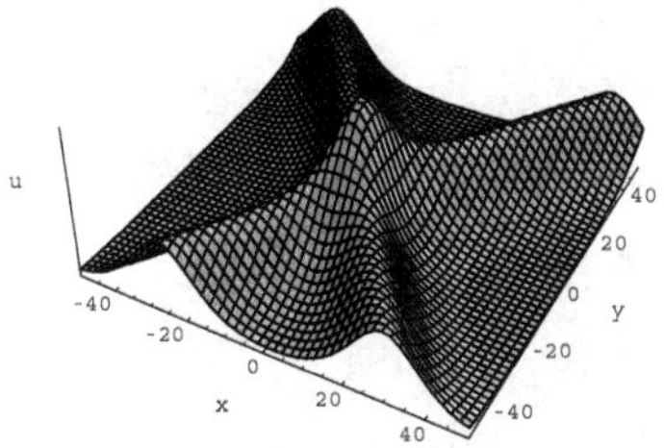

Figure 4. The two-soliton solution for KPI equation as $u_0 = -\frac{1}{4}$

KPII equation is given by

$$u_{xt} - u_{xxxx} - 3(u^2)_{xx} + u_{yy} = 0 \tag{54}$$

Making a variable transformation $\xi = x - t$ in Eq.(54), it can be transformed into the following form

$$u_{yy} - u_{\xi\xi} - 3(u^2)_{\xi\xi} - u_{\xi\xi\xi\xi} = 0 \tag{55}$$

Letting $u = u_0 + 2(\ln F)_{\xi\xi}$ and using a similar way dealing with KPI, we take

$$F = 1 + b_1(e^{ip\xi} + e^{-ip\xi})e^{\Omega y+\gamma} + b_2 e^{2\Omega y+2\gamma}$$

By computing, the exact solution of Eq.(55) is given by

$$u = u_0 - \frac{2p^2[4b_1^2 + b_1(e^{ip\xi} + e^{-ip\xi})(b_2 e^{\Omega y+\gamma} + e^{-\Omega y-\gamma})]}{[b_1(e^{ip\xi} + e^{-ip\xi}) + (b_2 e^{\Omega y+\gamma} + e^{-\Omega y-\gamma})]^2} \tag{56}$$

where parameters satisfy

$$\Omega^2 = p^4 - (1+6u_0)p^2 \qquad b_1^2 = \frac{\Omega^2 b_2}{\Omega^2 + 3p^4} \tag{57}$$

It is easily to see that $u_0 \geq -\frac{1}{6}$ is available as long as $p^2 \geq 1 + 6u_0$.

Taking $\xi = x - t$ into Eq.(56), the exact solution to KPII equation is expressed as

$$u_2(x,y,t) = u_0 - \frac{2p^2[4b_1^2 + b_1 cos(p(x-t))(b_2 e^{\Omega y+\gamma} + e^{-\Omega y - \gamma})]}{[b_1 cos(p(x-t) + (b_2 e^{\Omega y+\gamma} + e^{-\Omega y-\gamma})]^2} \tag{58}$$

where

$$\begin{cases} u_0 \geq -\frac{1}{6} \\ p^2 \geq 1 + 6u_0 \\ \Omega^2 = p^4 - (1+6u_0)p^2 \\ b_1^2 = \frac{\Omega^2 b_2}{\Omega^2+3p^4} \end{cases} \tag{59}$$

Obviously, $\cos p(x-t)$ is periodic, so the solution given by Eq.(58) is a periodic soliton solution with $x-t$-direction (see Fig.5).

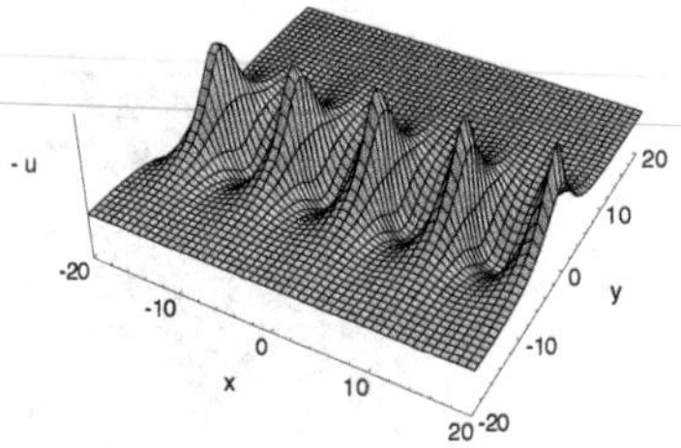

Figure 5. The $x-t$-periodic soliton solution for KPII equation as $u_0 = -\frac{1}{8}$

Comparing Eq.(47) with Eq.(55), it is easily to find that the Eq.(47) may be changed into Eq.(55) and vice versa by using the temporal and spatial transformation $(\xi, y) \to (i\xi, iy)$. Because the solutions of KP equation are real functions, it is naturally to take specially $b_2 = 1, \gamma = 0$. Making variable transformation $\xi \to i\xi, y \to iy, i^2 = -1$ in Eq.(50) and Eq.(56) yields

$$u_3(x,y,t) = u_0 - \frac{2p^2[b_1^2 + b_1 \cos(p(x-t))\cos(\Omega y)]}{[b_1 \cos(p(x-t)) + \cos(\Omega y)]^2} \tag{60}$$

It is noted that the solution given by Eq.(59) is a singular periodic solution to KPI equation. In order to avoid the singularity, we set $\cos(p(x-t)) > 0$ and $\cos(\Omega y) > 0$ (see Fig.6).

Besides, the y-periodic soliton solution to KPII is also given by

$$u_4(x,y,t) = u_0 + \frac{2p^2[4b_1^2 + b_1(e^{p(x-t)} + e^{-p(x-t)})(e^{i\Omega y} + e^{-i\Omega y})]}{[b_1(e^{p(x-t)} + e^{-p(x-t)}) + (e^{i\Omega y} + e^{-i\Omega y})]^2} \tag{61}$$

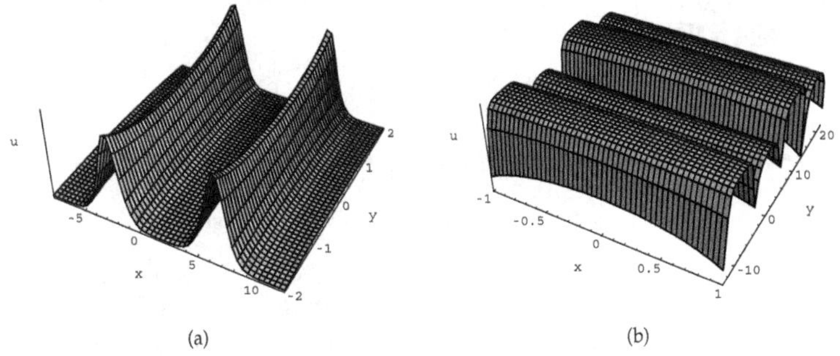

Figure 6. (a) The doubly periodic solution for KPI equation with $x-t$ direction as $u_0=-\frac{1}{8}$ (b) The doubly periodic solution for KPI equation with y direction as $u_0=-\frac{1}{8}$

where

$$\begin{cases} u_0 < -\frac{1}{6} \\ p^2 < -\frac{1+6u_0}{4} \\ \Omega^2 = -p^4-(1+6u_0)p^2 \\ b_1^2 = \frac{\Omega^2 b_2}{\Omega^2-3p^4} \end{cases} \tag{62}$$

The solution given by Eq.(60) represents periodic soliton with y-direction (see Fig.7).

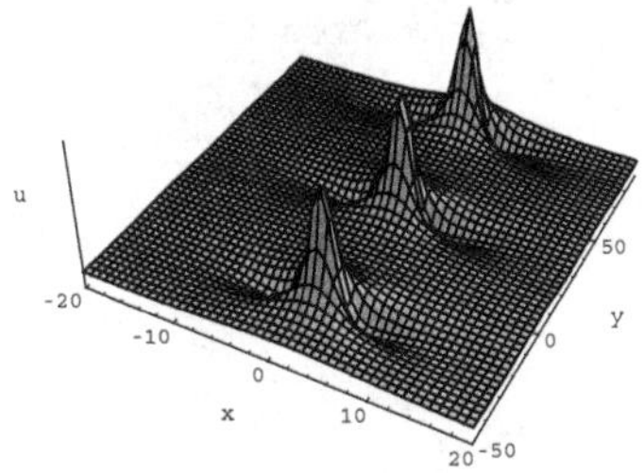

Figure 7. The y-periodic soliton solution for KPII equation as $u_0=-\frac{1}{4}$

According to above discussion, we get that $u_0=-\frac{1}{6}$ is a unique periodic bifurcation point for KPI and deflexion of soliton for KPII. Around the both sides at u_0, the property of solutions to KPI and KPII is all changed. As for KPI, when the equilibrium u_0 varies from one side of $-\frac{1}{6}$ to another side, two-soliton solution changes into doubly periodic solution. Whereas, the y-periodic soliton changes into x-periodic soliton for KPII. The double-soliton waves and doubly periodic soliton waves of KPI, periodic soliton waves on different spatial variable of KPII are interchanged around u_0.

4.2. Exact multi-wave solution

Let

$$u = 2(\ln f)_{xx} \tag{63}$$

in Eq.(45), where $f = f(x,y,t)$ is an unknown real function. Substituting Eq.(63) into Eq.(45), we can reduce Eq.(45) into the bilinear form

$$(D_x D_t - D_x^4 - p^2 D_y^2) f \cdot f = 0 \tag{64}$$

In order to obtain three wave solution of KP equation, we provide that

$$f = \cos(\gamma_1) + a_{-1}\exp(-\delta_1) + a_1\exp(\delta_1) + a_2\sinh(\xi_1) \tag{65}$$

where $\gamma_1 = p_1(x - \alpha_1 t), \delta_1 = p_3(x + \beta_3 y + \alpha_3 t)$ and $\xi_1 = p_2(x + \beta_2 y + \alpha_2 t)$. Substituting Eq.(65) into Eq.(64)and equating the coefficients of all powers of $\sinh(\xi_1)$, $\cos(\gamma_1)$, $\sinh(\xi_1)$, $\exp(\delta_1)$, $\sinh(\xi_1)\exp(-\delta_1)$, $\cosh(\xi_1)$, $\sin(\gamma_1)$, $\cosh(\xi_1)\exp(\delta_1)$, $\cosh(\xi_1)\exp(-\delta_1)$, $\sin(\gamma_1)$ $\exp(\delta_1)$, $\sin(\gamma_1)\exp(-\delta_1)$, $\cos(\gamma_1)\exp(\delta_1)$, $\cos(\gamma_1)\exp(-\delta_1)$ to zero, we can obtain a set of algebraic equations for $a_{-1}, a_2, a_1, \alpha_1, \alpha_2, \alpha_3, \beta_2, \beta_3, p_1, p_2$ and p_3, then by solving these set of algebraic equations and let $p_1 = p_2$, we obtain the breather two-solitary wave (three wave) solution of KPI as follows:

$$u_1(x,y,t) = 2\Big(\frac{a_2\sqrt{a_1 a_{-1}}((A_1 - p_1)^2\sinh(\delta_2 + \xi_2) - (A_1 + p_1)^2\sinh(\delta_2 - \xi_2))}{(\cos(\gamma_2) + 2\sqrt{a_1 a_{-1}}\cosh(\delta_2) + a_2\sinh(\xi_2))^2}$$

$$+\frac{4a_1 a_{-1}A_1^2 - a_2^2 p_1^2 + 2\sqrt{a_1 a_{-1}}(A_1^2 - p_1^2)\cosh(\delta_2)\cos(\gamma_2)}{(\cos(\gamma_2) + 2\sqrt{a_1 a_{-1}}\cosh(\delta_2) + a_2\sinh(\xi_2))^2}$$

$$+\frac{2p_1(a_2 p_1\cosh(\xi_2) + 2\sqrt{a_1 a_{-1}}A_1\sinh(\delta_2))\sin(\gamma_2) - p_1^2}{(\cos(\gamma_2) + 2\sqrt{a_1 a_{-1}}\cosh(\delta_2) + a_2\sinh(\xi_2))^2}\Big)$$

where

$$a_1 a_{-1} > 0, \qquad \theta_1 = \ln\Big(\sqrt{\frac{a_1}{a_{-1}}}\Big), \qquad \gamma_2 = p_1(x - C_1 t)$$

$$\delta_2 = A_1(x + D_1 y + B_1 t) + \theta_1, \qquad \xi_2 = p_1(x + \beta_2 y + C_1 t)$$

$$A_1 = \frac{\sqrt{2p_1^2\beta_2^2 + 12p_1^4}}{\beta_2}, \qquad B_1 = -\frac{(4p_1^2 + \beta_2^2)\beta_2^2 + 96p_1^4}{2\beta_2^2}$$

$$C_1 = -\frac{4p_1^2 - \beta_2^2}{2}, \qquad D_1 = -\frac{12p_1^4 - \beta_2^2}{2\beta_2}$$

Here, a_{-1}, a_2, p_1 and β_2 are free parameters.

In the case of $a_1 a_{-1} < 0$, the breather two-solitary wave solution of KPI can be expressed as

$$u_2(x,y,t) = 2\Big(\frac{a_2\sqrt{\gamma}((A_1 - p_1)^2\cosh(\delta_2 + \xi_2) - (A_1 + p_1)^2\cosh(\delta_2 - \xi_2)) + 4\gamma A_1^2 - a_2^2 p_1^2}{(\cos(\gamma_2) + \sqrt{\gamma}\sinh(\delta_2) + a_2\sinh(\xi_2))^2}$$

$$+\frac{2\sqrt{\gamma}(A_1^2 - p_1^2)\sinh(\delta_2)\cos(\gamma_2) + 2p_1(a_2 p_1\cosh(\xi_2) + 2\sqrt{\gamma}A_1\cosh(\delta_2))\sin(\gamma_2) - p_1^2}{(\cos(\gamma_2) + \sqrt{\gamma}\sinh(\delta_2) + a_2\sinh(\xi_2))^2}\Big)$$

where $\gamma = -a_1 a_{-1}$.

Similarity, the breather two-solitary wave solutions of KPII equation also obtained as follows

$$u_3(x,y,t) = 2(\frac{a_2\sqrt{a_1a_{-1}}((A_2-p_1)^2\sinh(\delta_3+\xi_3)-(A_2+p_1)^2\sinh(\delta_3-\xi_3))}{(\cos(\gamma_3)+2\sqrt{a_1a_{-1}}\cosh(\delta_3)+a_2\sinh(\xi_3))^2}$$

$$+\frac{a_1a_{-1}A_2^2-a_2^2p_1^2+2\sqrt{a_1a_{-1}}(A_2^2-p_1^2)\cosh(\delta_3)\cos(\gamma_3)}{(\cos(\gamma_3)+2\sqrt{a_1a_{-1}}\cosh(\delta_3)+a_2\sinh(\xi_3))^2}$$

$$+\frac{2p_1(a_2p_1\cosh(\xi_3)+2\sqrt{a_1a_{-1}}A_2\sinh(\delta_3))\sin(\gamma_3)-p_1^2}{(\cos(\gamma_3)+2\sqrt{a_1a_{-1}}\cosh(\delta_3)+a_2\sinh(\xi_3))^2})$$

where $\gamma_3 = p_1(x-C_2t), \delta_3 = A_2(x+D_2y+B_2t)+\theta_3$ and $\xi_3 = p_1(x+\beta_2y+C_2t)$ as $a_1a_{-1} > 0$. Here

$$A_2 = \frac{\sqrt{2p_1^2\beta_2^2-12p_1^4}}{\beta_2},\ B_2 = -\frac{(\beta_2^2-4p_1^2)\beta_2^2+96p_1^4}{2\beta_2^2},\ C_2 = -\frac{4p_1^2+\beta_2^2}{2},\ D_2 = \frac{12p_1^4+\beta_2^2}{2\beta_2}$$

a_{-1}, a_2, p_1 and β_2 are free parameters. And

$$u_4(x,y,t) = 2(\frac{a_2\sqrt{a_1a_{-1}}((p_3-p_1)^2\sinh(\delta_4+\xi_4)-(p_3+p_1)^2\sinh(\delta_4-\xi_4))}{(\cos(\gamma_4)+2\sqrt{a_1a_{-1}}\cosh(\delta_4)+a_2\sinh(\xi_4))^2}$$

$$+\frac{4a_1a_{-1}p_3^2-a_2^2p_1^2+2\sqrt{a_1a_{-1}}(p_3^2-p_1^2)\cosh(\delta_4)\cos(\gamma_4)}{(\cos(\gamma_4)+2\sqrt{a_1a_{-1}}\cosh(\delta_4)+a_2\sinh(\xi_4))^2}$$

$$+\frac{2p_1(a_2p_1\cosh(\xi_4)+2\sqrt{a_1a_{-1}}p_3\sinh(\delta_4))\sin(\gamma_4)-p_1^2}{(\cos(\gamma_4)+2\sqrt{a_1a_{-1}}\cosh(\delta_4)+a_2\sinh(\xi_4))^2})$$

where $\gamma_4 = p_1(x-C_3t), \delta_4 = p_3(x+D_3y+B_3t)+\theta_3, \xi_4 = p_1(x+A_3y+C_3t)$, with

$$A_3 = 2\sqrt{3}p_3,\ B_3 = -2(3p_1^2+p_3^2),\ C_3 = -2(3p_3^2+p_1^2),\ D_3 = \frac{\sqrt{3}(p_3^2+p_1^2)}{p_3},\ \theta_3 = \ln(\sqrt{\frac{a_1}{a_{-1}}})$$

a_{-1}, a_2, p_1 and p_3 are free parameters.

5. Kadomtsev-Petviashvili equation with positive dispersion

The purpose of this section is to investigate the fission and fusion interactions of the breather-type multi-solitary waves solutions to the KP equation with positive dispersion.

By transformation of independent variable $t \to -t, y \to \sqrt{3}y$ in (*), KP equation with positive dispersion can be written as

$$u_t + 6uu_x + u_{xxx} - 3\partial_x^{-1}u_{yy} = 0 \tag{66}$$

By the transformation of a dependent variable u

$$u = 2(\ln f)_{xx} \tag{67}$$

Eq.(66) can be transformed into the bilinear form

$$F(D_x, D_y, D_t)f\cdot f = (D_xD_t + D_x^4 - 3D_y^2)f\cdot f = 0 \tag{68}$$

where $f(x,y,t)$ is a real function.

5.1. Fission and fusion of multi-wave

In the following, by using generalized three-wave type of Ansätz approach, we study the interaction and spatiotemporal feature of three-wave of KP equation with positive dispersion. Now we suppose the solution of Eq.(68) as

$$f(x,y,t) = e^{\xi_1} + \delta_1 \cos \xi_2 + \delta_2 \cosh \xi_3 + \delta_3 e^{-\xi_1} \tag{69}$$

where $\xi_1 = a_1 x + b_1 y + c_1 t + \theta_1, \xi_2 = a_2 x + b_2 y + c_2 t + \theta_2, \xi_3 = a_3 x + b_3 y + c_3 t + \theta_3$ and $a_j, b_j, c_j, \delta_j, j = 1,2,3$ are some constants to be determined. Substituting the Ansätz Eq.(69) into Eq. (68) will produce the following relations:

$$a_3 = -\frac{b_2 a_1 - a_2 b_1}{{a_2}^2 + {a_1}^2}, \quad c_3 = \frac{\Delta_1}{\left({a_2}^2 + {a_1}^2\right)^3}$$
$$c_1 = -\frac{{a_1}^5 - 2\,{a_1}^3{a_2}^2 + 3\,{b_2}^2 a_1 - 3\,{b_1}^2 a_1 - 3\,{a_2}^4 a_1 - 6\,a_2 b_1 b_2}{{a_2}^2 + {a_1}^2}$$

$$\begin{aligned}
b_3 &= \frac{2\,{a_1}^3{a_2}^3 + {a_2}^2 b_1 b_2 + {a_1}^5 a_2 - {a_1}^2 b_1 b_2 + {a_2}^5 a_1 + {b_1}^2 a_1 a_2 - {b_2}^2 a_1 a_2}{\left({a_2}^2 + {a_1}^2\right)^2}\\
\delta_3 &= \frac{\Delta_2}{4\left({a_2}^2 + {a_1}^2\right)^3\left({a_1}^3 + a_2 b_1 + a_1 {a_2}^2 - b_2 a_1\right)\left({a_1}^3 + b_2 a_1 + a_1 {a_2}^2 - a_2 b_1\right)}\\
c_2 &= -\frac{3\,{a_1}^4 a_2 + 2\,{a_1}^2{a_2}^3 - 3\,{b_2}^2 a_2 + 3\,{b_1}^2 a_2 - {a_2}^5 - 6\,b_1 b_2 a_1}{{a_2}^2 + {a_1}^2}\\
\Delta_1 &= 3\,b_2{a_1}^7 + 3\,a_2 b_1{a_1}^6 + 9\,{a_2}^2 b_2{a_1}^5 + 9\,{a_2}^3 b_1{a_1}^4 + 9\,{a_1}^3{a_2}^4 b_2 - 3\,{a_1}^3{b_1}^2 b_2\\
&\quad + {a_1}^3{b_2}^3 + 3\,{a_1}^2 a_2{b_1}^3 - 9\,{a_1}^2 a_2 b_1{b_2}^2 + 9\,{a_1}^2{a_2}^5 b_1 + 3\,a_1{a_2}^6 b_2 - 3\,a_1{a_2}^2{b_2}^3\\
&\quad + 9\,a_1{a_2}^2{b_1}^2 b_2 + 3\,{a_2}^3 b_1{b_2}^2 - {a_2}^3{b_1}^3 + 3\,{a_2}^7 b_1\\
\Delta_2 &= -\left({b_2}^2{a_1}^2 - 2\,a_2 b_1 b_2 a_1 + {a_2}^2{b_1}^2 + {a_1}^4{a_2}^2 + 2\,{a_2}^4{a_1}^2 + {a_2}^6\right)\\
&\quad \left(\left({a_2}^2 + {a_1}^2\right)^3{\delta_1}^2 - \left(a_2 b_1 - b_2 a_1 + {a_1}^3 + a_1{a_2}^2\right)\left({a_1}^3 - a_2 b_1 + b_2 a_1 + a_1{a_2}^2\right){\delta_2}^2\right)
\end{aligned} \tag{70}$$

where $a_1, \delta_1, \delta_2, a_2, b_1, b_2$ are some free constants and $a_1 \neq 0, a_2 \neq 0$. Now we can explicitly write down the three-wave solutions using

$$\begin{aligned}
u(x,y,t) &= 2\,\frac{2{a_1}^2\sqrt{\delta_3}\cosh\left(\xi_1 - \ln\sqrt{\delta_3}\right) - \delta_1{a_2}^2\cos\xi_2 + \delta_2{a_3}^2\cosh\xi_3}{2\sqrt{\delta_3}\cosh\left(\xi_1 - \ln\sqrt{\delta_3}\right) + \delta_1\cos\xi_2 + \delta_2\cosh\xi_3}\\
&\quad -2\,\frac{\left(2\sqrt{\delta_3}a_1\sinh\left(\xi_1 - \ln\sqrt{\delta_3}\right) - \delta_1 a_2\sin\xi_2 + \delta_2 a_3\sinh\xi_3\right)^2}{\left(2\sqrt{\delta_3}\cosh\left(\xi_1 - \ln\sqrt{\delta_3}\right) + \delta_1\cos\xi_2 + \delta_2\cosh\xi_3\right)^2}
\end{aligned} \tag{71}$$

where a_j, b_j, c_j, δ_j $(j = 1,2,3)$ satisfy Eq.(70). It is called the breather type multi-solitary waves solutions. Fig.8 is the plot of the spatial structure with the parameters selected as

$$\begin{pmatrix} a_1 & b_1 & \delta_1 \\ a_2 & b_2 & \delta_2 \end{pmatrix} = \begin{pmatrix} 0.1 & 1 & 0.5 \\ 1.1 & 0.1 & 10^{-4} \end{pmatrix}$$

Fig.8 is the process of interaction for two solitons solutions with the evolution of time, where s_1 and s_2 are the different two solitons, and s_{12} represents the interaction between s_1, s_2 and the periodic wave $\cos(\xi_2)$. The phenomena of soliton interaction are clearly presented. It shows that the two soliton experience interaction, they will fusion with few oscillations and later travel ahead continuously. The value of δ_2 will determine the length of resonant soliton. Obviously, the solitons s_1 and s_2 can not approach each together closely from picture. They interact produce the breather wave.

Usually, the interactions between solitons for a lot of integrable or non-integrable system are considered to be completely elastic. That is to say, the amplitude, velocity and wave shape of a soliton do not change after the nonlinear collisions. However, for several nonlinear partial differential equations, completely nonelastic interactions will occur. On the other hand, two or more solitons will fusion to one soliton. These two types of phenomena was called soliton fission and soliton fusion, respectively. Now we demonstrate soliton fission and fusion of the Kadomtsev-Petviashvili equation with positive dispersion.

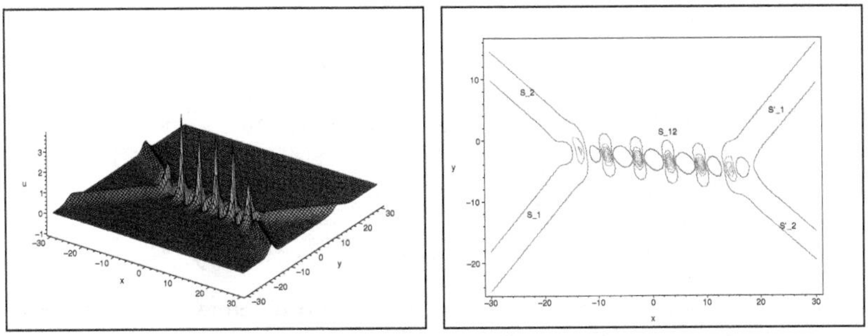

Figure 8. The plot of the space structure of the breather-type multi-solitary waves solutions and contour plot map of u in (x, y)-plane.

In Eq.(70), let $\delta_3 = 0$, then

$$\frac{\Delta\left((a_1{}^3 + a_2b_1 + a_1a_2{}^2 - a_1b_2)(a_1{}^3 + a_1b_2 + a_1a_2{}^2 - a_2b_1)\delta_2{}^2 - (a_1{}^2 + a_2{}^2)^3\delta_1{}^2\right)}{4(a_1{}^2 + a_2{}^2)^3(a_1{}^3 + a_1b_2 + a_1a_2{}^2 - a_2b_1)(a_1{}^3 + a_2b_1 + a_1a_2{}^2 - a_1b_2)} = 0$$

where $\Delta = a_2{}^2\left(a_2{}^2 + a_1{}^2\right)^2 + (a_2b_1 - b_2a_1)^2$, $a_1 \neq 0$ and $a_2 \neq 0$.

For $a_2{}^2\left(a_2{}^2 + a_1{}^2\right)^2 + (a_2b_1 - b_2a_1)^2$ can not be zero, then the condition must be satisfied:

$$(a_1(a_1^2 + a_2^2 - b_2) + a_2b_1)(a_1(a_1^2 + a_2^2 + b_2) - a_2b_1)\delta_2^2 - (a_1^2 + a_2^2)^3\delta_1^2 = 0$$

So, we obtain the following solution:

$$u(x,y,t) = \frac{2(a_1{}^2e^{\xi_1} - \delta_1a_2{}^2\cos\xi_2 + \delta_2a_3{}^2\cosh\xi_3)}{e^{\xi_1} + \delta_1\cos\xi_2 + \delta_2\cosh\xi_3} - \frac{2\left(a_1e^{\xi_1} - \delta_1a_2\sin\xi_2 + \delta_2a_3\sinh\xi_3\right)^2}{\left(e^{\xi_1} + \delta_1\cos\xi_2 + \delta_2\cosh\xi_3\right)^2} \tag{72}$$

where $\xi_1 = a_1x + b_1y + c_1t, \xi_2 = a_2x + b_2y + c_2t, \xi_3 = a_3x + b_3y + c_3t$ and $a_j, b_j, c_j, j = 1,2,3,$ and δ_1, δ_2 with the following relations:

$$c_2 = -\frac{3\,b_1{}^2a_2 - a_2{}^5 - 3\,b_2{}^2a_2 + 2\,a_1{}^2a_2{}^3 + 3\,a_1{}^4a_2 - 6\,a_1b_1b_2}{a_1{}^2 + a_2{}^2}$$

$$c_1 = -\frac{a_1{}^5 - 3\,a_2{}^4a_1 - 6\,a_2b_1b_2 - 2\,a_1{}^3a_2{}^2 - 3\,b_1{}^2a_1 + 3\,b_2{}^2a_1}{a_1{}^2 + a_2{}^2}$$

$$\begin{aligned} c_3 = {} & (3\,b_2a_1{}^7 + 3\,a_2b_1a_1{}^6 + 9\,a_2{}^2b_2a_1{}^5 + 9\,a_2{}^3b_1a_1{}^4 + 9\,a_1{}^3a_2{}^4b_2 + a_1{}^3b_2{}^3 \\ ] \quad & - 3\,a_1{}^3b_1{}^2b_2 - 9\,a_1{}^2a_2b_1b_2{}^2 + 9\,a_1{}^2a_2{}^5b_1 + 3\,a_1{}^2b_1{}^3a_2 + 3\,a_1a_2{}^6b_2 - 3\,a_1a_2{}^2b_2{}^3 \\ & + 9\,a_1a_2{}^2b_1{}^2b_2 + 3\,a_2{}^3b_1b_2{}^2 - b_1{}^3a_2{}^3 + 3\,a_2{}^7b_1)\left(\left(a_1{}^2 + a_2{}^2\right)^3\right)^{-1} \end{aligned}$$

$$a_3 = \frac{a_2b_1 - b_2a_1}{a_1{}^2 + a_2{}^2}$$

$$b_3 = \frac{a_2a_1{}^5 + 2\,a_2{}^3a_1{}^3 - a_1{}^2b_1b_2 + a_1b_1{}^2a_2 + a_1a_2{}^5 - a_1b_2{}^2a_2 + b_1b_2a_2{}^2}{\left(a_1{}^2 + a_2{}^2\right)^2}$$

$$\delta_2{}^2 = \frac{\left(a_1{}^2 + a_2{}^2\right)^3}{\left(a_1{}^3 + a_2b_1 + a_1a_2{}^2 - b_2a_1\right)\left(a_1{}^3 + b_2a_1 + a_1a_2{}^2 - a_2b_1\right)}\delta_1{}^2$$

Fig.9 shows the plot of two kinds of interaction behavior between two single solitons with different parameters and a breather wave, where

$$\begin{pmatrix} a_1 & b_1 & \delta_1 \\ a_2 & b_2 & \delta_2 \end{pmatrix} = \begin{pmatrix} -1.5 & 1 & 0.05 \\ -1.1 & 1.2 & 2 \end{pmatrix} \text{and} \begin{pmatrix} a_1 & b_1 & \delta_1 \\ a_2 & b_2 & \delta_2 \end{pmatrix} = \begin{pmatrix} 1.5 & -1 & 0.05 \\ -1.1 & 1.2 & 2 \end{pmatrix}$$

respectively. From the first picture of Fig.9, we can see that two single solitons interact strongly to make a resonance breather-wave solution from a point at which two incident solitons meet

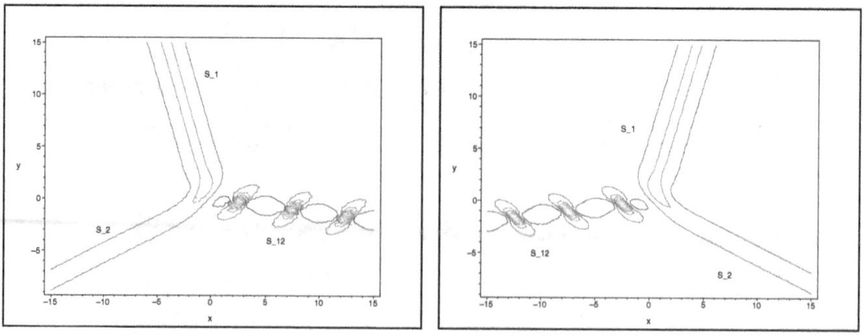

Figure 9. Contourplot of the breather-type multi-solitary waves solutions with the different parameters.

together. This phenomena is called soliton fusion. However, in the right figure it is found that the breather wave with the period oscillation can split up into two smaller line solitons with different directions. This phenomena is called the soliton fission.

6. Conclusion

Using Homoclinic test approach, Extend Homoclinic test approach, Three-wave method and Introducing parameters and small perturbation method, we obtain novel solutions of Potential Kadomtsev-Petviashvili equation and Kadomtsev-Petviashvili equation such as periodic solitary wave,breather solitary wave, breather homoclinic wave, breather heteroclinic wave, cross kink wave,kinky kink wave, periodic kink wave, two-solitary wave, doubly periodic wave, doubly breather solitary wave. Moreover, we observed that there were differently spatiotemporal features in two-wave and multi-wave propagations including the degeneracy of soliton, periodic bifurcation and soliton deflexion of two-wave, fission and fusion of breather two-wave and so on. In future, we intend to study the stability and the interactions patterns of N-wave solutions in KP equation. What's more, can we obtain similar results to another integrable or non-integrable system? How can one use the soliton fission and fusion of models to study the practically observed soliton fission and fusion in the experiments?

Acknowledgement

The work was supported by the National Natural Science Foundation of China (No. 10661028, 10801037 and 11161055).

Author details

Zhengde Dai
School of Mathematics and Physics, Yunnan University, Kunming 650091, P.R.China

Jun Liu and Gui Mu
College of Mathematics and Information Science, Qujing Normal University, Qujing 655000, P.R.China

Murong Jiang
School of Information Sciences and Engineering, Yunnan University, Kunming 650091, P.R.China

7. References

[1] Zabusky N, Kruskal M (1965) Interaction of "Solitons" in a Collisionless Plasma and the Recurrence of Initial States. Phys.Rev.lett. 15:240-243.

[2] Ablowitz M, Clarkson P (1991) Soliton, nonlinear evolution equations and inverse scattering. Cambridge University Press.

[3] Weiss J, Tabor M, Carnevale G (1983) The Painlevé property for partial differential equations. J. Math. Phys. 24:522-527.

[4] Kadomtsev B, Petviashvili V (1970) On the stability of solitary waves in weakly dispersive media. Soviet.Phys.Dokl.15:539-541.

[5] Chow K (2001) Periodic solutions for a system of four coupled nonlinear Schrödinger equations. Phys. Lett. A 285:319.

[6] Zeng X, Dai Z, Li D, Han S(2008) Some exact periodic soliton solkutions and resonance for the potential Kadomtsev-Petviashvili erquation. J. of Physics. Conference series 96: 012149.

[7] Dai Z, Li S, Dai Q, Huang J(2007) Singular periodic soliton solutions and resonance for the Kadomtsev-Petviashvili equation. Chaos, Solitons and Fractals :1148-1153.

[8] Dai Z, Liu J, Liu Z (2010) Exact periodic kink-wave and degenerative soliton solutions for potential Kadomtsev -petviashvili equation. Commun. Nonlinear. Sci. Numer. Simulat. 15: 2331-2336.

[9] Dai Z, Li S, Li D, Zhu A (2007) Periodic Bifurcation and Soliton deflexion for Kadomtsev-Petviashvili Equation, Chin. Phys. Lett. 24 :1420-1424.

[10] Dai Z, Liu J, Li D (2009) Applications of HTA and EHTA to YTSF Equation. Applied Mathematics and Computation. 207:360-364.

[11] Dai Z, Lin S, Fu H, Zeng X (2010) Exact three-wave solutions for the KP equation. Applied Mathematics and Computation. 216:1599-1604.

[12] Dai Z, Wang C, Lin S, Li D, Mu G (2010) The Three-wave Method for Nonlinear Evolution Equations. Nonl. Sci. Lett. A, 1:77-82.

[13] Wang C, Dai Z, Fission and fusion of breather multi-solitary waves to the Kadomtsev-Petviashvili equation with positive dispersion(to appear).

[14] Li D, Zhang H (2003) New soliton-like solutions to the potential Kadomtsev-Petviashvili equation. Appl. Math. Comput. 146:381-384.

[15] Li D, Zhang H (2003) Symbolic computation and various exact solutions of potential Kadomtsev- Petviashvili equation. Appl. Math. Comput. 145:351-359.

[16] Kaya D, EI-Sayed S (2003) Numerical soliton-like solutions of the potential Kadomtsev-Petviashvili equation by the decomposition method. Phys. Lett. A 320:192-199.

[17] Inan E, Kaya D (2003) Some exact solutions to the potential Kadomtsev-Petviashvili equation and to a system of shallow water equation. Phys. Lett. A 355:314-318.

[18] Fu H, Dai Z (2009) Double Exp-function method and application. Int. J. Nonlinear Sci. Num. Simul. 10(7):927-933.

[19] Ukai S (1989) Local solutions of the Kadomtsev-Petviashvili equation. J. Fac. Sci. Univ. Tokyo, Sect. IA Math. 36:193-209.

[20] Cheng Y, Li Y(1991) The constraint of the Kadomtsev-Petviashvili equation and its special solutions. Phys. Lett. A 157:22-26.

[21] Isaza P, Mejia J, Stallbohm V (1995) local solution for the Kadomtsev-Petviashvili equation in R^2. Math. Anal. Appl. 196:566-587.

[22] Tajiri M, Watanabe Y (1997) Periodic soliton solutions as imbricate series of rational solitons:solutions to the Kadomtsev-Petviashvili equation with positive dispersion. Nonlinear Math. Phys. 4:350-357.

[23] Zhu J (2007) Multisoliton excitations for the Kadomtsev-Petviashvili equation and the coupled Burgers equation. Chaos, Solitons and Fractals. 31: 648-657.

[24] Yomba E (2004) Construction of new soliton-like solutions for the (2 + 1)dimensional Kadomtsev-Petviashvili equation. Chaos, Solitons and Fractals. 22:321-325.

[25] Tagumi Y (1989) Soliton-like solutions to a (2+1) dimensional generalization of the KdV equation. Phys. Lett. A 141: 116-120.

[26] Miles J (1977) Resonantly interacting solitary waves. J. Fluid Mech. 79:171-179.

[27] Kodama Y (2004) Young diagrams and N-soliton solutions of the KP equation. J. Phys. A: Math. Gen. 37:11169-11190.

Chapter 6

Wave Propagation Methods for Determining Stiffness of Geomaterials

Auckpath Sawangsuriya

Additional information is available at the end of the chapter

1. Introduction

In many geoengineering design and analysis, the laboratory and field investigations are generally required to identify and classify geomaterials as well as to assess their engineering properties. One of the important engineering properties commonly used in geomechanical design and analysis is the stiffness of geomaterial. Such stiffness primarily describes the deformation characteristic of a geomaterial used to support engineered structures. Understanding the stiffness behaviour of geomaterials is therefore essential for improving design and analysis of structural behavior under varying loading and environmental conditions.

The importance of accurate stiffness measurements from small to large strains has gained increased recognition in both static and dynamic analyses over the past 20 years. In the static triaxial test, the local displacement transducer has been used to measure local axial strains (Goto et al. 1991). The resonant column and torsional shear devices have been widely used for many years to study cyclic and dynamic properties of geomaterials at strains ranging from 10^{-4} to 0.1% (Drnevich 1985, Saada 1988). The American Association of State Highway and Transportation Officials (AASHTO) has adopted the use of resilient modulus (M_R) in pavement design, which is customarily used by the pavement community. The precise measurement of M_R values of pavement materials is typically in the strain range from 10^{-2} to 0.1% (Kim and Stokoe 1992).

Based on the typical variation of shear moduli with shear strain levels shown in Fig. 1 (after Atkinson and Sallförs 1991, Mair 1993, Ishihara 1996, Sawangsuriya et al. 2005), three strain ranges are defined: (1) very small strain, (2) small strain, and (3) large strain. The very small strain range corresponds to the range of strain less than the elastic threshold strain (γ_{et}), approximately between 10^{-3}% and 10^{-2}%, depending on plasticity index (I_p) for plastic soils (Vucetic and Dobry 1991) and on confining pressure (σ_o) for non-plastic soils (Ishibashi and

Zhang 1993). Within this very small strain region, the geomaterial exhibits linear-elastic behaviour and the shear modulus is independent of strain amplitude, approaching a nearly constant limiting value of the maximum shear modulus (G_{max}). The small strain range starts from elastic threshold strain to 1% where the shear modulus is highly non-linear and strain-dependent. The large strain range corresponds to strain generally larger than 1%. In the large strains, the geomaterial is approaching failure and the shear modulus is substantially decreased. In many geoengineering applications, e.g. foundations, retaining walls, tunnels, pavements etc., the stress-strain behavior of geomaterial is highly non-linear, resulting in shear modulus degradation with strain by orders of magnitude. The variation of shear moduli and other properties of geomaterial with respect to shear strain levels for different geotechnical applications as measured by in situ and laboratory tests are also shown in Fig. 1.

Estimation of stiffness has traditionally been made in a triaxial apparatus using precise displacement transducers or resonant column devices (Lo Presti et al. 2001). Although several methods become commercially available to determine the stiffness of geomaterials both in the laboratory and in the field, the wave propagation techniques are widely accepted for their rapid, non-destructive, and low-cost evaluation methods. By knowing the elastic wave velocities as measured with the wave-based techniques and total mass density of the media, the stiffness of the geomaterials can be determined. In particular, a shear or S-wave velocity is a keystone for calculating the shear modulus of geomaterial. Such S-wave measurement has been researched extensively using shear plates (e.g. Lawrence 1963, 1965), resonant column tests (e.g. Hardin and Drnevich 1972), and bender elements (e.g. Shirley 1978, Shirley and Hampton 1978). The use of the shear plates is limited due to their large size and their need for a high excitation voltage (Ismail et al. 2005) while complexities and high cost of test equipments are disadvantages of resonant column tests. In contrast, bender elements have gained reputation particularly in research on geoengineering because of their smaller size, and lower voltage required leading to easier operation. Such method also provides cost-effectiveness and realistic design parameter which in turn becomes most valuable tool for mechanistic-based design and analysis, long-term performance monitoring, quality control process during construction etc.

2. Wave propagation methods for determining stiffness of geomaterials

Wave propagation methods become popular techniques in the evaluation of stiffness of geomaterials and process monitoring both in the laboratory and in the field (Richart et al. 1970, Matthews et al. 2000, Santamarina et al. 2001, LoPresti et al. 2001). Table 1 summarizes the existing wave propagation methods used by the geoengineering community in the U.S. Some of them have already been approved by the American Society for Testing and Materials (ASTM). Wave propagation methods for determining stiffness of geomaterials have several advantages:

1. Most wave propagation methods are relatively simple, rapid, repeatable, and nondestructive.

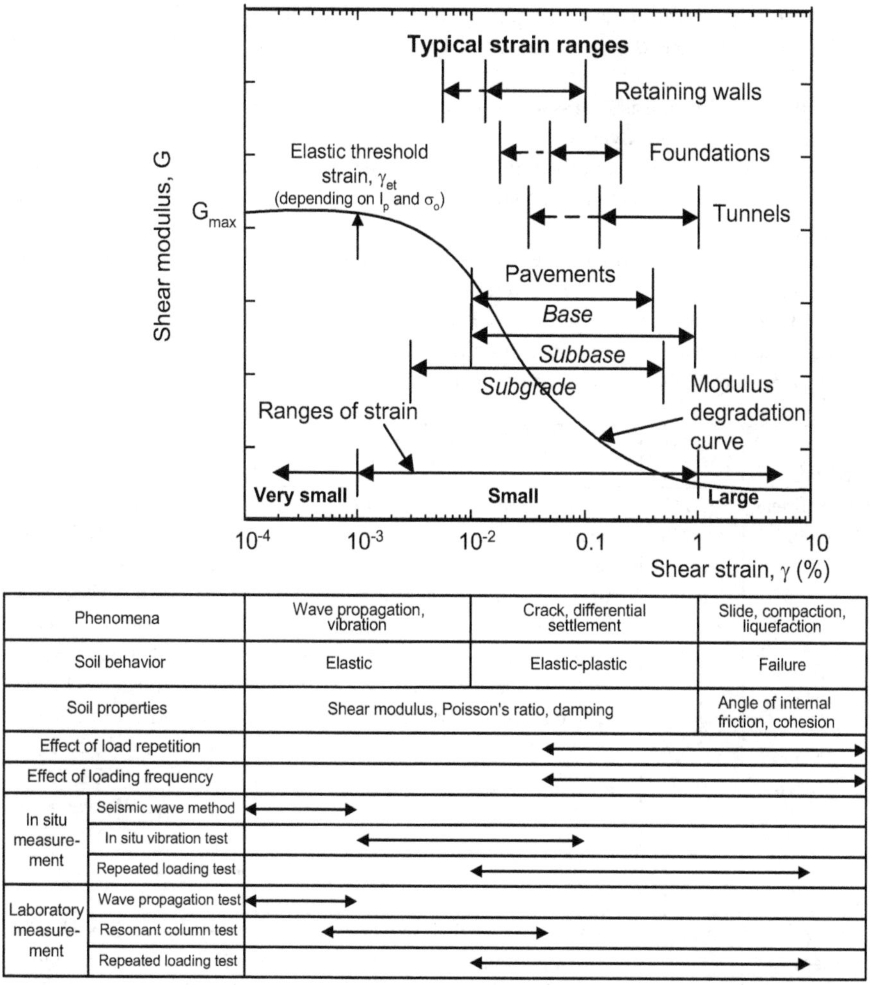

Figure 1. Variation in shear modulus with different shear strain levels for different geoengineering applications, in-situ tests, and laboratory tests (after Atkinson and Sallfors 1991, Mair 1993, Ishihara, 1996, Sawangsuriya et al. 2005).

2. Good agreement between stiffness measured in the laboratory and in the field is made when the laboratory specimens are at the same conditions as those in the field (Anderson and Woods 1976, Viggiani and Atkinson 1995a, Nazarian et al. 1999, Atkinson 2000).
3. Load repetition, strain rate, and loading frequency have only minor influence in the small-strain range (Iwasaki et al. 1978, Ni 1987, Bolton and Wilson 1989, Tatsuoka and Shibuya 1991, Jardine 1992, Shibuya et al. 1992, 1995, Ishihara 1996).

4. Stiffness of geomaterials is unique for both static (monotonic) and dynamic (cyclic) loading conditions (Georgiannou et al. 1991, Jamiolkowski et al. 1994, Tatsouka et al. 1997).
5. Little or no hysteresis (stress-strain loop) exists in both slow repetitive and dynamic cyclic loading tests (Silvestri 1991).
6. Volumetric and shear deformations (or strains) are fully recoverable and the tendency of geomaterials to dilate or to contract during drained shear does not occur (Ishihara 1996).
7. Stiffness is independent of drainage since the induced strain levels are too small to cause pore water pressure to build up during undrained shear test (Ohara and Matsuda 1988, Dobry 1989, Georgiannou et al. 1991, Silvestri 1991). Pore water pressure does not build up if the shear strain amplitude is smaller than 10^{-2}% for sands (Dobry 1989) and 0.1% for clays (Ohara and Matsuda 1988).

Test Methods	Standard	Test Principle	References
Soil Stiffness Gauge (SSG)	ASTM D 6758	A small dynamic force generated inside the device is applied through a ring-shaped foot resting on the ground surface and a deflection is measured using velocity sensors. The near-surface stiffness of geomaterials is then determined as the ratio of the applied force to the measured deflection.	Wu et al. (1998), Humboldt (1999, 2000a, 2000b), Fiedler et al. (1998, 2000), Nelson and Sondag (1999), Chen et al. (1999), Siekmeir et al. (1999), Hill et al. (1999), Sargand et al. (2000), Weaver et al. (2001), Lenke et al. (2001, 2003), Sargand (2001), Peterson et al. (2002), Sawangsuriya et al. (2002, 2003, 2004)
Bender Element	None	Shear wave velocity is determined by measuring the travel time of shear wave and the tip-to-tip distance of piezoceramic bender elements. The corresponding shear stiffness is calculated by knowing the shear wave velocity and mass density of geomaterial.	Dyvik and Madshus (1985), Thomann and Hryciw (1990), Hryciw and Thomann (1993), Souto et al. (1994), Fam and Santamarina (1995), Nakagawa et al. (1996), Viggiani and Atkinson (1995b, 1997), Jovicic and Coop (1998), Zeng and Ni (1998), Fioravante and Capoferri (2001), Santamarina et al. (2001)

Resonant Column	ASTM D 4015	Resonant frequency is measured and is related to the shear wave velocity and the corresponding shear stiffness.	Wilson and Dietrich (1960), Hardin and Music (1965), Hardin (1970), Drnevich (1977), Drnevich et al. (1978), Edil and Luh (1978), Isenhower (1980), Drnevich (1985), Ray and Woods (1988), Morris (1990), Lewis (1990), Cascante et al. (1998)
Pulse Transmission (Ultrasonic Pulse)	ASTM C 597	Elastic wave velocity is determined by measuring a travel time of either compressional wave or shear wave arrivals and the distance between ultrasonic transducers made of piezoelectric materials. The stiffness of geomaterial is calculated based on an elastic theory.	Lawrence (1963), Nacci and Taylor (1967), Sheeran et al. (1967), Woods (1978), Nakagawa et al. (1996), Yesiller et al. (2000)
Seismic Reflection	None	Travel time of seismic waves reflected from subsurface interfaces following the law of reflection is measured so that the elastic wave propagation velocity and the corresponding stiffness of geomaterial are determined.	Kramer (1996), Sharma (1997), Frost and Burns (2003)
Seismic Refraction	ASTM D 5777	Travel time of seismic refracted waves when they encounter a stiffer material (higher shear wave velocity) in the subsurface interface following the law of refraction (Snell's law) is measured so that the elastic wave propagation velocity and the corresponding stiffness of geomaterial are determined.	Kramer (1996), Sharma (1997), Frost and Burns (2003)
Spectral Analysis of Surface Waves (SASW)	None	Surface (Rayleigh) wave velocity varied with frequency is measured by utilizing the dispersion characteristics of surface wave and the fact that surface waves propagate to depths that are proportional to their wavelengths or frequencies in order to determine the stiffness of subsurface profiles.	Nazarian and Stokoe (1987), Sanchez-Salinero et al. (1987), Rix and Stokoe (1989), Campanella (1994), Nazarian et al. (1994), Wright et al. (1994), Mayne et al. (2001)

Seismic Cross-Hole	ASTM D 4428	Measurement of wave propagation velocity either compressional or shear wave from one subsurface boring to other adjacent subsurface borings in a linear array. The seismic wave is generated by various means so that the elastic waves propagate in the horizontal direction through the geomaterial and are detected by the geophones located in the other hole.	Stokoe and Woods (1972), Stokoe and Richart (1973), Anderson and Woods (1975), Hoar and Stokoe (1978), Campanella (1994), Mayne et al. (2001), Frost and Burns (2003)
Seismic Down-Hole or Up-Hole	None	Compressional and/or shear waves propagating vertically in a single borehole are monitored. The travel time of compressional and/or shear waves from the source to receiver(s) is measured. The wave propagation velocity at any depths is obtained from a plot of travel time versus depth.	Richart (1977), Campanella (1994), Ishihara (1996), Mayne et al. (2001), Frost and Burns (2003)
Seismic Cone Penetration	None	Similar to the seismic down-hole test, except that no borehole is required. The profile of shear wave velocity is obtained in a same manner as the seismic down-hole test. The receiver is located in the cone	Campanella et al. (1986), Robertson et al. (1986), Baldi et al. (1988), Campanella (1994), Kramer (1996), Mayne (2001), Frost and Burns (2003)

Table 1. Summary of wave propagation methods for determining stiffness of geomaterials.

In particular, pulse transmission method using piezoelectric transducers have been commonly used to monitor P- and S-wave propagation in many different types of geomaterials (Lawrence 1963, Sheeran et al. 1967, Woods 1978, Yesiller et al. 2000). In spite of its common use, such method has several shortcomings. For instance, weak transmitted wave signals, poor coupling between transducer and medium, near field effects, and high operating frequencies reduce the signal-to-noise ratio of the collected signals exacerbating the difficulties in interpreting the waveforms generated with these transducers (Sanchez-Salinero et al. 1986, Brignoli et al. 1996, Nakagawa et al. 1996, Ismail and Rammah 2005). A number of studies on the use of bender elements in wave-based technique has overcome many of the aforementioned problems and recently become a very popular method to measure the S-wave velocity and small-strain shear modulus (G_o) and their evolution with changes in effective stresses, water content, and cementation (Dyvik and Madshus 1985, Thomann and Hryciw 1990, Souto et al. 1994, Fam and Santamarina 1995, Cho and Santamarina 2001, Pennington et al. 2001, Mancuso et al. 2002, Zeng et al. 2002, Lee and Santamarina 2005).

A bender element test utilizes a pair of two-layer piezoelectric ceramic materials for transmitting and receiving S-waves (Shirley 1978). Bender elements are widely used to determine the wave velocity as S-wave pass through a geomaterial. Because the movement of a bender element is relatively small, when it is energised, the resulting sample displacements are tiny. Thus, the shear modulus obtained from this device can be defined as the shear modulus in the small-strain region. The first application of bender elements in laboratory geotechnical testing was by Shirley and Hampton (1978). Dyvik and Madshus (1985) also measured G_o in laboratory tests using bender elements which were installed in a top cap and a pedestal of an oedometer at the Norwegian Geotechnical Institute. Since then much research has been carried out using bender elements in triaxial apparatus, e.g. Viggiani 1992, Jovicic 1997, and Pennington 1999, along with their interpretation techniques of the testing results, e.g. Viggiani and Atkinson 1995 and Jovicic et al. 1996. Nowadays, bender elements are widely used to determine the wave velocity as shear wave pass through a sample.

3. Bender element measurement and instrumentation system

A pair of bender elements (i.e., one is a transmitter and another is a receiver) is utilized in the shear wave (S-wave) measurement. Bender elements act both as actuators and sensors that they distort or bend when subjected to a change in voltage and generate a voltage when are distorted or bent. Mounted as cantilever beams, bender elements are protruded a small distance into a specimen to provide robust coupling and induce elastic disturbances. During the excitation of bender elements, two types of mechanical waves are generated: compression (P) and shear (S) waves (Santamarina et al. 2001).

Typical configuration and electrical wiring for transmitting and receiving elements are illustrated in Fig. 2. A bender element can be tailor made in accordance with the testing apparatus and requirement (Sawangsuriya et al. 2008b). Fig. 3 illustrates an example of a series-connected piezoelectric ceramic bender element with the dimensions of 6.4-mm wide, 11.0-mm long, and 0.6-mm thick. Thin coaxial cables are soldered to the conductive bender element surfaces and their exposed electric wiring is completely coated with thinned polyurethane in order to provide electrical insulation. Typically after insulating with polyurethane, the bender elements can be coated with a conductive silver painting that creates an electric shield. The shield must be grounded to minimize electrical noise and to avoid electrical cross-talk between source and receiver. As also illustrated in Fig. 3, a custom made aluminum bolt-clamp anchoring system can be used to create a rigid cantilever system for supporting and accommodating the bender element. By using this anchoring system, one third of the bender element length (~4 mm) is anchored in the housing and is rigidly clamped by screws. The bender element-anchoring system can be directly mounted on opposite ends of the specimen such that the S-wave propagating longitudinally was sent and received (Sawangsuriya et al. 2008b).

The transmitting bender element produces an S-wave which propagates through a medium when it is excited by an applied voltage signal. This S-wave impinges on the receiving bender element, causing it to bend, which in turn produces a very small voltage signal. Fig. 4 illustrates typical electrical input and output signals from the transmitting and receiving bender elements. Fig. 5 illustrates a modeled input electrical step function, a modeled source bender element response and the receiver bender element response. The true input signal is somewhat significantly different from the signal generated by the signal generator. This is due to the fact that the response of the source bender element is controlled by the elastic properties of the bimorph, the cantilever length, the support properties (i.e., fixity), and the elastic properties of the surrounding medium (Lee and Santamarina 2005). Furthermore, the receiver bender element's response is governed not only by the stiffness, attenuation and dispersive properties of the medium but also by the distance between source and receiver bender elements, the wavelength, the distance to other boundaries, and the generation of reflected waves (Sawangsuriya et al. 2006, Arroyo et al. 2006).

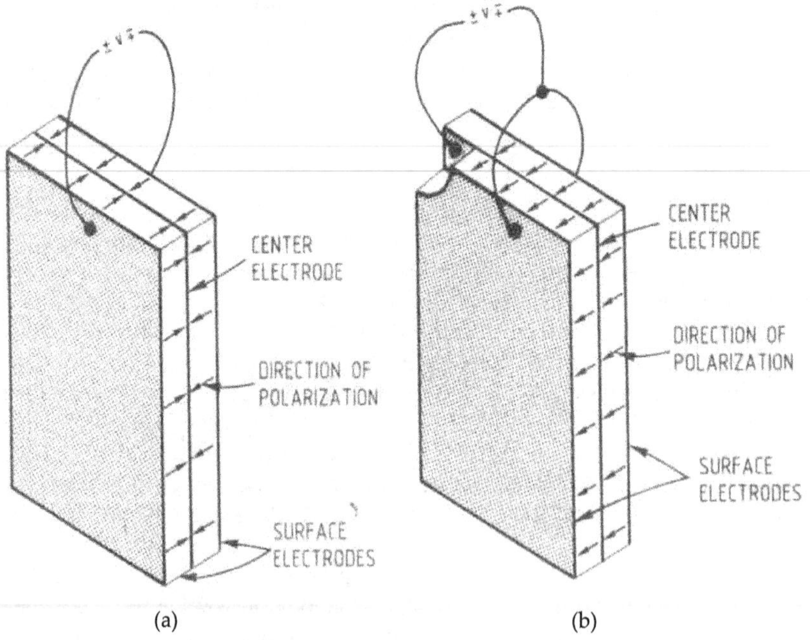

Figure 2. Two types of piezoceramic bender elements: (a) series connected and (b) paralleled connected (Dyvik and Madshus 1985).

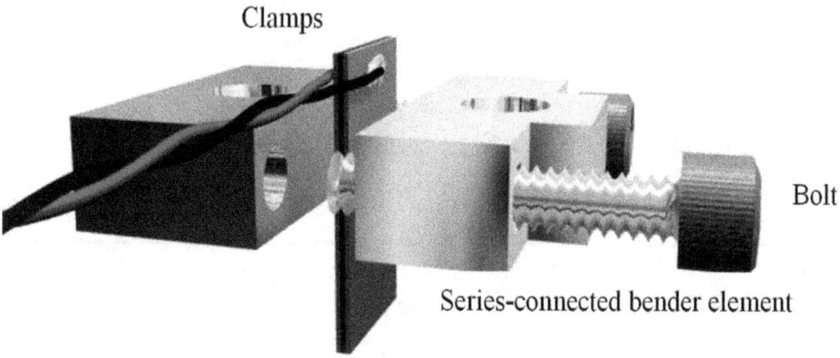

Figure 3. A 3-D drawing of bender element housing (Sawangsuriya et al. 2008b).

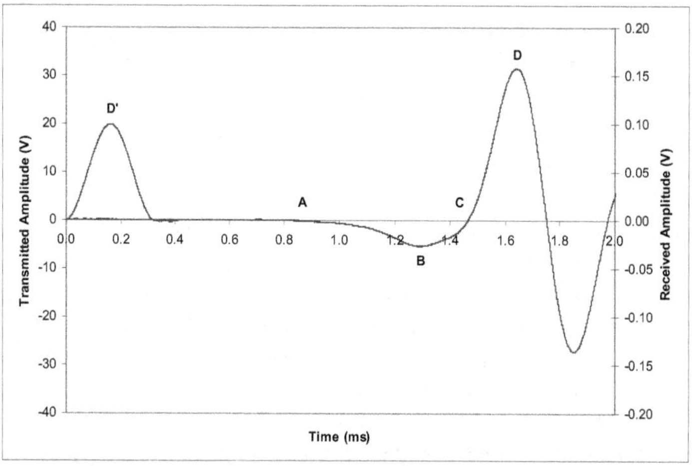

Figure 4. Typical input and output signals from the transmitting and receiving bender elements (Sukolrat 2007).

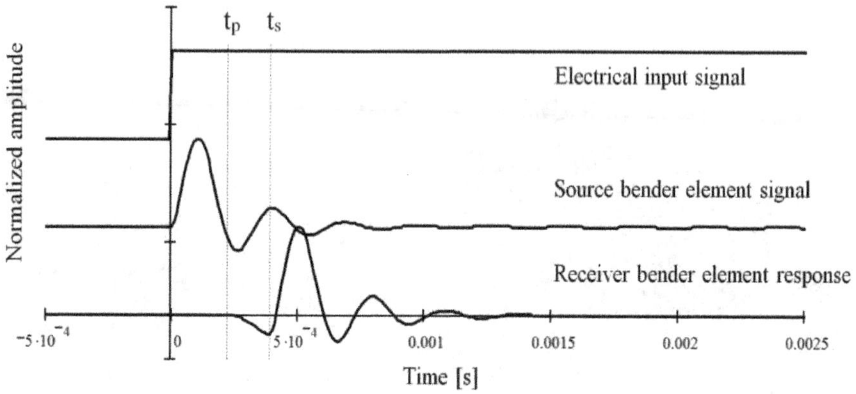

Figure 5. Modeled input electrical step function, signal generated by the source bender element and output signals from the source and receiving bender elements (Received signal modeled using Sanchez-Salinero et al. 1986– Model parameters: wave velocities V_s=200 m/s and V_p=310 m/s, and damping D=0.005) (Sawangsuriya et al. 2006).

Once these boundary and scale effects are evaluated and their effects are considered, the travel time between source and receiver bender elements can be determined. The recorded traces provide a means to measure the S-wave travel time, calculate the S-wave velocity, and evaluate the corresponding shear modulus (if the density is known). By measuring the travel time of the S-wave (t_s) and the tip-to-tip distance between transmitting and receiving bender element (L'), the S-wave velocity of the specimen (V_s) is obtained as:

$$V_s = \frac{L'}{t_s} \tag{1}$$

The instrumentation system for the bender element test consists of a pair of bender elements, function generator, signal amplifier, voltage divider for the input signals and digital oscilloscope, signal amplifier/filter and digital oscilloscope. Fig. 6 illustrates a schematic diagram of the bender element instrumentation system and that of an automatic multiplex bender element acquisition system when more than three pairs of bender elements are installed to a sample. A signal generator is commonly used to generate an input signal to the transmitting bender element. It is also noteworthy that the square-wave input signal gives the clearest response regardless of sample's stiffness because it includes all frequencies, which is advantageous when the resonant frequency of the bender element-medium system is unknown (Kawaguchi et al. 2001, Lee and Santamarina 2005).

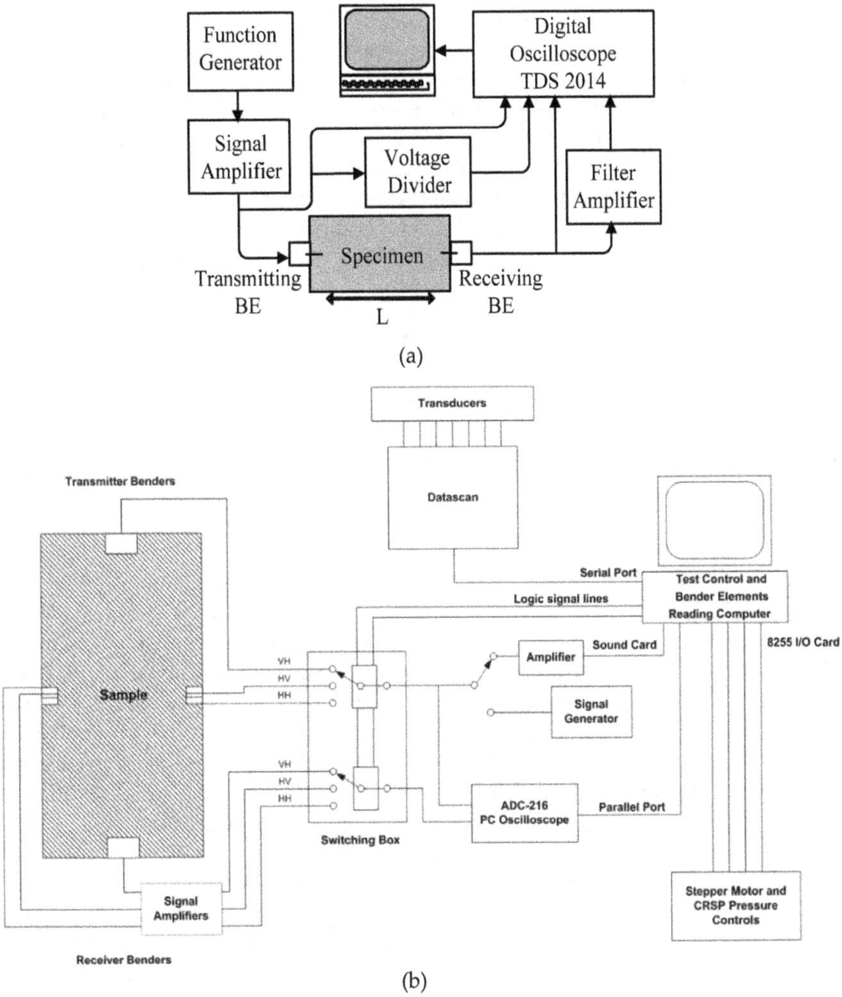

Figure 6. Schematic diagram of the bender element instrumentation system: (a) manual system for one direction of wave propagation (Sawangsuriya et al. 2006), (b) automatic multiplexing system for triaxial sample (Sukolrat 2007).

4. Collection and interpretation of bender element measurement data in geomaterials

Although S-wave measurements using bender elements are promising, the convenience of bender element tests is limited by subjectivity associated with identifying wave travel time arrivals. The effect of distances to boundaries and between source and receiver plays an important role in the evaluation of wave propagation test data. For example when S-waves are created in a bender element wave propagation test, reflection and refraction from boundaries may produce waves other than direct S-waves; near-field effect may obscure the S-wave arrival; or wave attenuation and dispersion may prevent the proper identification of multiple reflections (Sanchez-Salinero et al. 1986, Fratta and Santamarina 1996). Sawangsuriya et al. (2006) performed a series of experimental studies to explore and to establish the limits to minimize the boundary effects in the wave propagation experiments. The experimental results are presented as simple dimensionless ratios that can provide guidelines for the design and interpretation of the wave propagation experiments as the following:

Near-field effects

Distinguishing S-wave arrivals in the near-field is difficult since the arrival of the S-wave may be obscured by other wavefronts that propagate at higher velocities (e.g. P-wave velocity - Sanchez-Salinero et al. 1986, Lee and Santamarina 2005). This effect becomes significant especially at closer distances between sources and receivers. Sanchez-Salinero et al. (1986) developed a closed-form solution for longitudinal and transverse motions in linear-elastic and homogeneous media. Fig. 7a shows the arrivals for distances between source and receiver (L) equal to 1, 2, 3, 4, 6, and 8 shear-wave wavelengths (λ_s). Note that for greater relative source-receiver distances (L/λ_s), the separation between P- and S-wave arrivals increases, the amplitude of the transverse displacement caused by the P-wave decreases and the S-wave arrival is better evaluated. To clearly separate the P- and S-wave arrivals, L/λ_s must be greater than 4. However, this ratio also depends on the material damping. Understanding of the near-field effects may help interpreting both P- and S-waves arrival times which can also be used to evaluate the elastic modulus and Poisson's ratio. According to the wave propagation experiments by Sawangsuriya et al. (2006), wave propagation results were obtained using long cylindrical Kaolinite specimens (77.5 mm diameter) that were trimmed to different heights and hence different source-receiver distances. Since the specimens were identical, the slope of the measured travel time versus distance was used to calculate the P- and S-wave velocities (see Fig. 7b). The calculated Poison's ratio (ν) is 0.17. As the source-receiver distance increases, the P-wave arrival is more difficult to determine (Fig. 7a), for this reason there are less P-wave arrival time data points at greater source-receiver distances in Fig. 7b. It is therefore recommended that the minimum source-receiver distance (L) must be at least three times the wavelength (λ_s) (greater for large damping ratios) to facilitate the interpretation of the S-wave arrivals.

Boundary effects - reflection

When a bender element deflects, it generates S-waves in the direction of the bender element plane and P-waves in the direction normal to the bender element plane (Santamarina et al. 2001). Therefore, the presence of lateral boundaries may cause the reflected P-waves to arrive faster than direct S-waves hindering the proper evaluation of S-wave velocities. The difference in travel time arrivals depends on the Poisson's ratio (ν) of the geomaterial (Lee and Santamarina 2005). In 3-D arrangements of sources and receivers, the limiting separations can be expressed as a function of dimensionless ratios H/L and R/L and the Poisson's ratio as illustrated in Fig.8. Note that the normalized distances below the curves are the acceptable distances where the S-wave arrives earlier than the reflected P-wave. It is therefore important that depending on the Poisson's ratio of the geomaterial, the relative distance between sources and receivers and to the boundaries (H/L and R/L) must be limited to avoid possible P-wave reflection interferences.

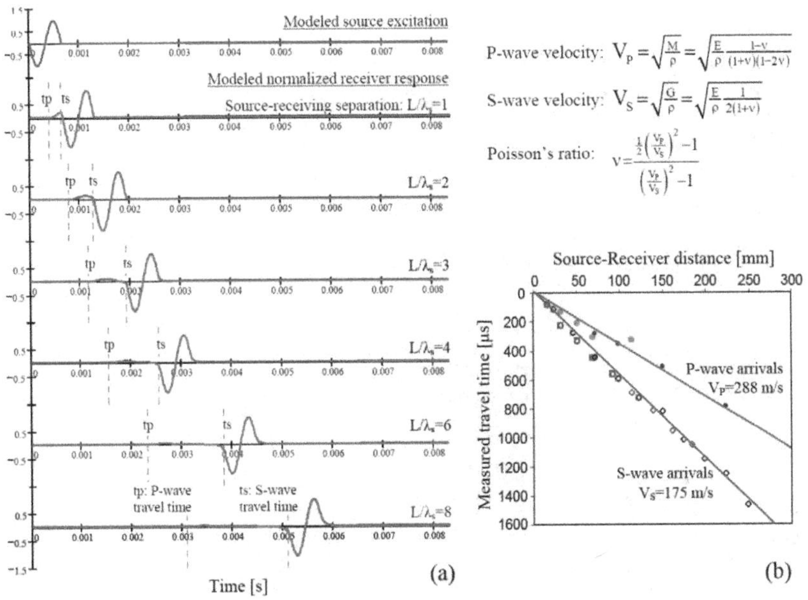

Figure 7. Near-field effects: (a) normalized receiver response (transverse particle motion) with increasing source-receiver separation (model parameters: V_P=288 m/s, V_s=175 m/s, damping D=0.01 – closed-form solution by Sanchez-Salinero et al. 1986), and (b) measured P- and S-waves travel times in Kaolinite specimens (Sawangsuriya et al. 2006). Note: M = constraint modulus, G = shear modulus, E = Young's modulus, ρ = mass density.

Boundary effects - refraction

As the separation between source and receiver increases, refracted waves may travel along rigid boundaries (having higher velocities) masking the S-wave arrivals. This phenomenon can be observed when the test specimen is confined within a rigid wall container. When monitoring S-waves, the velocity is calculated by an assumed straight ray travel length and the first arrival time. However, the presence of the rigid boundary may mask the arrival of the direct S-wave if a refracted wave arrives first. This phenomenon was observed by using specimens prepared inside the cylindrical rigid wall container where the source-receiver distance and the distance to the wall varied (Sawangsuriya et al. 2006). Results matched well with a simple refraction model used to determine the shortest travel time from the two possible wave paths. A sketch of the model and a summary of experimental results are shown in Fig. 9. Results suggest that for a given medium and PVC wall having finite wave velocities, the bender elements need to be placed at a minimum distance (d) greater than 0.4 L to avoid receiving refracted signals. As a consequence, it is recommended that with the presence of rigid boundaries, the distance of bender elements to the boundary (d) shall be greater than approximately 0.4 L to avoid refracting wave from arriving first.

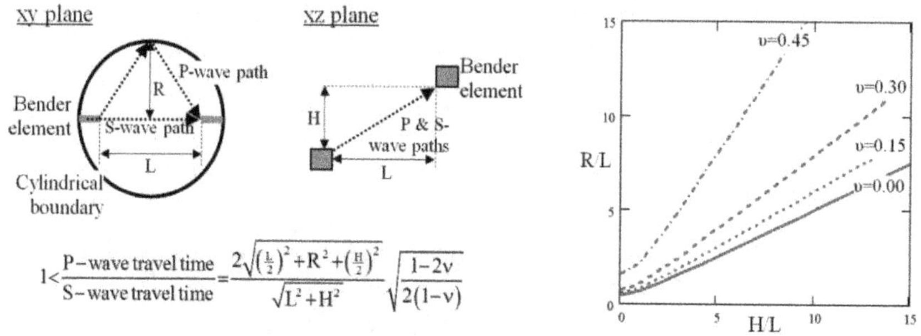

Figure 8. Limiting distances to avoid reflected P-wave arrivals when testing with boundaries. The lines in the plot represent the upper limits, the recommended test dimensions should fall under these lines (Sawangsuriya et al. 2006).

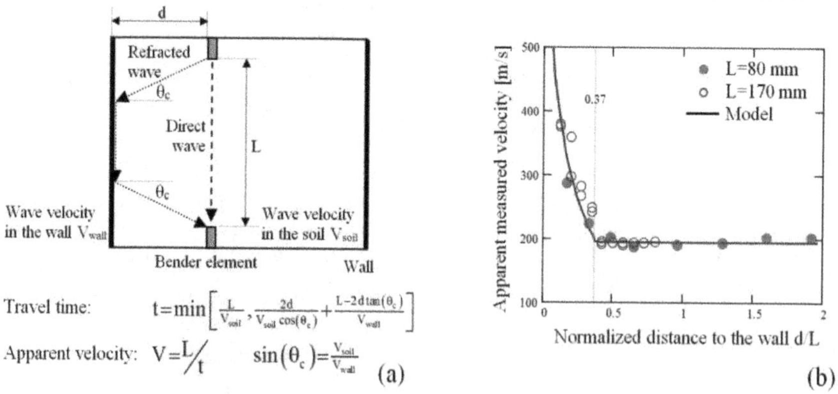

Figure 9. (a) Sketch of the test setup and the model for data evaluation and (b) experimental results from Kaolinite specimens (water content = 22%, unit weight = 9.7 kN/m^3) (Sawangsuriya et al. 2006).

Identifying wave travel time arrivals

A selection of the travel time for a wave travelling from a transmitter bender to a receiver bender has been controversial. Generally, the travel time can be determined either in the time domain or in the frequency domain. In the time-domain method, the travel time of the S-wave is monitored by plotting recorded signal with time. The S-wave velocity is then calculated if the tip-to-tip distance between the transmitting and receiving bender elements is known (Dyvik and Madshus 1985; Viggiani and Atkinson 1995). In general, the signals may be different from those in Figs. 4 and 5, possibly due to the stiffness of the sample, the boundary conditions, the test apparatus, the degree of fixity of the bender element into the platen or housing, the elastic properties of the bimorph, the cantilever length, the support properties (i.e., fixity), and the elastic properties of the surrounding medium as aforementioned. Once these boundary and scale effects are evaluated, and their effects are considered, the travel time between the transmitter and the receiver benders can be determined. The S-wave arrival corresponds to point C (Fig. 4) is typically chosen for the determination of the travel time in order to avoid the near-field effect caused by the arrival of energy with P-wave velocity (point A in Fig. 4 – Sanchez-Salinero et al. 1986; Kawaguchi et al. 2001; Lee and Santamarina 2005).

In the frequency-domain method, the transmitter sends a continuous sinusoidal wave at a single frequency through the sample to the receiver, and the phase shift between the two waves is measured. By varying the frequency, the phase delay of the transmitted and received signals can be observed, and the frequencies may be found when they are exactly in-phase and exactly out-of-phase when the phase differences are exact multiple N of π. The gradient of the plot between wave number (N) against frequency gives the group travel time. This is also called as π-point method. Similar results may be obtained using a sine sweep instead of a continuous wave of a single frequency. The Automatic Bender Elements Testing System (ABETS) program has been developed and uses phase-delay method to determine the shear wave velocities from bender element tests (Greening and Nash 2004). The principle of the ABETS is to send a sine wave pulse of continuously varying frequency generated by a soundcard in a computer through the sample. The signals from the transmitter and the receiver are collected and analysed by embedded Fast Fourier Transform (FFT) analysis. By selecting the range of frequencies for which the coherence is close to unity which indicates high quality signal that there is no interference at that transmitted frequency, a linear fit to the unwrapped phase data is used to determine the group's travel time within selected frequency range.

5. Evaluation of S-wave velocity

The S-wave velocity in particulate geomaterials depends on the state and history of the effective stresses, void ratio, degree of saturation, and type of particles. Based on experimental evidence and analytical studies, the velocity-effective stress relationship for granular geomaterials is expressed as a power function having two physically-meaningful parameters: a coefficient α and an exponent β. These parameters represent the S-wave velocity at a given state of stress and its variation with stress changes. The velocity-stress power relationship for granular media under isotropic loading is expressed as (Roesler 1979, White 1983):

$$V_s = \alpha \left(\frac{\sigma'_o}{p_r} \right)^{\beta} \tag{2}$$

where V_s is the S-wave velocity, σ'_o is the isotropic effective stress, p_r=1 kPa is a reference stress, α and β are experimentally determined parameters. Fig. 10 illustrates an example of the velocity-stress relationship as determined by S-wave velocity measurements at different stress states for dry medium sand prepared by different methods exhibits a power function (see Fig. 10) along with the experimentally determined coefficient α and exponent β (Sawangsuriya et al. 2007a). As shown in Fig. 10, for a dry medium sand subjected to isotropic loading, the velocity-stress power relation in Eq. (2) closely fits the data and as expected, the S-wave velocity increases as the confining pressure increases.

Several researchers have attempted to quantify the physical meaning of these experimentally determined parameters. Duffy and Mindlin (1957), Hardin and Black (1968), Santamarina et al. (2001), Fernandez and Santamarina (2001), and Fratta and Santamarina (2002) described the physical meaning of the parameters α and β: the coefficient α relates to the type of packing (i.e., void ratio or coordination number), the properties of the material, and fabric changes, while the exponent β relates to the effects of contact behavior. Both parameters indicate the effects of stress history, cementation and rock weathering in the formation. For example, dense sands, overconsolidated clays, and soft rocks have higher coefficient α and lower exponent β. In case of loose sands, normally consolidated clays, and clays with high plasticity, the coefficient α becomes lower while the exponent β becomes higher. Santamarina et al. (2001) suggested an inverse relationship between α and β values for various granular media, ranging from sands and clays to lead shot and steel spheres:

$$\beta \approx 0.36 - \frac{\alpha}{700\frac{m}{s}} \tag{3}$$

In addition, Hardin and Black (1968) suggested that the coefficient α can be separated into two coefficients: one accounts for the grain characteristic (or nature of grains), and the other accounts for the properties of the packing (i.e., void ratio or coordination number). The separation into these different components is justified on particulate material models. For example for random particles of uniform spheres with Hertzian contacts, the relationship between isotropic stresses and S-wave velocity is (Chang et al. 1991, Santamarina et al. 2001):

$$V_s = \underbrace{\left(\frac{\sqrt{\pi}}{4 \cdot \rho'} \cdot \frac{5-4\nu'}{2-\nu'} \right)^{1/2} \left(\frac{G'}{1-\nu'} \right)^{1/3}}_{\text{particle properties}} \cdot \underbrace{\left[\frac{1.7-e}{(1+e)^{5/2}} \right]^{1/3}}_{\text{packing properties}} \cdot \underbrace{(\sigma'_o)^{1/6}}_{\text{effective stress}} \tag{4}$$

where ρ', G', and ν' are the density, shear modulus and Poisson's ratio of the particles. This equation is derived assuming that the number of contacts per particle or coordination number C_n is related to the void ratio e by the following equation (Chang et al. 1991):

$$C_n = 13.28 - 8e \tag{5}$$

Experimentally, the velocity-stress power relation is re-written as:

$$V_s = A \cdot F(e) \cdot \left(\frac{\sigma_o^{'}}{p_r} \right)^{\beta} \tag{6}$$

where A is a fitting parameter which accounts for the properties of the grains and the function F(e) depends on the initial void ratio e of the geomaterial (Mitchell and Soga 2005). For round-grained particles, Hardin and Black (1968) proposed the following function:

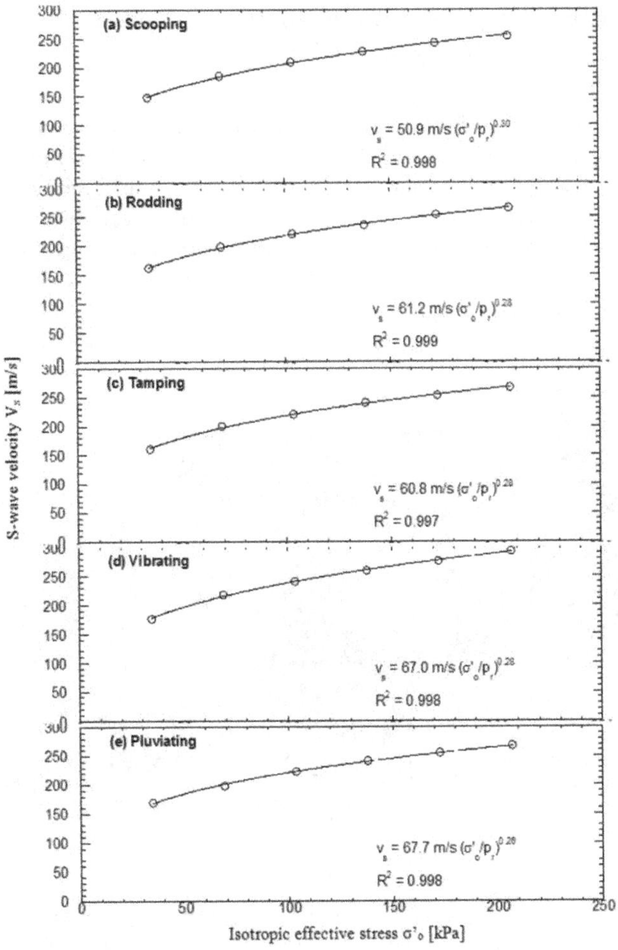

Figure 10. Velocity-stress response of specimen prepared with different methods (Sawangsuriya et al. 2007a).

$$F(e) = \frac{(2.17 - e)^2}{1 + e} \tag{7}$$

The evaluation of S-wave velocity using bender elements has been commonly performed on oedometer and standard triaxial samples. Normally, the top and bottom caps of an oedometer cell and a conventional triaxial cell are modified to incorporate bender elements and their electrical connections (see Fig. 11 – Sawangsuriya et al. 2007a). As shown in Fig. 11, the evaluation of S-wave velocity using bender elements is performed on standard triaxial specimens and subjected to isotropic loading. The top and bottom caps of a conventional triaxial cell are modified to incorporate bender elements and their electrical connections (see Fig. 11a). Examples of S-wave traces from bender element tests are shown in Fig. 12 along with the S-wave arrival as indicated by the arrows. As shown in Fig. 12, the S-wave arrival increases with confining pressure.

Laterally mounted benders were also used in triaxial apparatus (Pennington 1999). The horizontal pairs of bender elements are installed to study the anisotropy of the sample (see Fig. 13- Sukolrat 2007). In horizontal bender system, two similar bender elements were mounted orthogonally; one pair is used for S-wave transmitters and the other pair is for S-wave receivers. The bender elements are placed diagonally opposite one another and oriented so that the shear stiffness in both HV and HH directions are synchronized (depending on different polarization direction).

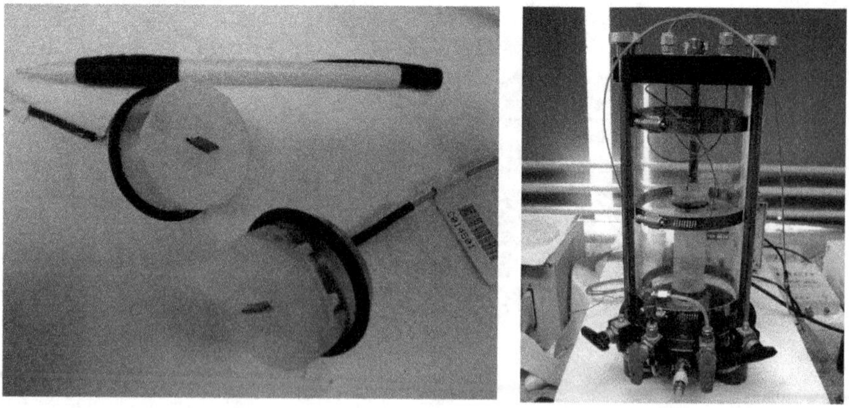

(a) (b)

Figure 11. Modified triaxial cell with bender element measurement system: (a) details of bender elements and (b) assembled modified triaxial cell (Sawangsuriya et al 2007a).

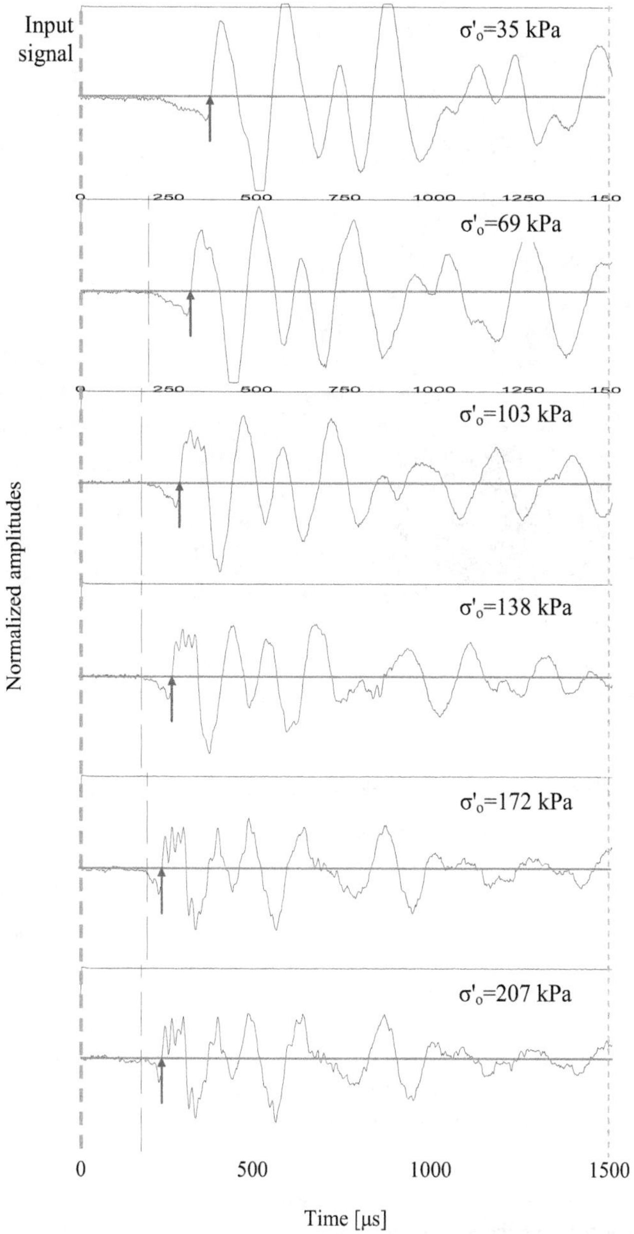

Figure 12. S-wave traces from bender element tests. The arrow corresponds to the estimated S-wave arrival. The first deflection corresponds to the near-field effect (Sawangsuriya et al. 2007a).

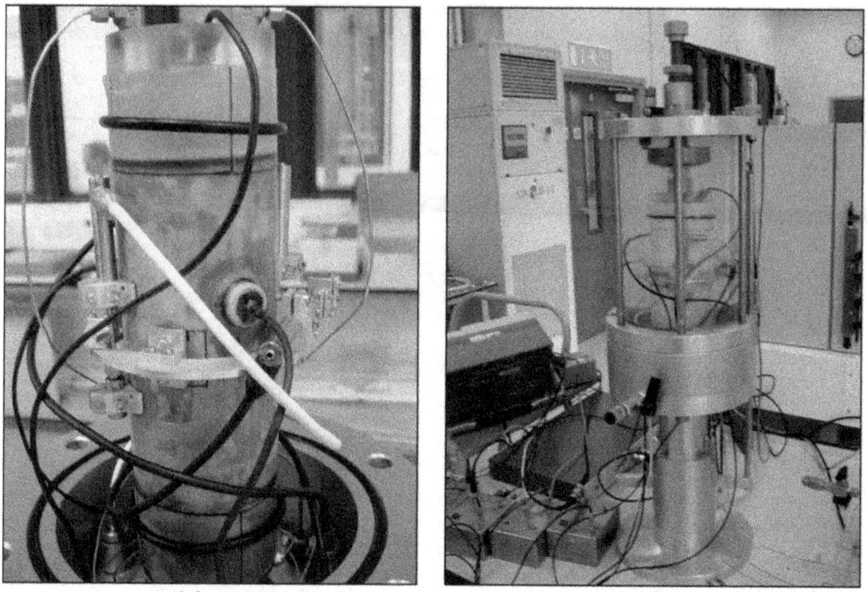

(a) (b)

Figure 13. Modified stress path cell to incorporate vertical and horizontal bender system and local strain measuring system: (a) Installation of measuring devices, (b) triaxial stress path cell during testing (Sukolrat, 2007).

6. Small-strain shear modulus based on bender element measurement

According to elastic theory using the measured S-wave velocity (V_s) and total mass density of the specimen ($\varrho = \gamma/g$), the small-strain shear modulus (G_o) can be calculated with the relationship $G_o=\varrho \cdot V_s^2$. G_o is an important and fundamental geomaterial property for a variety of geotechnical design applications and can be applied to all kinds of static (monotonic) and dynamic geotechnical problems at small strains (Richart et al. 1970, Jardine et al. 1986, Burland 1989). Note that the term "small-strain" is typically associated with the shear strain range below the elastic threshold strain (10^{-3}-10^{-2}%). Within the small strain range where the deformations or strains are purely elastic and fully recoverable, the shear modulus is independent of strain amplitude and reaches a nearly constant limiting value of the maximum shear modulus. In this strain region, most geomaterials exhibit linear-elastic behavior.

A number of factors affecting G_o have been extensively investigated and reported. These include the current state of the sample (e.g. stress state, overconsolidation ratio, density, void ratio, and microstructure), anisotropy, degree of saturation, aging, cementation, and temperature. Such factors can be briefly explained as the followings:

Current state

The current state of a sample relative to G_o is defined by: (i) existing normal stresses in the ground which is also known as the mean effective principle stress or confining pressure (σ_o'), (ii) the overconsolidation ratio (OCR), and (iii) the void ratio (e) or the density of the geomaterial (ρ). By taking all parameters into account, a general expression as proposed by different investigators for G_o of geomaterials is of the following form:

$$G_o = A(OCR)^k f(e) p_a^{(1-n)} (\sigma_o')^n \quad (8)$$

where A is a dimensionless material constant coefficient, k is a overconsolidation ratio exponent, f(e) is a void ratio function, p_a is the reference stress or atmospheric pressure (~100 kPa) expressed in the same units as G_o and σ_o', and n is a stress exponent. A number of studies have been conducted to estimate these parameters by relating with other physical geomaterial properties as summarized in Table 2 and Table 3.

Parameter	Dependency	Typical value	References
A	Grain characteristics or nature of grains, fabric or microstructure of geomaterial	Determined by regression analysis for individual test	See Table 2
k	Plasticity index (PI)	Vary from 0 to 0.5 (for PI<40, k=0; PI>40, k = 0.5)	Hardin and Black (1968), Hardin and Drnevich (1972)
f(e)	Properties of packing and density	See Table 3	See Table 3
N	Contact between particles and strain amplitude	Approximately 0.5 at small strains	Hardin and Richart (1963), Hardin and Black (1966, 1968), Drnevich et al. (1967), Seed and Idriss (1970), Silver and Seed (1971), Hardin and Drnevich (1972), Kuribayashi et al. (1975), Kokusho (1980)

Table 2. Parameters describing a current state of sample for G_o

$$G_o = A \cdot f(e) \cdot (\sigma_o')^n$$

Type of geomaterials	A	f(e)	n	References
Round-grained Ottawa sand	6,900	$(2.17-e)^2/(1+e)$	0.5	Hardin and Black (1968)
Angular-grained crushed quartz	3,270	$(2.97-e)^2/(1+e)$	0.5	
Clean sand	41,600	$0.67-e/(1+e)$	0.5	Shibata and Soelarno (1975)
Clean sand (C_u < 1.8)	14,100	$(2.17-e)^2/(1+e)$	0.4	Iwasaki and Tatsuoka (1977)
Clean sand	9,000	$(2.17-e)^2/(1+e)$	0.4	Iwasaki et al. (1978)
Toyoura sand	8,400	$(2.17-e)^2/(1+e)$	0.5	Kokusho (1980)
Clean sand	7,000	$(2.17-e)^2/(1+e)$	0.5	Yu and Richart (1984)
Ticino sand	7,100	$(2.27-e)^2/(1+e)$	0.4	Lo Presti et al. (1993)
Clean sand	9,300	$1/e^{1.3}$	0.45	Lo Presti et al. (1997)
Reconstituted NC Kaolinite (PI = 20) and undisturbed NC clays	3,270	$(2.97-e)^2/(1+e)$	0.5	Hardin and Black (1968)
Reconstituted NC Kaolinite (PI = 35)	4,500	$(2.97-e)^2/(1+e)$	0.5	Marcuson and Wahls (1972)
Reconstituted NC Bentonite (PI = 60)	445	$(4.4-e)^2/(1+e)$	0.5	
Remolded clay (PI = 0~50)	2,000~4,000	$(2.97-e)^2/(1+e)$	0.5	Zen et al. (1978)
Undisturbed NC clay (PI = 40~85)	90	$(7.32-e)^2/(1+e)$	0.6	Kokusho et al. (1982)
Clay deposits (PI = 20~150)	5,000	$1/e^{1.5}$	0.5	Shibuya and Tanaka (1996)†
Remolded clay (PI = 20~60)	24,000	$1/(1+e)^{2.4}$	0.5	Shibuya et al. (1997)†
Sand and clay	6,250	$1/(0.3+0.7e^2)$	0.5	Hardin (1978)
Several soils	5,700	$1/e$	0.5	Biarez and Hicher (1994)

Note: G_o and σ_o' are in kPa, † using effective vertical stress (σ_v') instead of σ_o' , Cu = Coefficient of Uniformity

Table 3. Function and constants in proposed empirical equations on G_o

Anisotropy

Anisotropic G_o of geomaterials is generally described in terms of stress-induced anisotropy and inherent anisotropy (Stokoe et al. 1985).The stress-induced anisotropy results from the anisotropy of the current stress condition and is independent of the stress and strain history of the geomaterial. The inherent or fabric anisotropy results from structure or fabric of the geomaterial that reflects the deposition or forming process (e.g. aging, cementation). Both stress-induced anisotropy and fabric anisotropy of G_o depend on the direction of loading (Mitchell and Soga 2005). Such fabric anisotropy of G_o can be evaluated based on S-wave measurements using bender elements (Sawangsuriya et al. 2007b). For example, G_o is a function of the principal effective stresses in the directions of wave propagation and particle motion and is independent of the out-of-plane principal stress (Stokoe et al. 1995). The inherent anisotropy can be evaluated by measuring body wave velocities propagating through the specimen subjected to isotropic states of stress (i.e., mean effective stress). For the stress-induced anisotropy, the measurements are taken from specimen subjected to anisotropy states of stress (i.e., changes in vertical stress while maintaining average principal stresses).

Under anisotropic states of stress, the representative stiffness values can be different, depending on the measurement conditions and the sample preparation procedures. The anisotropy of the stress state induces anisotropy of small-strain stiffness. An empirical equation for G_o under anisotropic stress condition is expressed as (Roesler 1979, Stokoe et al. 1985):

$$G_o = A(OCR)^k F(e) p_a^{1-n} \sigma'^{n_i}_i \sigma'^{n_j}_j \quad (9)$$

where σ_i' is the effective normal stress in the direction of wave propagation, σ_j' is the effective normal stress in the direction of particle motion, and $n = n_i + n_j$.

Degree of saturation

Early studies on the influence of the degree of saturation on G_o described a coupled motion of the solid particles and the fluid (Biot 1956, Hardin and Richart 1963, Richart et al. 1970). According to Biot's theory, no structural coupling exists between the solid particle and the fluid (the fluid has no shearing stiffness), the coupling in the shearing mode is only developed by the relative motions of the solid and fluid as indicated by the term involving the apparent additional mass density and thus G_o can be expressed as:

$$G_o = v_s^2 \left(\rho + \frac{\rho_f \rho_a}{\rho_f + \rho_a} \right) \quad (10)$$

where ρ is the mass density of the solid particles, ρ_f is the mass density of fluid, and ρ_a is the mass density of an additional apparent mass. In a real geomaterial, ρ_a varies with the grain size and permeability; however, the total mass density of the saturated geomaterial could be substituted into the mass density term of Eq. (10) to take into account the coupling effect of the mass of the fluid. The shear wave velocity of saturated geomaterial is therefore less than that of dry geomaterial because the added apparent mass of water moving along with the geomaterial skeleton (i.e., the drag of the water in the pores). Recent studies by Santamarina et al. (2001) and Inci et al. (2003) indicated that the response of G_o by varying the degree of saturation demonstrates three phases of behavior and is attributed to contact-level capillary forces or suction. A sharp increase in G_o is observed at the beginning of the drying process, followed by a period of gradual increase in measured G_o, and a final sharp increase in G_o at the end of the drying period.

Aging

A time-dependent nature of G_o of geomaterials has been reported by several investigators (Afifi and Woods 1971, Marcuson and Wahls 1972, Afifi and Richart 1973, Stokoe and Richart 1973, Trudeau et al. 1974, Anderson and Woods 1975, Anderson and Woods 1976, Anderson and Stokoe 1978, Isenhower and Stokoe 1981, Athanasopoulos 1981, Kokusho 1987). Results of these investigations indicate that G_o tends to increase with the duration of time under a constant confining pressure after the primary consolidation is complete due to a time effect results from strengthening of particle bonding. The time dependency of G_o increase can be characterized by two phases: (i) an initial phase due to primary consolidation and (ii) a second phase in which G_o increases about linearly with the logarithm of time and occurs after completion of primary consolidation, also referred as the long-term time effect (Fig. 14). The second phase of secondary consolidation occurs after primary phase when G_o increases continuously with time. The rate of secondary increase in G_o is related to thixotropic changes in the clay structure and is determined to be linear when plotted versus the logarithm of time. To incorporate this long-term time effect, the change in G_o with time can be expressed by:

$$\Delta G = N_G G_{1000} \tag{11}$$

and

$$N_G = \left(\frac{1}{G_{1000}}\right)\left(\frac{\Delta G}{\log_{10}(t_2 / t_1)}\right) \tag{12}$$

where ΔG is the increase in G_o over one logarithm cycle of time, G_{1000} is the value of G_o measured after 1,000 minutes of application of constant confining pressure following the primary consolidation, and N_G is the aging increment coefficient, which indicates an increase of G_o within one logarithmic cycle of time.

The duration of primary consolidation and the magnitude of the secondary increase, as defined by change in G_o per logarithmic cycle of time, vary with geomaterial types and stress conditions (i.e., confining pressure). For sands, the rate of increase in G_o is relatively small (1 to 3% per log cycle of time) but for clays the effect is quite remarkable as illustrated in Fig. 15.

Cementation

Cementation occurs either naturally due to the precipitation or formation of salts, calcite, alumina, iron oxides, silicates, and aluminates or artificial stabilization processes produced by adding lime, cement, asphalt, fly ash, or other bonding agents to geomaterials. The effect of cementation on G_o have been evaluated by Clough et al. (1981), Acar and El-Tahir (1986), Saxena et al. (1988), Lade and Overton (1989), Baig et al. (1997), Fernandez and Santamarina (2001), Yun and Santamarina (2005). G_o of cemented materials increases with increasing cement content and confining pressure (Fig. 16). Additionally at low confinement, the stiffness behavior of cemented materials is controlled by the cementation and the materials become brittle, whereas at high confinement the behavior is controlled by the state of stress and resembles an uncemented material, which becomes more ductile.

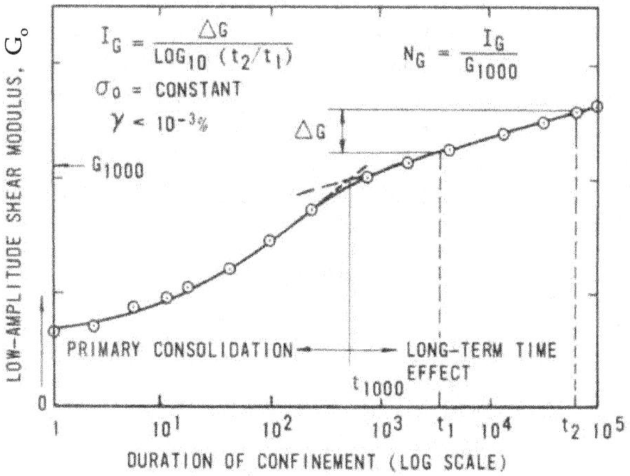

Figure 14. Phases of Go versus confinement time (Anderson and Stokoe 1978).

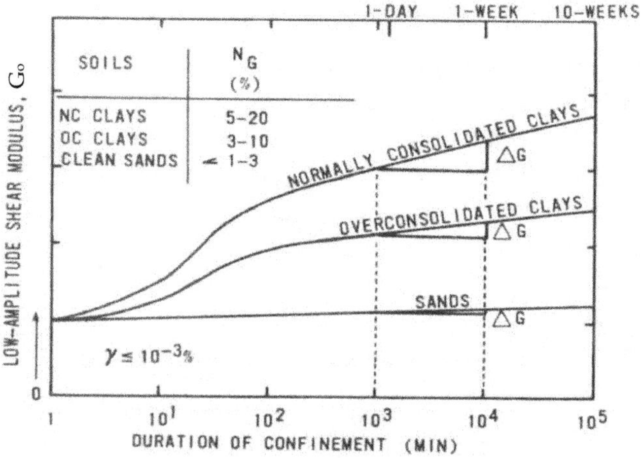

Figure 15. Effect of aging on Go (Anderson and Stokoe 1978).

Temperature

Effect of temperature on time-dependent changes in G_o was reported in Bosscher and Nelson (1987), Fam et al. (1998). The dependency of G_o on temperature suggests that higher temperatures cause the stiffness increase with time. Fam et al. (1998) presented the evolution in velocity with time for coarse-grained granular salt specimen under a constant effective stress and subjected to a temperature step (heating-cooling cycle) as illustrated in Fig. 17. The rate of increase in velocity with time increases at higher temperatures (Fam et al. 1998). Bosscher and Nelson (1987) studied G_o of frozen Ottawa 20-30 sand as a function of the confining pressure, the degree of ice saturation, the relative density, and the temperature. They found that G_o of frozen sand is higher than that of non-frozen state. At temperatures near the melting point of ice, G_o can be significantly influenced by the confining pressure, the degree of ice saturation, and the relative density (Bosscher and Nelson 1987).

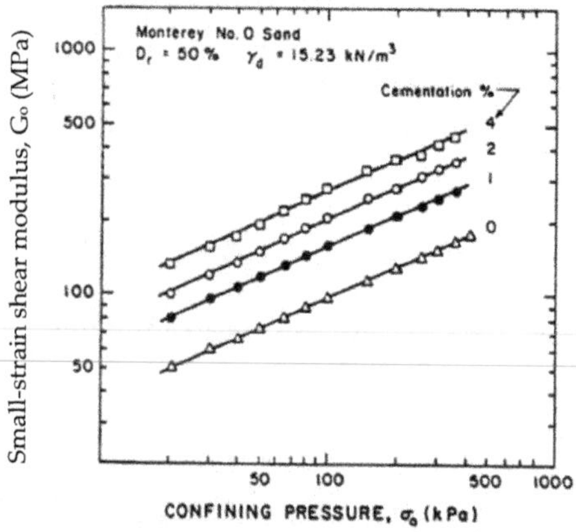

Figure 16. Effect of cementation on G_o (Acar and El-Tahir 1986).

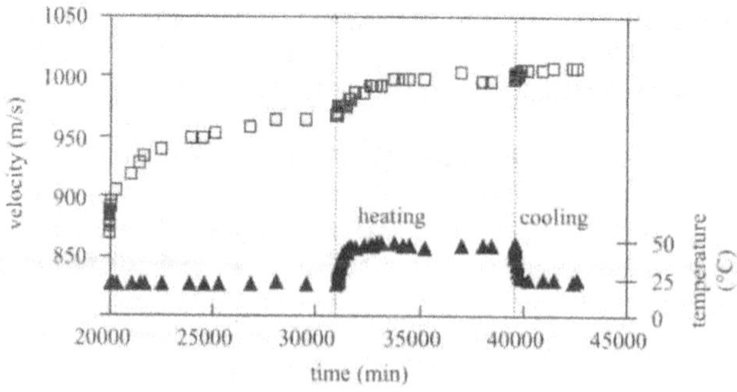

Figure 17. Effect of temperature on time-dependent changes in velocity for a coarse-grained granular salt specimen under a constant vertical load (Fam et al. 1998).

7. Application of wave propagation methods in geoengineering quality control process

The quality of the engineered earth fills depends on the suitability and compaction of the materials used. Earthwork compaction acceptance criteria are typically based on adequate dry density of the placed earthen materials achieved through proper moisture content and compaction energy. For instance, compaction specifications often require achievement of an in-situ dry density of 90-95% of the maximum value obtained from laboratory standard or modified Proctor test. According to this approach, by achieving a certain dry density using an acceptable level of compaction energy assures attainment of an optimum available level of structural properties and also minimises the available pore space and thus future moisture changes.

The question of the achieved structural property, which is the ultimate objective quality control, however remains unfulfilled. Dry density is only a quality index used to judge compaction acceptability and is not the design parameter or relevant property for engineering purposes. For compacted highway, railroad, airfield, parking lot, and mat foundation subgrades and support fills, the ultimate engineering parameter of interest is often the shear modulus of geomaterials, which is a direct structural property for determining load support capacity and deformation characteristic in engineering design.

Shear modulus of compacted geomaterials depends on density and moisture but also on fabric and texture of geomaterials, which varies along the roadway route. The conventional approach of moisture-density control, however, does not reflect the variability of the texture and microstructure of geomaterials and thus their shear modulus. Even if the structural layers satisfy a compaction quality control requirement based on density testing, a large variability in shear modulus can still be observed (Sargand et al. 2000; Nazarian and Yuan, 2000). Additionally, the comparison between density and modulus tests suggests that conventional density testing cannot be used to define subtle changes in the stiffness of the compacted earth fills (Fiedler et al. 1998). Shear modulus is a more sensitive measure of the texture and fabric uniformity than density. The stiffness non-uniformity is directly related to progressive failures and life-cycle cost, direct stiffness testing which can be conducted independently and in conjunction with conventional moisture-density testing is anticipated to reduce variability and substantially enhance quality control of the earthwork construction.

Ismail and Rammah (2006) proposed a test setup and procedure by which the small-strain shear modulus can be measured accurately by propagating elastic shear wave through the compacted geomaterial during laboratory compaction test. They designed a test setup in such a way that it can be readily incorporated into the conventional compaction mould as shown in Fig. 18. In addition, their procedures can be adopted in compaction works e.g. road construction, embankments, and earth structures. Consequently, incorporating the shear modulus into laboratory compaction tests, will guarantee fulfilment of the design criteria in term of dry density-moisture-modulus relationship.

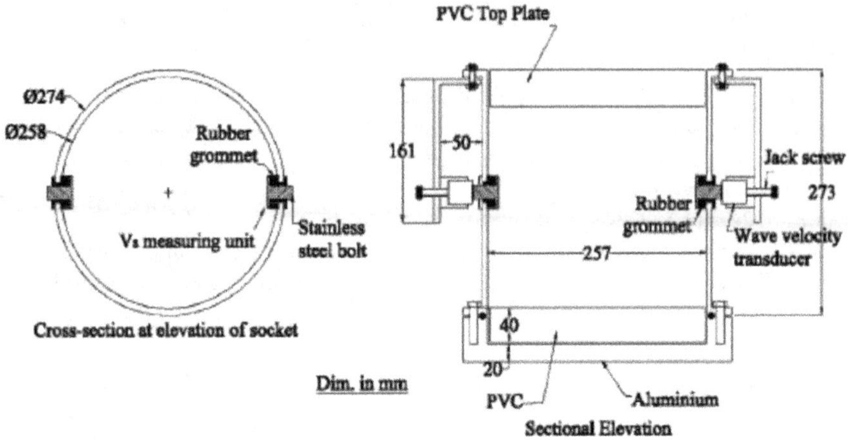

Figure 18. Seismic compaction mold (Ismail and Rammah 2006).

Sawangsuriya et al. (2008a) presented the experimental investigation of the shear modulus-matric suction-moisture content-dry unit weight relationship using three compacted subgrade soils. Compacted specimens were prepared over a range of molding water contents from dry to wet of optimum using enhanced, standard, and reduced Proctor efforts. A S-wave propagation technique known as bender elements was utilized to assess the shear wave velocity and corresponding G_o of the compacted specimens. S-wave traces of the clayey sand compacted with enhanced Proctor, standard Proctor, and reduced Proctor efforts are shown in Fig. 19. S-wave traces of the silt and lean clay compacted with standard Proctor effort are shown in Fig. 20.

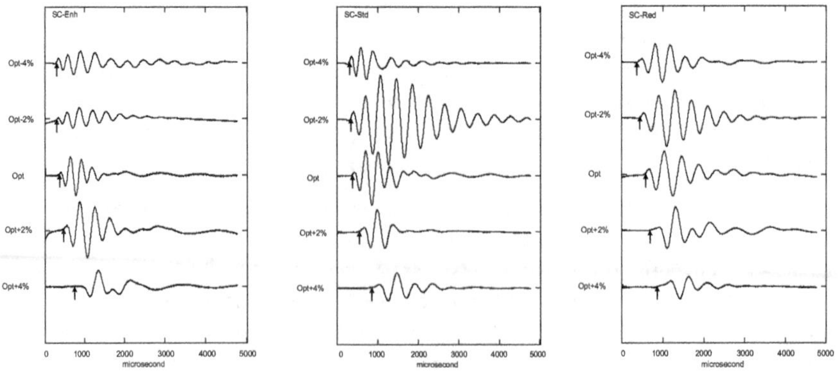

Figure 19. S-wave traces of the clayey sand compacted with enhanced Proctor (a), standard Proctor (b), and reduced Proctor (c) efforts (Sawangsuriya et al. 2008a).

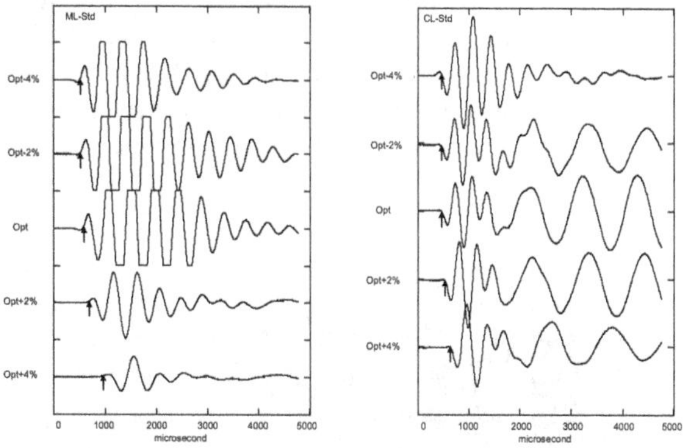

Figure 20. S-wave traces of the silt and lean clay compacted with standard Proctor effort (Sawangsuriya et al. 2008a).

8. Application of wave propagation methods for ground monitoring

In the past 20 years, the use of wave propagation methods (e.g. seismic. etc.) seismic reflection and refraction, seismic down-hole, up-hole and cross-hole etc. for ground monitoring has been promoted intensively because it is non-destructive and is conducted under in situ condition. The wave propagation methods are commonly used to determine stiffness-depth profile. Typically, the wave propagation methods can be classified into subsurface and near-surface methods as schematised in Fig. 21.

The subsurface methods including seismic down-hole, seismic up-hole, and seismic cross-hole are normally employed to monitor ground stiffness when the depth of interest is greater than 15 meters (Hooker 1998). In these methods, one or more boreholes are usually required in order that the source and/or the receiver can be installed. While the seismic cone penetration test can provide both stiffness and strength properties of the geomaterials during its penetration. The disadvantage of the subsurface method is that the cost of measurement is relatively high due to the cost of borehole casting. An alternative method is the near-surface method which provides simpler and more cost-effective approaches. The near-surface method is performed on the basis of the ability of wave propagation through the ground strata. When waves propagate through soil layers having different properties, they refract and/or reflect at different time. Once the arrival time is known, wave velocities and stiffness of each layer can be determined.

With the near-surface method, the Spectral Analysis of Surface Waves (SASW) using surface waves is another technique that can monitor both ground stiffness at shallow depth and layer thicknesses of subsurface profiles. The surface wave method utilises the dispersive characteristic of Rayleigh waves, which are elastic waves that propagate along the ground surface. Surface (Rayleigh) wave velocity varied with frequency is measured by utilizing the dispersion characteristics of surface wave and the fact that surface waves propagate to depths that are proportional to their wavelengths or frequencies in order to determine the stiffness of subsurface profiles. The objective in SASW testing is to make field measurements of surface wave dispersion (i.e., measurements of surface wave velocity at various wavelengths) and to determine the shear wave velocities of the layers in the profile based on the fact that surface waves propagate to depths that are proportional to their wavelengths (Mayne et al. 2001).

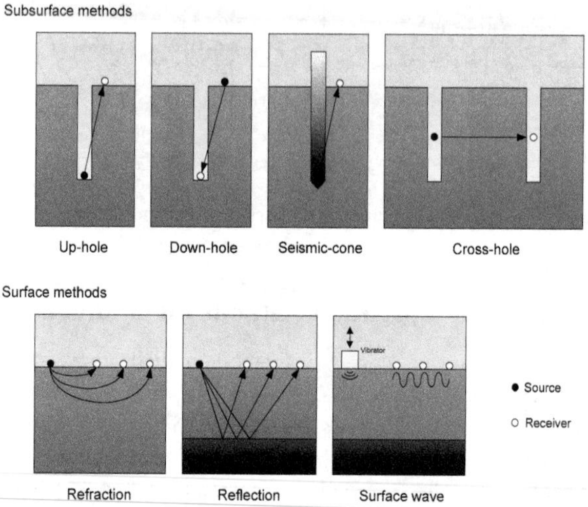

Figure 21. Wave propagation methods in ground monitoring (Sawangsuriya et al. 2008c).

The seismic methods have been employed to monitor ground improvement such as dynamic compaction and vibrofloatation methods (Hooker 1998). The measurement was made at the site before and after the ground improvement to establish stiffness profile in order that the performance of ground improvement can be monitored. Moreover, in some ground treatments e.g. grouted backfill and stone column, the near-surface wave propagating method can be employed to monitor their performance (Cuellar 1997).

Since ground moisture is sensitive to rise in water table, infiltration, and evaporation. Changes in ground moisture and hence in its modulus can occur over the service life of an engineered structure irrespective of the initial moisture conditions imposed during construction. The variation in ground modulus with moisture should be monitored systematically to reflect mechanical behavior of compacted engineered structures after construction (i.e., during post-compaction state). The influences of ground moisture should be also considered in performance assessment in such a way that the anticipated in-service conditions are taken into account. Sawangsuriya et al. (2009b) developed a laboratory method for identifying and examining the variation of modulus with moisture and suction in order to improve the ground performance assessment. Their study evaluated the G_o of compacted specimens using bender elements under the suction and moisture variation in the laboratory as shown in Fig. 22. Such method allows monitoring the modulus change with respect to the suction and moisture on the same specimen without disturbance. Fig. 23 illustrates shear wave time series during desaturation of lean clay specimen compacted near optimum.

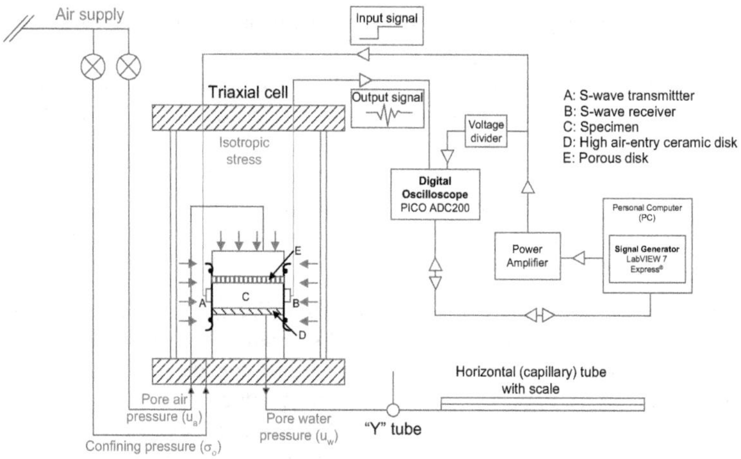

Figure 22. Schematic diagram of G₀ test using bender elements under the suction and moisture variation (Sawangsuriya et al. 2009b).

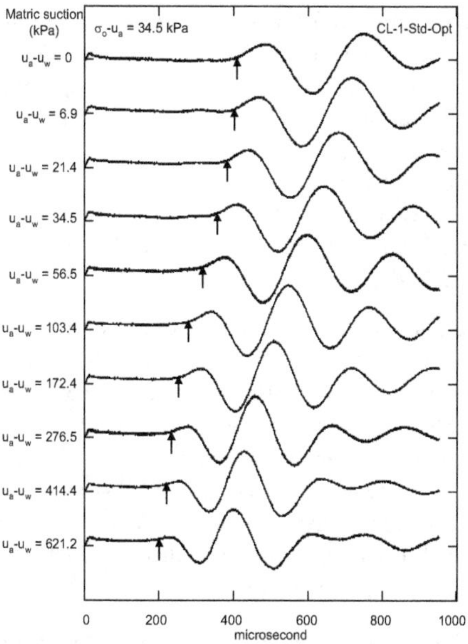

Figure 23. Shear wave time series during desaturation of lean clay specimen compacted near optimum (Sawangsuriya et al. 2009b).

9. Application of wave propagation methods in pavement design and analysis

The new AASHTO 2002 highway design relies heavily on mechanistic pavement properties. Current mechanistic-empirical design procedures for structural design of flexible pavements consider the mechanical properties of pavement material such as stiffness and strength. Several countries are currently either implementing or favoring the development of performance-based specifications. To successfully implement a mechanistic-empirical design procedure and to move toward performance-based specifications, a cost-effective, reliable, and practical means to assess the stiffness and strength of highway materials is necessary since both stiffness and strength of pavement layers plays a key role on the overall quality and performance of pavements.

One of the potential approaches to directly assess stiffness of geomaterials both in the laboratory and in the field is to employ G_o tests. Generally, G_o of geomaterials is routinely measured in earthquake engineering. In highway engineering, the application of G_o tests to assess the stiffness of highway materials and structural variability for pavement performance has increased dramatically (Kim and Stokoe 1992, Souto et al. 1994, Kim et al. 1997, Chen et al. 1999, Nazarian et al. 1999, Fiedler et al. 2000, Yesiller et al. 2000, Zeng et al. 2002, Nazarian et al. 2003). The key advantage of G_o tests over conventional modulus tests is their ability to rapidly and non-destructively assess the shear modulus of highway materials at the surface or under a free-field condition (i.e., zero confining pressure). Laboratory test methods are also available for G_o tests that can reproduce similar results to those measured in the field. Moreover in the field, the modulus changes in response to temporal variations in moisture content. The modulus changes in response to changes in moisture can be simply incorporated in G_o test in order to provide a model to estimate changes in modulus due to changes in moisture content for the pavement design and analysis. The relationship can be quantified empirically using a modulus ratio defined as the ratio of G_o at a specified moisture equilibrium to G_o at optimum compaction conditions. An example of modulus ratio is plotted against the degree of saturation for types of geomaterials as illustrated in Fig. 24.

Typically, the associated strain levels corresponding to many proposed geotechnical engineering structures such as foundations, retaining walls, tunnels, and pavements are however much larger (Mair 1993, Sawangsuriya et al. 2005). For example, the strain levels of the bender element are below $5x10^{-3}\%$, whereas the strain levels of the resilient modulus commonly used in the design of flexible pavement structures ranges from 0.01 to 0.1% (Sawangsuriya et al. 2005). In order to correct the small-strain modulus measurements to such relevant levels of strain amplitude imposed by the proposed structure, the modulus-strain relationship, or called strain-dependent modulus degradation curve, can be employed for a given operating stress level and soil type.

For pavement bases, subbases and subgrades, the stress-strain behaviour of soil is highly nonlinear and soil modulus may decay with strain by orders of magnitude. A relationship between the small-strain modulus (strains less than 10^{-2} %) and non-linear behaviour exhibited by soils at large strains (above 10^{-2} %) must be established. Generally, strains in base and subbase courses vary from 0.01 to 1%, whereas those in subgrades may vary from 0.003 to 0.6% (see Fig. 1). Therefore, the pavement base, subbase, and subgrade layers involve strains at higher levels, i.e., typical strain range of 10^{-2} to 1%, and soil exhibits non-linear properties. Fig. 1 also shows that the resilient modulus (M_R) test operates within these strain range. However, the decay curve of measured soil moduli from the small-strain tests must be evaluated and corrected to the modulus corresponding to these strain levels.

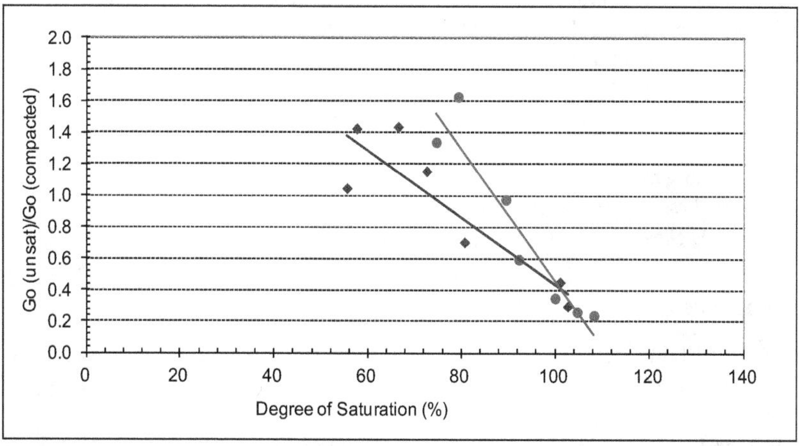

Figure 24. G_o at a specified moisture equilibrium normalized with G_o at optimum compaction condition vs. degree of saturation (Gupta et al. 2007).

10. Summary

Stiffness of geomaterials is an important engineering property, commonly used in geomechanical design and analysis. The deformation performance of geomaterials used to support engineered structures depends primarily on their stiffness. Therefore, it is important that the laboratory and field investigation should be performed to identify and classify geomaterials as well as to assess their mechanical properties for the engineering design and analysis. This chapter summarizes the wave propagation methods that permit the determination of the stiffness of the geomaterials. By knowing the elastic wave velocities as measured with the wave propagation method and total mass density of the medium, the corresponding stiffness of the geomaterials can be determined. One of the wave propagation methods based on the S-wave measurement is called a bender element test. It is a rapid, non-destructive, and low cost method, which can provide a realistic engineering design parameter (e.g. shear modulus). Moreover, elastic waves permit monitoring the elastic (stiffness) and inertial (mass density) properties of geomaterials while evaluating other coupled processes, i.e., the changes in state of effective stresses, the stress-induced anisotropy, the generation of geo-structure in soft sediments, etc.

The use of bender elements to generate and receive shear waves in geomaterials has become a very robust technique in geoengineering design and analysis and has been widely adopted for determining and monitoring stiffness of geomaterials both in the laboratory and field. However as with any other wave propagation techniques, the interpretation of bender element-collected data is controlled by wave characteristics, boundary conditions, and properties of the medium. Guidelines for designing test geometries and interpretation measured data from the wave propagation experiments are summarized herein. A bender element test has been employed for a variety of geoengineering applications, i.e., the mechanistic based design development, the long-term performance monitoring as well as quality control process during construction.

Author details

Auckpath Sawangsuriya
Bureau of Road Research and Development
Department of Highways, Thailand

11. References

Acar, Y. B. and EL-Tahir, E. A. (1986), "Low Strain Dynamic Properties of Artificially Cemented Sand," *Journal of Geotechnical Engineering*, ASCE, Vol. 112, No. 11, pp. 1001-1015.

Afifi, S. S. and Richart, F. E., Jr. (1973), "Stress-History Effects on Shear Modulus of Soils," *Soils and Foundations*, Vol. 13, No. 1, pp. 77-95.

Afifi, S. S. and Woods, R. D. (1971), "Long-Term Pressure Effects on Shear Modulus of Soils," *Journal of the Soil Mechanics and Foundations Division*, ASCE, Vol. 97, No. SM10, pp. 1445-1460.

Anderson, D. G. and Stokoe, K. H., II (1978), "Shear Modulus: A Time-Dependent Soil Property," *Dynamic Geotechnical Testing, ASTM STP 654*, Philadelphia, PA, pp. 66-90.

Anderson, D. G. and Woods, R. D. (1975), "Comparison of Field and Laboratory Shear Moduli," *Proceedings of the Conference on In Situ Measurement of Soil Properties*, Raleigh, NC, Vol. 1, pp. 69-92.

Anderson, D. G. and Woods, R. D. (1976), "Time-Dependent Increase in Shear Modulus of Clay," *Journal of the Geotechnical Engineering Division*, ASCE, Vol. 102, No. GT5, pp. 525-537.

Arroyo, M., Wood, D.M., Greening, P.D., Medina, L., and Rio, J. (2006). "Effect of sample size on bender-based axial G_o measurements" *Geotechnique*, Vol.56, No.1, 39-52.

Athanasopoulos, G. A. (1981), "Time Effects on Low-Amplitude Shear Modulus of Cohesive Soils," *Report UMEE 81R1*, The University of Michigan, MI.

Atkinson, J. H. (2000), "Non-Linear Soil Stiffness in Routine Design," *Geotechnique*, 40th Rankine Lecture, pp. 487-508.

Atkinson, J. H. and Sallfors, G. (1991). "Experimental determination of stress-strain-time characteristics in laboratory and in situ tests" *Proceedings of the 10th European Conference on Soil Mechanics and Foundation Engineering*, Florence, Italy, 915-956.

Baig, S., Picornell, M., and Nazarian, S. (1997), "Low Strain Shear Moduli of Cemented Sands," *Journal of Geotechnical Engineering*, ASCE, Vol. 123, pp. 540-545.

Baldi, G., Bruzzi, D., and Superbo, S. (1988), "Seismic Cone in Po River Sand," *Proceedings of the 1st International Symposium on Penetration Testing*, Balkema, Vol. 1, pp. 643-650.

Biarez, J. and Hicher, P.-Y. (1994), *Elementary Mechanics of Soil Behaviour-Saturated Remoulded Soils*, Balkema.

Biot, M. A. (1956), "Theory of Propagation of Elastic Waves in a Fluid-Saturated Porous Solid," *Journal of the Acoustic Society of America*, Vol. 28, No. 2, pp. 168-191.

Bolton, M. D. and Wilson, J. M. R. (1989), "An Experimental and Theoretical Comparison between Static and Dynamic Torsional Soil Tests," *Geotechnique*, Vol. 39, No. 4, pp. 585-599.

Bosscher, P. J. and Nelson, D. L. (1987), "Resonant Column Testing of Frozen Ottawa Sand," *Geotechnical Testing Journal*, ASTM, Vol. 10, No. 3, pp. 123-134.

Brignoli, E. G. M, Gotti, M. and Stokoe, K. H. (1996). "Measurement of shear waves in laboratory specimens by means of piezoelectric transducers" *Geotechnical Testing Journal*, Vol.19, No.4, 384-397.

Burland, J. B. (1989), "The Ninth Lauritis Bjerrum Memorial Lecture: 'Small is Beautiful'- the Stiffness of Soils at Small Strains," *Canadian Geotechnical Journal*, Vol. 26, pp. 499-516.

Campanella, R. G. (1994), "Field Methods for Dynamic Geotechnical Testing: An Overview of Capabilities and Needs," *Dynamic Geotechnical Testing II, ASTM STP 1213*, Philadelphia, PA, pp. 3-23.

Campanella, R. G., Robertson, P. K., and Gillespie (1986), "Seismic Cone Penetration Test," *Proceedings of Sessions on In Situ 86, ASCE Specialty Conference on Use of In Situ Tests in Geotechnical Engineering*, Blackburg, VA, pp. 116-130.

Cascante, G., Santamarina, J. C., and Yassir, N., (1998), "Flexural Excitation in a Standard Resonant Column Device," *Canadian Geotechnical Journal*, Vol. 35, No. 3, pp. 488-490.

Chang, C. S., Misra, A., and Sundaram, S. S. (1991), "Properties of Granular Packings under Low Amplitude Cyclic Loading," *Soil Dynamics and Earthquake Engineering*, Vol. 10, No. 4, pp. 201-211.

Chen, D.-H., Wu, W., He, R., Bilyeu, J., and Arrelano, M. (1999), "Evaluation of In-Situ Resilient Modulus Testing Techniques," *Recent Advances in the Characterization of Transportation Geo-Materials*, Geotechnical Special Publication, No. 86, ASCE, Reston, VA., pp. 1-11.

Cho, G. C. and Santamarina, J. C. (2001), "Unsaturated Particulate Materials-Particle-Level Studies," *Journal of Geotechnical and Geoenvironmental Engineering*, ASCE, Vol. 127, No. 1, pp. 84-96.

Clough, G. W., Sitar, N., Bachus, R. C., and Rad, N. S. (1981), "Cemented Sands under Static Loading," *Journal of the Geotechnical Engineering Division*, ASCE, Vol. 107, pp. 799-817.

Dobry, R. (1989), "Some Basic Aspects of Soil Liquefaction during Earthquakes," *Earthquake Hazards and the Design of Constructed Facilities in the Eastern United States*, Annual of the New York Academy of Sciences, No. 558, pp. 172-182.

Drnevich, V. P. (1977), "The Resonant Column Test," *Report to the U.S. Army Corps of Engineers*, Soil Mechanics Series, No. 23, Department of Civil Engineering, The University of Kentucky.

Drnevich, V. P. (1985), "Recent Developments in Resonant Column Testing," *Richart Commemorative Lectures, Proceedings of the ASCE Specialty Session*, Detroit, MI, pp. 79-107.

Drnevich, V. P., Hall, J. R., and Richart, F. E. (1967), "Effects of Amplitude of Vibration on the Shear Modulus of Sand," *Proceedings of the International Symposium on Wave Propagation and Dynamic Properties of Earth Materials*, Albuquerque, NM, pp. 189-199.

Drnevich, V. P., Hardin, B. O., and Shippy, D. J. (1978), "Modulus and Damping of Soils by the Resonant-Column Method," *Dynamic Geotechnical Testing, ASTM STP 654*, Philadelphia, PA, pp. 91-125.

Duffy, J. and Mindlin, R. D. (1957), "Stress-Strain Relations of a Granular Medium," *Journal of the Applied Mechanics*, Vol. 24, No. 4, pp. 585-593.

Dyvik, R. and Madshus, C. (1985), "Lab Measurements of G_{max} Using Bender Elements," *Proceedings of the Geotechnical Engineering Division: Advances in the Art of Testing Soil Under Cyclic Conditions*, ASCE, Detroit, MI, pp. 186-196.

Edil, T. B. and Luh, G.-F. (1978), "Dynamic Modulus and Damping Relationships for Sands," *Proceedings on the Specialty Conference on Earthquake Engineering and Soil Dynamics*, ASCE, Pasadena, CA, pp. 394-409.

Fam, M. and Santamarina, J. C. (1995), "Study of Geoprocesses with Complementary Wave Measurements in an Oedometer," *Geotechnical Testing Journal*, ASTM, Vol. 18, No. 3, pp. 307-314.

Fam, M., Santamarina, J. C., and Dusseault, M. (1998), "Wave-Based Monitoring Processes in Granular Salt," *Journal of the Environmental and Engineering Geophysics*, Vol. 3, pp. 15-26.

Fernandez, A. and Santamarina, J. C. (2001), "Effect of Cementation on the Small-Strain Parameters of Sands," *Canadian Geotechnical Journal*, Vol. 38, No. 1, pp. 191-199.

Fiedler, S. A., Main, M., and DiMillio, A. F. (2000), "In-Place Stiffness and Modulus Measurements," *Proceedings of Sessions of ASCE Specialty Conference on Performance Confirmation of Constructed Geotechnical Facilities*, Geotechnical Special Publication, No. 94, ASCE, Amherst, MA, pp. 365-376.

Fiedler, S. A., Nelson, C. R., Berkman, E. F. and DiMillio, A. F. (1998), "Soil Stiffness Gauge for Soil Compaction Control," *Public Roads*, U.S. Department of Transportation, Federal Highway Administration, Vol. 61, No. 4.

Fioravante, V. and Capoferri, R. (2001), "On the Use of Multi-Directional Piezoelectric Transducers in Triaxial Testings," *Geotechnical Testing Journal*, ASTM, Vol. 24, No. 3, pp. 243-255.

Frost J. D. and Burns, S. E. (2003), "In Situ Subsurface Characterization," *The Civil Engineering Handbook*, 2nd Edition by W. F. Chen and J. Y. Richard, CRC Press LLC.

Georgiannou, V. N., Rampello, S., and Silvestri, F. (1991), "Static and Dynamic Measurement of Undrained Stiffness of National Overconsolidated Clays," *Proceedings of the 10th European Conference Soil Mechanics*, Florence, Vol. 1, pp. 91-96.

Goto, S., Tatsuoka, F., Shibuya, S., and Sato, T. (1991), "A Simple Gauge for Local Strain Measurements in the Laboratory," *Soils and Foundations*, Vol. 31, No. 1, pp. 169-180.

Greening, P.D. and Nash, D.F.T. (2004). "Frequency domain determination of G_0 using bender elements" *Geotechnical Testing Journal*, Vol.27, No.3, 288-294.

Gupta, S., Ranaivoson, A., Edil, T.B., Benson, C.H., and Sawangsuriya, A. (2007), *Pavement design using unsaturated soil technology*, Report No. MN/RC-2007-11, Minnesota Department of Transportation, MN.

Hardin, B .O. and Black, W. L. (1966), "Sand Stiffness under Various Triaxial Stress," *Journal of the Soil Mechanics and Foundations Division*, ASCE, Vol. 92, No. SM2, pp. 27-42.

Hardin, B .O. and Richart, F. E., Jr. (1963), "Elastic Wave Velocities in Granular Soils," *Journal of the Soil Mechanics and Foundations Division*, ASCE, Vol. 89, No. SM1, pp. 33-65.

Hardin, B. O. (1978), "The Nature of Stress-Strain Behavior of Soils," *Proceedings of the Geotechnical Engineering Division Specialty Conference on Earthquake Engineering and Soil Dynamics*, ASCE, Pasadena, CA, Vol. 1, pp. 1-90.

Hardin, B. O. and Black, W. L. (1968), "Vibration Modulus of Normally Consolidated Clay," *Journal of the Soil Mechanics and Foundations Division*, ASCE, Vol. 94, No. SM2, pp. 353-369.

Hardin, B. O. and Drnevich, V. P. (1972), "Shear Modulus and Damping in Soils: Design Equations and Curves," *Journal of the Soil Mechanics and Foundations Division*, ASCE, Vol. 98, No. SM7, pp. 667-692.

Hardin, B. O. and Music, J. (1965), "Apparatus for Vibration during the Triaxial Test," *Symposium on Instrumentation and Apparatus for Soils and Rocks, ASTM STP 392*, pp. 55-74.

Hardin, B. O., (1970), "Suggested Methods of Test for Shear Modulus and Damping of Soils by the Resonant Column," *ASTM STP 479*, Philadelphia, PA, pp. 516-529.

Hill, J. J., Kurdziel, J. M., Nelson, C. R., Nystrom, J. A., and Sondag, M. (1999), "MnDOT Overload Field Tests of Standard and SIDD RCP Installations," *Transportation Research Board 78th Annual Meeting*, Washington, D.C. (CD-ROM)

Hoar, R. J. and Stokoe, K. H, II (1978), "Generation and Measurement of Shear Waves In Situ," *Dynamic Geotechnical Testing, ASTM STP 654*, Philadelphia, PA, pp. 3-29.

Hryciw, R. D. and Thomann, T. G. (1993), "Stress-History-Based Model for G^e of Cohesionless Soils," *Journal of Geotechnical Engineering*, ASCE, Vol. 119, No. 7, pp. 1073-1093.

Humboldt Mfg. Co. (1999), *Humboldt Soil Stiffness Gauge (GeoGauge) User Guide: Version 3.3*, Norridge, IL.

Humboldt Mfg. Co. (2000a), *Test Results: Evaluation of the Humboldt GeoGauge on Soil-Fly Ash-Cement Mixtures*, Norridge, IL.

Humboldt Mfg. Co. (2000b), *Test Results: Evaluation of the Humboldt GeoGauge on New Mexico Route 44*, Norridge, IL.

Inci, G., Yesiller, N., and Kagawa, T. (2003), "Experimental Investigation of Dynamic Response of Compacted Clayey Soils," *Geotechnical Testing Journal*, ASTM, Vol. 26, No. 2, pp. 125-141.

Isenhower, W. M. (1980), "Torsional Simple Shear/Resonant Column Properties of San Francisco Bay Mud," *M.S. Thesis*, The University of Texas at Austin, Austin, TX.

Isenhower, W. M. and Stokoe, K. H., II (1981), "Strain-Rate Dependent Shear Modulus of San Francisco Bay Mud," *Proceedings of the International Conference on Recent Advances in Geotechnical Earthquake Engineering and Soil Dynamic*, St. Louis, MO, Vol. 2, pp. 597-602.

Ishibashi, I., and Zhang, X. (1993). "Unified dynamic shear moduli and damping ratios of sand and clay" *Soils and Foundations*, Vol.33, No.1, 182-191.

Ishihara, K. (1996), *Soil Behavior in Earthquake Geotechnics*, Oxford University Press, Inc., NY.

Ismail and Rammah (2005). "A new setup for measuring G_o during laboratory compaction" *Geotechnical Testing Journal*, Vol.29, No.4, 1-9.

Ismail, M.A., Sharma, S.S., and Fahey, M. (2005). "A small true triaxial apparatus with wave velocity measurement" *Geotechnical Testing Journal*, Vol.28, No.2, 1-10.

Iwasaki, T. and Tatsuoka, F. (1977), "Effects of Grain Size and Grading on Dynamic Shear Moduli of Sands," *Soils and Foundations*, Vol. 17. No. 3, pp. 19-35.

Iwasaki, T., Tatsuoka, F., and Takagi, Y. (1978), "Shear Moduli of Sands under Cyclic Torsional Shear Loading," *Soils and Foundations*, Vol. 18, No. 1, pp. 39-50.

Jamiolkowski, M., Lancellotta, R., Lo Presti, D. C. F., and Pallera, O. (1994), "Stiffness of Toyoura Sand at Small and Intermediate Strain," *Proceedings of the 13th International Conference on Soil Mechanics and Foundation Engineering*, New Delhi, pp. 169-172.

Jardine, R. J. (1992), "Some Observations on the Kinetic Nature of Soil Stiffness," *Soils and Foundations*, Vol. 32, No. 2, pp. 111-124.

Jardine, R. J., Potts, D. M., Fourie, A. B., and Burland, J. B. (1986), "Studies of the Influence of Non-Linear Stress-Strain Characteristics in Soil-Structure Interaction," *Geotechnique*, Vol. 36, No. 3, pp. 377-396.

Jovicic, V. (1997). The measurement and interpretation on small strain stiffness of soils. Ph.D. thesis, City University.

Jovičić, V. and Coop, M. R. (1998), "The Measurement of Stiffness Anisotropy in Clays with Bender Element Tests in the Triaxial Apparatus," *Geotechnical Testing Journal*, ASTM, Vol. 21, No. 1, pp. 3-10.

Jovicic, V., Coop, M.R., and Simic, M. (1996). "Objective criteria for determining Gmax from bender elements tests" *Geotechnique*, Vol.46, No.2, 357-362.

Kawaguchi, T., Mitachi, T., Shibuya, S. (2001). "Evaluation of shear wave travel time in laboratory bender element test" *Proceedings of the 15th International Conference on Soil Mechanics and Geotechnical Engineering*, Istanbul, Turkey, 155-158.

Kim, D.-S. and Stokoe, K. H., II (1992). "Characterization of resilient modulus of compacted subgrade soils using resonant column and torsional shear tests" *Transportation Research Record*, No.1369, Washington, D.C., 83-91.

Kokusho, T. (1980), "Cyclic Triaxial Test of Dynamic Soil Properties for Wide Strain Range," *Soils and Foundations*, Vol. 20, pp. 45-60.

Kokusho, T. (1987), "In Situ Dynamic Soil Properties and their Evaluation," *Proceedings of the 8th Asian Regional Conference on Soil Mechanics and Foundation Engineering*, Kyoto, Japan, Vol. 2, pp. 215-435.

Kokusho, T., Yoshida, Y., and Esashi, Y. (1982), "Dynamic Properties of Soft Clays for Wide Strain Range," *Soils and Foundations*, Vol. 22, No. 4, pp. 1-18.

Kramer, S. L. (1996), *Geotechnical Earthquake Engineering*, Prentice-Hall, Inc., Upper Saddle River, NJ.

Kuribayashi, E., Iwasaki, T., Tatsuoka, F., and Horiuchi, S. (1975), "Effects of Particle Characteristics on Dynamic Deformational Properties of Soils," *Proceedings of the 5th Asian Regional Conference on Soil Mechanics and Foundation Engineering*, Bangalore, India, pp. 361-367.

Lade, P. V. and Overton, D. D. (1989), "Cementation Effects in Frictional Materials," *Journal of Geotechnical Engineering*, ASCE, Vol. 115, pp. 1373-1387.

Lawrence, F. V., Jr. (1963), "Propagation Velocity of Ultrasonic Wave Through Sand," *MIT Research Report R63-8*, Massachusetts Institute of Technology, Cambridge, MA.

Lawrence, F.V. (1963). Propagation of ultrasonic waves through sand, Research report R63-8, MIT, Cambridge.

Lawrence, F.V. (1965). Ultrasonic shear wave velocity in sand and clay, Research report R65-05, Soil publication No.17, MIT, Cambridge.

Lee, J.-S. and Santamarina, J. C. (2005). "Bender elements: performance and signal interpretation" *Journal of Geotechnical and Geoenvironmental Engineering*, ASCE, Vol.131, No.9, 1063-1070.

Lenke, L. R., McKeen, R. G., and Grush, M. (2001), "Evaluation of a Mechanical Stiffness Gauge for Compaction Control of Granular Media," *Report NM99MSC-07.2*, New Mexico State Highway and Transportation Department, Albuquerque, NM.

Lenke, L. R., McKeen, R. G., and Grush, M. (2003), "Laboratory Evaluation of the GeoGauge for Compaction Control," *Transportation Research Record*, No. 1849, Washington, D.C., pp. 20-30.

Lewis, M. D. (1990), "A Laboratory Study of the Effect of Stress State on the Elastic Moduli of Sand," *Ph.D. Thesis*, The University of Texas at Austin, Austin, TX.

Lo Presti, D. C. F., Jamiolkowski, M., Pallara, O., Cavallararo, A., and Pedroni, S. (1997), "Shear Modulus and Damping of Soils," *Geotechnique*, Vol. 47, No. 3, pp. 603-617.

Lo Presti, D. C. F., Pallara, O., Lancellota, R., Armandi, M., and Maniscalco, R. (1993), "Monotonic and Cyclic Loading Behavior of Two Sands at Small Strains," *Geotechnical Testing Journal*, ASTM, Vol. 16, No. 4, pp. 409-424.

Lo Presti, D. C. F., Shibuya, S., and Rix, G. J. (2001). "Innovative in soil testing" *Pre-failure Deformation Characteristics of Geomaterials*. Jamiolkowski, H., Lancellota, R. H., and LoPresti, D. C. F., Eds. Swets and Zeitlinger, Lisse, 1027-1076.

Mair, R. J. (1993), "Developments in Geotechnical Engineering Research: Application to Tunnels and Deep Excavations," Unwin Memorial Lecture 1992, *Proceedings of the Institution of Civil Engineers-Civil Engineering*, Vol. 97, No. 1, pp. 27-41.

Mancuso, C., Vassallo, R., and d'Onofrio, A. (2002), "Small Strain Behavior of a Silty Sand in Controlled-Suction Resonant Column-Torsional Shear Tests," *Canadian Geotechnical Journal*, Vol. 39, No. 1, pp. 22-31.

Matthews, M. C., Clayton, C. R. I., and Own, Y. (2000) "The use of field geophysical techniques to determine geotechnical stiffness parameters" *Geotechnical Engineering*, Vol.143, 31-42.

Mayne, P. W. (2001), "Stress-Strain-Strength-Flow Parameters from Enhanced In Situ Tests," *Proceedings of the International Conference on In Situ Measurement of Soil Properties and Case Histories*, Bali, Indonesia, pp. 27-48.

Mayne, P. W., Christopher, B. R., and DeJong, J. (2001), *Manual on Subsurface Investigations*, National Highway Institute Publication No. FHWA NHI-01-031, FHWA, Washington, D.C.

Mitchell, J. K. and Soga, K. (2005), *Fundamentals of Soil Behavior*, John Wiley and Sons, Inc., NJ.

Morris, D. V. (1990), "Automatic Feedback System for Resonant Column Testing," *Geotechnical Testing Journal*, ASTM, Vol. 13, No. 1, pp. 16-23.

Nacci, V. A. and Taylor, K. J. (1967), "Influence of Clay Structure on Elastic Wave Velocities," *Proceedings of the International Symposium on Wave Propagation and Dynamic Properties of Earth Materials*, Albuquerque, New Mexico, pp. 491-502.

Nakagawa, K., Soga, K., and Mitchell, J. K. (1996), "Pulse Transmission System for Measuring Wave Propagation in Soils," *Journal of Geotechnical Engineering*, ASCE, Vol. 122, No. 4, pp. 302-308.

Nazarian, S. and Stokoe, K. H., II (1987), "In Situ Determination of Elastic Moduli of Pavements Systems by Special-Analysis-of-Surface-Waves Method (Theoretical Aspects)," *Research Report 437-2*, Center for Transportation Research, The University of Texas at Austin, Austin, TX.

Nazarian, S., Yuan, D., and Arellano, M. (2002), "Quality Management of Base and Subgrade Materials with Seismic Methods," *Transportation Research Record*, No. 1786, Washington, D.C., pp. 3-10.

Nazarian, S., Yuan, D., and Baker, M. R. (1994), "Automation of SASW Method," *Dynamic Geotechnical Testing II, ASTM STP 1213*, Philadelphia, PA, pp. 88-100.

Nazarian, S., Yuan, D., and Tandon, V. (1999), "Structural Field Testing of Flexible Pavement Layers with Seismic Methods for Quality Control," *Transportation Research Record*, No. 1654, Washington, D.C., pp. 50-60.

Nelson, C. R. and Sondag, M. (1999), "Comparison of the Humboldt GeoGauge with In-Place Quasi-Static Plate Load Tests," *CNA Consulting Engineers Report*, Minneapolis, MN.

Ni, S. H. (1987), "Dynamic Properties of Sand under True Triaxial Stress States from Resonant Column and Torsion Shear Tests," *Ph.D. Thesis*, The University of Texas at Austin, TX.

Ohara, S. and Matsuda, H. (1988), "Study on the Settlement of Saturated Clay Layer Induced by Cyclic Shear," *Soils and Foundations*, Vol. 28, pp. 103-113.

Pennington, D.S. (1999). The anisotropic small strain stiffness of Cambridge Gault clay. Ph.D. Thesis, University of Bristol.

Ray, R. P. and Woods, R. D. (1988), "Modulus and Damping Due to Uniform and Variable Cyclic Loading," *Journal of Geotechnical and Geoenvironmental Engineering*, ASCE, Vol. 114, No. 8, pp. 861-876.

Richart, F. E., Hall, J. R., and Woods, R. D. (1970). *Vibrations of soils and Foundations*. Prentice Hall, Englewood Cliffs.

Richart, F.E. (1977), "Dynamic Stress-Strain Relations for Soils," *Proceedings of the 9th International Conference on Soil Mechanics and Foundation Engineering*, Tokyo, Vol. 2, pp. 605-612.

Rix, G. J. and Stokoe, K. H., II (1989), "Stiffness Profiling of Pavement Subgrades," *Transportation Research Record*, No. 1235, Washington, D.C., pp. 1-9.

Robertson, P. K., Campanella, R. G., Gillespie, D., and Rice, A. (1986), "Seismic CPT to Measure In Situ Shear Wave Velocity," *Journal of Geotechnical Engineering*, ASCE, Vol. 112, No. 8, pp. 71-803.

Roesler, S. K. (1979), "Anisotropic Shear Modulus due to Stress Anisotropy," *Journal of the Geotechnical Engineering Division*, ASCE, Vol. 105, No. GT7, pp. 871-880.

Saada, A. S. (1988), "Hollow Cylinder Torsional Devices: Their Advantages and Limitation," *Advanced Triaxial Testing of Soil and Rock, ASTM STP 977*, Philadelphia, PA. pp. 766-779.

Sanchez-Salinero, I., Roesset, J. M., and Stokoe, K. H., II (1986). Analytical studies of body wave propagation and attenuation. Report GR 86-15, University of Texas, Austin, TX.

Sanchez-Salinero, I., Roesset, J. M., Shao, K., Stokoe, K. H., II, and Rix, G. J. (1987), "Analytical Evaluation of Variables Affecting Surface Wave Testing of Pavements," *Transportation Research Record 1136*, Washington, D.C., pp. 132-144.

Santamarina, J. C., Klein, K. A., and Fam, M. A. (2001), *Soils and Waves*, John Wiley & Sons, Chichester, UK.

Sargand, S. M. (2001), "Direct Measurement of Backfill Stiffness for Installation and Design of Buried Pipes," *Ohio Research Institute for Transportation and the Environment*, Ohio University, OH.

Sargand, S. M., Edwards, W. F., and Salimath, S. (2000), "Evaluation of Soil Stiffness via Non-Destructive Testing," *Ohio Research Institute for Transportation and the Environment*, Ohio University, OH.

Sawangsuriya, A., Biringen, E., Fratta, D., Bosscher, P. J. and Edil, T. B. (2006). "Dimensionless limits for the collection and interpretation of wave propagation data in soils" *GeoShanghai Conference, Site and Geomaterial Characterization*, ASCE, Geotechnical Special Publication, No.149, Shanghai, China, 160-166.

Sawangsuriya, A., Bosscher, P. J., and Edil, T. B. (2002), "Laboratory Evaluation of the Soil Stiffness Gauge," *Transportation Research Record*, No. 1808, Washington, D.C., pp. 30-37.

Sawangsuriya, A., Bosscher, P. J., and Edil, T. B. (2005). Alternative testing techniques for modulus of pavement bases and subgrades. *Proceedings of the 13th Annual Great Lakes Geotechnical and Geoenvironmental Engineering Conference, Geotechnical Applications for Transportation Infrastructure*, ASCE, Geotechnical Practice Publication, No.3, Milwaukee, WI, 108-121.

Sawangsuriya, A., Edil, T. B., and Benson, C. H. (2009a). "Effect of suction on resilient modulus of compacted fine-grained subgrade soils" *Transportation Research Record 2101*, TRB, National Research Council, Washington, D.C., 82-87.

Sawangsuriya, A., Edil, T. B., and Bosscher, P. J. (2003), "Relationship between Soil Stiffness Gauge Modulus and Other Test Moduli for Granular Soils," *Transportation Research Record*, No. 1849, Washington, D.C., pp. 3-10.

Sawangsuriya, A., Edil, T. B., and Bosscher, P. J. (2004), "Assessing Small-Strain Stiffness of Soils Using the SSG," *Proceedings of the 15th Southeast Asia Geotechnical Conference*, Bangkok, Thailand, pp. 101-106.

Sawangsuriya, A., Edil, T. B., and Bosscher, P. J. (2008a). "Modulus-suction-moisture relationship for compacted soils" *Canadian Geotechnical Journal*, Vol.45, No.7, 973-983.

Sawangsuriya, A., Edil, T. B., and Bosscher, P. J. (2009b). "Modulus-suction-moisture relationship for compacted soils in postcompaction state" *Journal of Geotechnical and Geoenvironmental Engineering*, Vol.135, No.10, 1390-1403.

Sawangsuriya, A., Fall, M., and Fratta, D. (2008b). "Wave-based techniques for evaluating elastic modulus and Poisson's ratio of laboratory compacted lateritic soils" *Geotechnical and Geological Engineering*, Vol.26, No.5, 567-578.

Sawangsuriya, A., Fratta, D., Bosscher, P. J., and Edil, T. B. (2007a). "S-wave velocity-stress power relationship: packing and contact behavior of sand specimens" *Geo-Denver Conference, Advances in Measurement and Modeling of Soil Behavior*, ASCE, Geotechnical Special Publication, No.173, Denver, CO, 1-10.

Sawangsuriya, A., Fratta, D., Edil, T. B., and Bosscher, P. J. (2007b). "Small-strain shear modulus anisotropy of compacted soils using bender elements" *60th Canadian Geotechnical Conference & 8th Joint CGS/IAH-CNC Groundwater Conference*, Ottawa, Canada

Sawangsuriya, A., Sukolrat, J., and Jotisankasa, A. (2008c), "Innovative testing methods for determining stiffness of geomaterials," *Seminar of Highway Engineering: Best Practices in Highway Engineering*, Department of Highways, Bangkok, Thailand, pp. 589-609.

Saxena, S. K., Avramidis, A. S., and Reddy, K. R. (1988), "Dynamic Moduli and Damping Ratios for Cemented Sands at Low Strains," *Canadian Geotechnical Journal*, Vol. 25, No. 2, pp. 353-368.

Seed, H. B. and Idriss, I. M. (1970), "Soil Moduli and Damping Factors for Dynamic Response Analyses," *Report EERC 70-10*, Earthquake Engineering Research Center, University of California, Berkeley, CA.

Sharma, P. V. (1997), *Environmental and Engineering Geophysics*, Cambridge University Press, Cambridge, UK.

Sheeran, D. E., Baker, W. H., and Krizek, R. J. (1967), "Experimental Study of Pulse Velocities in Compacted Soils," *Highway Research Record*, No. 177, Washington, D.C., pp. 226-238.

Shibata, T. and Soelarno, D. S. (1975)," Stress-Strain Characteristics of Sands under Cyclic Loading," *Proceedings of the Japan Society of Civil Engineering*, No. 239, pp. 57-65. (in Japanese)

Shibuya, S. and Tanaka, H. (1996), "Estimate of Elastic Shear Modulus in Holocene Soil Deposits," *Soils and Foundations*, Vol. 36, No. 4, pp. 45-55.

Shibuya, S., Hwang, S. C., and Mitachi, T. (1997), "Elastic Shear Modulus of Soft Clays from Shear Wave Velocity Measurement," *Geotechnique*, Vol. 47, No. 3, pp. 593-601.

Shibuya, S., Mitachi, T., Fukuda, F., and Degoshi, T. (1995), "Strain Rate Effect on Modulus and Damping of Normally Consolidated Clay," *Geotechnical Testing Journal*, ASTM, Vol. 18, No. 3, pp. 365-375.

Shibuya, S., Tatsuoka, S., Teachavorasinskun, S., Kong, X. J., Abe, F., Kim, Y. S., and Park, C. S. (1992), "Elastic Deformation Properties of Geomaterials," *Soils and Foundations*, Vol. 32, No. 3, pp. 26-46.

Shirley, D.J. (1978). "An improved shear wave transducer" J. Acoust. Soc. Am., 63, No.5, 1643-1645.

Shirley, D.J. and Hampton, L.D. (1978). "Shear–wave measurements in laboratory sediments" J. Acoust. Soc. Am., Vol.63, No.2, 607-613.

Silver, M. L. and Seed, H. B. (1971), "Deformation Characteristics of Sands under Cyclic Loading," *Journal of the Soil Mechanics and Foundations Division*, ASCE, Vol. 97, No. SM8, pp. 1081-1098.

Silvestri, F. (1991), "Stress-Strain Behaviour of Natural Soils by Means of Cyclic/Dynamic Torsional Shear Tests," *Experimental Characterization and Modelling of Soils and Soft Rocks*, The University of Napoli Federico, pp. 7-73.

Souto, A., Hartikainen, J., and Özüdoğru, K. (1994), "Measurement of Dynamic Parameters of Road Pavement Materials by the Bender Element and Resonant Column Tests," *Geotechnique*, Vol. 44, No. 3, pp. 519-526.

Stokoe, K. H., II and Richart, F. E., Jr. (1973), "In Situ and Laboratory Shear Wave Velocities," *Proceedings of the 8th International Conference on Soil Mechanics and Foundation Engineering*, Vol. 1, Part 2, Moscow, U.S.S.R., pp. 403-409.

Stokoe, K. H., II and Woods, R. D. (1972), "In Situ Shear Wave Velocity by Cross-Hole Method," *Journal of the Soil Mechanics and Foundations Division*, ASCE, Vol. 98, No. SM5, pp. 443-460.

Stokoe, K. H., II, Hwang, S. K., Lee, J. N.-K., and Andrus, R. D. (1995), "Effects of Various Parameters on the Stiffness and Damping of Soils at Small to Medium Strains," *Pre-failure Deformation of Geomaterials*, Balkema, Rotterdam, Vol. 2, pp. 785-816.

Stokoe, K. H., II, Lee, S. H. H., and Knox, D. P. (1985), "Shear Moduli Measurement under True Triaxial Stresses," *Proceedings of the Geotechnical Engineering Division: Advances in the Art of Testing Soil Under Cyclic Conditions*, ASCE, Detroit, MI, pp. 166-185.

Sukolrat, J. (2007). Destructuration of Bothkennar clay, Ph.D. Thesis, Department of Civil Engineering, University of Bristol, UK.

Tatsuoka, F. and Shibuya, S. (1991), "Deformation Characteristics of Soils and Rocks from Field and Laboratory Tests," *Proceedings of the 9th Asian Regional Conference on Soil Mechanics and Foundation Engineering*, Bangkok, Thailand, Vol. 2, pp. 101-170.

Thomann, T. G. and Hryciw, R. D. (1990), "Laboratory Measurement of Small Strain Shear Modulus under K_o Conditions," *Geotechnical Testing Journal*, ASTM, Vol. 13, No. 2, pp. 97-105.

Trudeau, P. J., Whitman, R. V., and Christian, J. T. (1974), "Shear Wave Velocity and Modulus of a Marine Clay," *Journal of the Boston Society of Civil Engineers*, Vol. 61, No. 1, pp. 12-25.

Viggiani, G. (1992). Small strain stiffness of fine grained soils. Ph.D. Thesis, City University.

Viggiani, G. and Atkinson, J. H. (1995a), "Stiffness of Fine-Grained Soil at Very Small Strains," *Geotechnique*, Vol. 45, No. 2, pp. 249-265.

Viggiani, G. and Atkinson, J. H. (1995b), "Interpretation of Bender Element Tests," *Geotechnique*, Vol. 45, No. 1, pp. 149-154.

Viggiani, G. and Atkinson, J. H. (1997), Discussion: "Interpretation of Bender Element Tests," *Geotechnique*, Vol. 47, No. 4, pp. 873-877.

Vucetic, M. and Dobry, R. (1991), "Effect of Soil Plasticity on Cyclic Response," *Journal of Geotechnical Engineering*, ASCE, Vol. 117, No. 1, pp. 89-107.

Weaver, E., Ormsby, W. C., and Zhao, X. (2001), *Pavement Variability Study: Phase I Interim Report*, National Pooled Fund Study 2(212), Non-Nuclear Testing of Soils and Granular Bases Using the GeoGauge (Soil Stiffness Gauge) and Other Similar Devices, U.S. Department of Transportation, Federal Highway Administration.

Wilson, S. D. and Dietrich, R. J. (1960), "Effect of Consolidation Pressure on Elastic and Strength Properties of Clay," *Proceedings of the Soil Mechanics and Foundations Division, Research Conference on Shear Strength of Cohesive Soils*, ASCE, Boulder, CO, pp. 419-435.

Woods, R. D. (1978), "Measurement of Dynamic Soil Properties," *Proceedings of the Earthquake Engineering and Soil Dynamics Specialty Conference*, ASCE, Pasadena, CA, Vol. 1, pp. 91-178.

Wright, S. G., Stokoe, K. E., II, and Roesset, J. M. (1994), "SASW Measurements at Geotechnical Sites Overlaid by Water," *Dynamic Geotechnical Testing II, ASTM STP 1213*, Philadelphia, PA, pp. 39-57.

Wu, W., Arellano, M., Chen, D.-H., Bilyeu, J., and He., R. (1998), "Using a Stiffness Gauge as an Alternative Quality Control Device in Pavement Construction," *Texas Department of Transportation Report*, Austin, TX.

Yesiller, N., Inci, G., and Miller, C. J. (2000), "Ultrasonic Testing for Compacted Clayey Soils," *Proceedings of Sessions of Geo-Denver 2000, Advances in Unsaturated Geotechnics*, ASCE, Geotechnical Special Publication, No. 99, Denver, CO, pp. 54-68.

Yu, P. and Richart, F. E., Jr. (1984), "Stress Ratio Effects on Shear Modulus of Dry Sands," *Journal of the Geotechnical Engineering Division*, ASCE, Vol. 110, No. GT3, pp. 331-345.

Yun, T. S. and Santamarina, J. C. (2005), "Decementation, Softening, and Collapse: Changes in Small-Strain Shear Stiffness in K_o Loading," *Journal of Geotechnical and Geoenvironmental Engineering Division*, ASCE, Vol. 131, No. 3, pp. 350-358.

Zen, K., Umehara, Y., and Hamada, K. (1978), "Laboratory Tests and In Situ Seismic Survey on Vibratory Shear Modulus of Clayey Soils with Various Plasticities," *Proceedings of the 5th Japanese Earthquake Engineering Symposium*, pp. 721-728.

Zeng, X. and Ni, B. (1998), "Application of Bender Elements in Measuring G_{max} of Sand under K_o Condition," *Geotechnical Testing Journal*, ASTM, Vol. 21, No. 3, pp. 251-263.

Zeng, X., Figueroa, J. L., and Fu, L. (2002). Measurement of base and subgrade layer stiffness using bender element technique. Proc. of the 15th ASCE Engineering Mechanics Conference, Columbia University, New York, NY, 1-10.

Chapter 7

An Efficient Approach for Seismoacoustic Scattering Simulation Based on Boundary Element Method

Zheng-Hua Qian

Additional information is available at the end of the chapter

1. Introduction

A seismogram synthesis is the theoretically calculated ground motion that would be recorded for a given crust structure and seismic source, which is an important tool that can be applied to the study of earthquake source from strong motion records. At the early stage of seismology, the reflectivity method [1] and the generalized ray method [2] were widely used for simulating seismic wave excitation and propagation, where a laterally homogeneous model is considered. The real Earth's crust, however, is a laterally inhomogeneous structure indeed. To study the site effect of strong ground motion, a stratified structure with irregular interfaces is an eligible model for simulating seismic wave propagation. For the study of seismic waves propagating through a sediment filled basin in the case of rigid grains, the work [3-4] is noteworthy, where a model was proposed to reproduct a nonlinear effect experimentally observed for real seismic waves: site amplification decreases as the amplitude of the incident wave increases.

In the last three decades, the study of seismic wave excitation and propagation in laterally inhomogeneous media has become extensively important in seismology and geophysics, including both analytical solutions and non-analytical results. The analytical solutions, however, only exist for a few special cases [5], such as the semi-cylindrical canyon and alluvial valleys, semi-elliptical canyon, and hemispherical canyon. The methods based on high-frequency approximation [6](such as asymptotic ray theory, Gaussian beam method, and so on) can give a proper solution by including the contribution from the neighboring rays but are only appropriate for handling body wave propagation under relatively high frequency. The methods based on plane wave decomposition (such as discrete wave number method [7-10]) can provide a complete wave field for any finite frequency but usually involve the truncation of matrices and vectors having an infinite dimension. Nowadays, the major approaches to study a laterally inhomogeneous model mainly rely on numerical methods which include domain methods and boundary methods. The domain numerical methods (such as finite difference [11], finite element [12], pseudo-spectral method [13], and spectral element method [14]) have become the standard tools for seismic wave modeling, however, these methods do not explicitly consider the boundary continuity conditions between different formations. That disables the methods to sufficient accuracy for modeling the reflection/transmission across irregular interfaces. Furthermore, they have difficulties in dealing with a large-scaled model because they require the subdivision of the domain into elements, especially the case when the domain needs to be extended to infinity. Boundary numerical methods have emerged as good alternatives to the domain numerical methods in dealing with an infinite continuum model because the radiation condition at infinity is easily fulfilled. They also have an obvious advantage over the domain numerical methods: the dimensionality of the problem under consideration is reduced by one order, because only a boundary instead of a domain discretization is required. Among this type of methods, the boundary element method (BEM) has been extensively used in simulating seismic wave excitation and propagation [15]. However, for a stratified model with irregular interfaces, the dimension of the final system of simultaneous equations in the traditional boundary element metod is proportional to the production of the element number and the interface number, which leads to an exponential increase in computing time and memory requirement. In order to overcome that problem, different approaches have been tried, for example, Fu [16] proposed an improved block Gaussian elimination scheme and Bouchon et al. [17] introduced a sparse method. Recently, global matrix propagators were introduced to improve the efficiency of the traditional BEM for modeling seismic wave excitation and propagation in multilayered solids [18-19], which expects great savings in computing time and memory requirement.

On the other hand, the seismoacoustic scattering due to an irregular fluid-solid interface must be considered when we model the seismic wave propagation in oceanic regions or gulf areas. This kind of problem is related to a wide range of seismic research conducted at or close to oceanic regions or gulf areas [20], such as deep ocean acoustic experiments, ocean bottom seismic observations, or the interpretation of the effects of water layer on the observed surface wave traveling through gulf areas. Many metropolitan cities all over the world are located at or close to seaside, so it is important to take into account the effect of the sea water when conducting the earthquake resistant design for high-rise buildings in such big gulf cities. Also, the underground structures beneath sea bottom are known to be strongly inhomogeneous. Therefore, it is necessary to simulate the seismoacoustic scattering in irregularly multilayered elastic media overlain by a fluid layer due to some scenario earthquake event. Furthermore, ocean bottom observations are a key feature in the modern world-wide standardized seismograph network [21]. Correct seismic wave modeling can help to obtain valid interpretation of the observed waveforms. In the past three decades, many techniques, classified into different categories [7, 18], have been developed for calculating seismic wave excitation and propagation in laterally inhomogeneous media. Among those methods, the finite difference method (FDM) [22-23] and the finite element method (FEM) [24] become very popular in modelling seismoacoustic scattering due to irregular fluid-solid interface because of their flexibility in modeling complex media.

However, there are still difficuilties in using these methods to simulate wave propagation in models with highly irregular topography since these methods do not explicitly consider the boundary and continuity conditions between different formations. That disables the methods to sufficient accuracy for modeling the wave reflection/transmission across irregular interfaces. Furthermore, the two methods usually require very large computational resources (i.e. a large amount of memory and long calculation time).

To overcome the problem, the reflection/transmission matrices for the fluid-solid irregular interface [20] were developed based on the discrete wavenumber representation of the wave field, i.e. the Aki-Larner method [25]. But their method is only suitable for a long-wavelength irregularity and becomes unstable at very steep interfaces (e.g. a vertical interface), which is due to the intrinsic feature of the method itself. On the other hand, BEM is more suitable than the other methods to model the seismoacoustic scattering at very steep irregular interfaces[26], especially in the cases when better accuracy is required near boundaries and sources. However, the traditional BEM requires a tremendous increase in memory capacity and computational time when it's used to model the wave propagation in multilayered media. Motivated by the work in [19-20], we develop a method based on the combination of the traditional BEM and a global matrix propagator to simulate the seismoacoustic scattering due to an irregular fluid-solid interface. This method takes the advantage of the global matrix propagator to suppress the tremendous increase in computational resources required for simulating wave propagation in multilayered media, but also retains the merits of the boundary element method. This implies that higher-frequency seismoacoustic scattering in complex structures can be calculated with reduced computer resources.

For simplicity, the Chapter is organized as follows: the mathematical formulation for an SH-wave (Shear wave polarized in the horizontal plane) propagation in a multilayered solid is first presented in an easily-understood way in Section 2, where some simple examples are calculated to test the formulation in solid layers; then we gratually go deep into the full formulation for the fluid-solid scattering simulation in Section 3, where two irregular fluid-solid models are used to test the validity of the fluid-solid formulation; based on the formulation in Section3, one of the two fluid-solid models is calculated to show the effects of water layer and the water reverberation in water layer in Section 4; finally summary is drawn in Section 5.

2. SH-wave propagation modeling in mutilayered solids

2.1. Methodology statement

The problem to be studied is illustrated in Figure 1. In this model, there are L homogeneous layers $\Omega^{(i)}$ (i=1, 2, ..., L) over a half-space, among which the ith layer is bounded by two irregular interfaces $\Gamma_{(i-1)}$ and $\Gamma_{(i)}$ (Throughout the section, the superscript with round brackets indicates layer index and the subscript with round brackets interface index). The uppermost interface is a free surface, and an arbitrary source is embedded in the sth layer. Assume each individual layer to be isotropic, linearly elastic material, so the tensor of elastic constants is determined by two independent Lame constants because of the rotational and translational symmetry. For the two-dimensional *SH* case under consideration, the physical properties of each layer can only be described by the shear modulus $\mu^{(i)}$ and mass density $\rho^{(i)}$. This problem will be solved by a boundary element method with a global matrix propagator. In this section, we will present the formulations of the method in the solid layers.

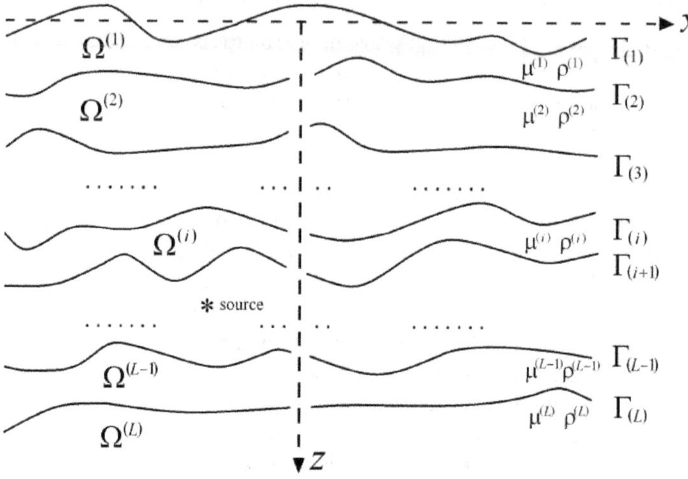

Figure 1. A multilayered 2D model with irregular interfaces

Consider the boundary element method problem in the *i*th layer, the scalar wave equation in frequency domain is given by (the details on the derivation of the SH-wave equation (1) and PSv-wave equation (29), which are suppressed here for brevity, refer to [27])

$$\mu^{(i)}\nabla^2 u(\mathbf{r},\omega)+\rho^{(i)}\omega^2 u(\mathbf{r},\omega)=-F(\mathbf{r},\omega)\delta_{si}, \tag{1}$$

where $u(\mathbf{r}, \omega)$ is the seismic response at a position vector $\mathbf{r}$ in frequency domain, $\delta_{si} = 1$ for $i = s$ and vanishes else, and $F(\mathbf{r}, \omega)$ is the seismic source in the *s*th layer.

Suppose the seismic source distribution consists of a simple point source at a position vector $\mathbf{s}$, i.e.,

$$F(\mathbf{r},\omega)=F(\omega)\delta(\mathbf{r}-\mathbf{s}). \tag{2}$$

With the aid of the free-space Green's function, Eq. (1) can thus be transformed into the following boundary integral equation for the displacement u on the boundary $\Gamma = \Gamma_{(i)}+\Gamma_{(i+1)}$

$$Cu(\mathbf{r})+\int_{\Gamma} T^*(\mathbf{r},\mathbf{r}')u(\mathbf{r}')d\Gamma(\mathbf{r}')=\int_{\Gamma} U^*(\mathbf{r},\mathbf{r}')t(\mathbf{r}')d\Gamma(\mathbf{r}')+\delta_{si}U^*(\mathbf{r},\mathbf{s})F(\mathbf{r},\omega), \tag{3}$$

where $\mathbf{r}$ and $\mathbf{r}'$ are the position vectors of "field point" and "source point" on the boundary Γ, respectively. The coefficient C generally depends on the local geometry at $\mathbf{r}$, $t(\mathbf{r}')$ are the traction components, and $U^*(\mathbf{r}, \mathbf{r}')$ and $T^*(\mathbf{r}, \mathbf{r}')$ are the fundamental solutions for displacements and tractions, respectively,

$$\begin{aligned} U^*(\mathbf{r},\mathbf{r}') &= -\frac{\mathrm{i}}{4\mu^{(i)}} H_0^2\left(k^{(i)}\left|\mathbf{r}-\mathbf{r}'\right|\right) \\ T^*(\mathbf{r},\mathbf{r}') &= \frac{\mathrm{i}k^{(i)}}{4} H_1^2\left(k^{(i)}\left|\mathbf{r}-\mathbf{r}'\right|\right)(\mathbf{r}'-\mathbf{r})\mathbf{n}, \end{aligned} \tag{4}$$

and they satisfy the following equations

$$\begin{aligned} &\mu^{(i)}\nabla^2 U^*(\mathbf{r},\mathbf{r}') + \rho^{(i)}\omega^2 U^*(\mathbf{r},\mathbf{r}') = -\Delta(\mathbf{r}'-\mathbf{r}) \\ &T^*(\mathbf{r},\mathbf{r}') = \mu^{(i)}\frac{\partial}{\partial \mathbf{n}} U^*(\mathbf{r},\mathbf{r}'), \end{aligned} \tag{5}$$

where i is the imaginary unit different from the italic i being index, $k^{(i)}=\omega/c^{(i)}$ with $c^{(i)}$ being the shear wave velocity in the ith layer, Δ represents a Dirac delta function which goes to infinity at the point $\mathbf{r}' = \mathbf{r}$ and is equal to zero elsewhere, and $\mathbf{n}$ denotes the outward normal vector of the surface on which the traction is acting. Please note that the position vectors $\mathbf{r}$ and $\mathbf{r}'$ are now separately defined as "field point" and "source point" in the whole domain $\Omega^{(i)}$ bounded by the boundary Γ. Here, $U^*(\mathbf{r}, \mathbf{r}')$ represents the displacement field at the "field point" $\mathbf{r}$ generated by a concentrated unit source acting at the "source point" $\mathbf{r}'$. The symbol H^2 means the Hankel function of the second kind and its subscript indicates the order.

To solve Eq. (3), we need to do discretization on the boundary. For simplicity, we discretize the boundary $\Gamma_{(i)}$ and $\Gamma_{(i+1)}$ each into N constant elements (the results in the case of linear element can be deduced by analog), so that the coefficient C is taken as 1/2 always. For a given element j (taken as the "source point"), Eq. (3) can be discretized as follows

$$\frac{1}{2}u_j + \sum_{k=1}^{2N} H_{jk}u_k = \sum_{k=1}^{2N} G_{jk}t_k + F_j\delta_{si}, \tag{6}$$

where H_{jk} is an integral of $T(\mathbf{r}^j, \mathbf{r}^k)$ over field element k, G_{jk} is an integral of $U(\mathbf{r}^j, \mathbf{r}^k)$ over field element k, and F_j is the source term. When the source element goes from j = 1 to 2N, one can obtain a system of equations which can be expressed in matrix form

$$\mathbf{H}\mathbf{u}^{(i)} = \mathbf{G}\mathbf{t}^{(i)} + \mathbf{F}\delta_{si}, \tag{7}$$

where the superscript (i) denotes the layer index. Obviously, the column vector $\mathbf{u}^{(i)}$ of the dimension 2N×1 consists of the element displacements on both the boundary $\Gamma_{(i)}$ and the boundary $\Gamma_{(i+1)}$, the same situation applies to the column vector $\mathbf{t}^{(i)}$. And the coefficient matrices $\mathbf{H}$ and $\mathbf{G}$ are of the dimension 2N×2N.

The boundary and continuity conditions for the whole model under consideration are:

1. the traction-free condition on the free surface,

$$\mathbf{t}_{(1)}^{(1)} = 0, \tag{8}$$

2. the continuity conditions of displacement and traction at the inner interfaces,

$$\left.\begin{aligned}\mathbf{u}_{(i+1)}^{(i)} &= \mathbf{u}_{(i+1)}^{(i+1)} \\ \mathbf{t}_{(i+1)}^{(i)} &= -\mathbf{t}_{(i+1)}^{(i+1)}\end{aligned}\right\}, \tag{9}$$

3. the radiation boundary conditions imposed on the far-field behavior at infinity,

$$\left.\begin{aligned}\mathbf{u}_{(L+1)}^{(L)} &= 0 \\ \mathbf{t}_{(L+1)}^{(L)} &= 0\end{aligned}\right\}, \tag{10}$$

where the superscript denotes the layer index and the subscript denotes the interface index, which is followed as the same hereafter. And the minus sign in equation $(9)_2$ is due to the definition of outward normal vector in boundary element method. Please note that the displacements and tractions in equations (8-10) are different from those in equation (7). The former are of the dimension $N\times1$ while the latter $2N\times1$.

In order to apply those boundary conditions to equation (7), we need to rewrite equation (7) into

$$\begin{bmatrix}\mathbf{H}_{(i)}^{(i)} & \mathbf{H}_{(i+1)}^{(i)}\end{bmatrix}\begin{bmatrix}\mathbf{u}_{(i)}^{(i)} \\ \mathbf{u}_{(i+1)}^{(i)}\end{bmatrix} = \begin{bmatrix}\mathbf{G}_{(i)}^{(i)} & \mathbf{G}_{(i+1)}^{(i)}\end{bmatrix}\begin{bmatrix}\mathbf{t}_{(i)}^{(i)} \\ \mathbf{t}_{(i+1)}^{(i)}\end{bmatrix} + \mathbf{F}\delta_{si}, \tag{11}$$

where $\mathbf{u}_{(i)}^{(i)}$ and $\mathbf{u}_{(i+1)}^{(i)}$ are the element displacements separately corresponding to the upper interface and lower interface of the ith layer, similarly $\mathbf{t}_{(i)}^{(i)}$ and $\mathbf{t}_{(i+1)}^{(i)}$ are the element tractions corresponding to the upper interface and lower interface of the ith layer. Obviously, they are column vectors of the dimension $N\times1$. It should be pointed out that for one inner interface we should number the element displacements and tractions in the same order for the upper neighboring layer domain as that for the lower neighboring layer domain. Although this is a simple numbering, it is a common bug in computer programs. Look out for it! More attention should be paid to the coefficient matrices $\mathbf{H}_{(i)}^{(i)}$ and $\mathbf{H}_{(i+1)}^{(i)}$, $\mathbf{G}_{(i)}^{(i)}$ and $\mathbf{G}_{(i+1)}^{(i)}$ which are all of the dimension $2N\times N$ (except the case when $i = L$) and called matrix propagators. They will be used to form the global matrix propagators for the hybrid boundary element method being discussed in this paper.

For the sake of convenient derivation, we first define the upward direction as the positive direction of the tractions in the whole model. Following this, we just need to make a minor change to Eq. (11)

$$\begin{bmatrix}\mathbf{H}_{(i)}^{(i)} & \mathbf{H}_{(i+1)}^{(i)}\end{bmatrix}\begin{bmatrix}\mathbf{u}_{(i)} \\ \mathbf{u}_{(i+1)}\end{bmatrix} = \begin{bmatrix}\mathbf{G}_{(i)}^{(i)} & -\mathbf{G}_{(i+1)}^{(i)}\end{bmatrix}\begin{bmatrix}\mathbf{t}_{(i)} \\ \mathbf{t}_{(i+1)}\end{bmatrix} + \mathbf{F}\delta_{si}, \tag{12}$$

where only one subscript index is needed to indicate the element displacement and traction vector at different interfaces, and the minus sign in the front of the matrix propagator $\mathbf{G}_{(i+1)}^{(i)}$

is due to the above-defined positive direction. Equation (12) gives the relationship between the element tractions and the element displacements at the interfaces $\Gamma_{(i)}$ and $\Gamma_{(i+1)}$ which enclose the ith layer.

From now on, we introduce the global matrix propagators which is the key to the introduced method under discussion. For completeness, we cite some formula directly from the work by Ge and Chen [18] with minor modification.

The global matrix propagators $\mathbf{D}_{uu}^{(i)}$ and $\mathbf{D}_{tu}^{(i)}$ are separately defined as [18]

$$\begin{cases} \mathbf{u}_{(i)} = \mathbf{D}_{uu}^{(i)} \mathbf{u}_{(i+1)} \\ \mathbf{t}_{(i+1)} = \mathbf{D}_{tu}^{(i)} \mathbf{u}_{(i+1)} \end{cases}, i = 1, 2, \cdots, s-1, \tag{13}$$

for the layers above the source, and

$$\begin{cases} \mathbf{u}_{(i+1)} = \mathbf{D}_{uu}^{(i)} \mathbf{u}_{(i)} \\ \mathbf{t}_{(i)} = \mathbf{D}_{tu}^{(i)} \mathbf{u}_{(i)} \end{cases}, i = s+1, s+2, \cdots, L, \tag{14}$$

for the layers below the source.

In Eqs. (13) and (14), the superscripts in the global matrix propagators denote the layer index, and the subscripts in the element displacements and tractions denote the interface index. Each layer has two global matrix propagators which function in different ways for the layers above the source and the layers below the source. For one arbitrary layer above the source, $\mathbf{D}_{uu}$ transfers information of the element displacements from the upper interface to the lower interface in the layer whilst $\mathbf{D}_{tu}$ bridges the element tractions and the element displacements at the lower interface of the layer. However, for one arbitrary layer below the source, $\mathbf{D}_{uu}$ transfers information of the element displacements from the lower interface to the upper interface in the layer whilst $\mathbf{D}_{tu}$ bridges the element tractions and the element displacements at the upper interface of the layer. That's the main idea of the global matrix propagators which propagate information from "top and bottom" where the boundary conditions are known to "middle" where the source is located by the displacement and traction continuity conditions at inner interfaces. Both of the two matrices are of the dimension $N \times N$.

In the following part, how to obtain the global matrix propagators recursively will be briefly reviewed. Applying boundary condition (8) and Eq. (13) with $i = 1$ to Eq. (12) gives

$$\mathbf{H}_{(1)}^{(1)} \mathbf{D}_{uu}^{(1)} \mathbf{u}_{(2)} + \mathbf{H}_{(2)}^{(1)} \mathbf{u}_{(2)} + \mathbf{G}_{(2)}^{(1)} \mathbf{D}_{tu}^{(1)} \mathbf{u}_{(2)} = 0. \tag{15}$$

Because $\mathbf{u}_{(2)}$ could be the displacement at the second interface excited by an arbitrary source, the necessary condition of Eq. (15) is that its coefficient matrix equals zero, which gives

$$\begin{bmatrix} \mathbf{D}_{uu}^{(1)} \\ \mathbf{D}_{tu}^{(1)} \end{bmatrix} = -\begin{bmatrix} \mathbf{H}_{(1)}^{(1)} & \mathbf{G}_{(2)}^{(1)} \end{bmatrix}^{-1} \mathbf{H}_{(2)}^{(1)}. \tag{16}$$

Similarly, for the other layers above the source, i.e., $i = 2, \ldots, s-1$, we have

$$\begin{bmatrix} \mathbf{D}_{uu}^{(i)} \\ \mathbf{D}_{tu}^{(i)} \end{bmatrix} = -\begin{bmatrix} \mathbf{H}_{(i)}^{(i)} - \mathbf{G}_{(i)}^{(i)}\mathbf{D}_{tu}^{(i-1)} & \mathbf{G}_{(i+1)}^{(i)} \end{bmatrix}^{-1} \mathbf{H}_{(i+1)}^{(i)}. \tag{17}$$

Substituting boundary condition (10) and Eq. (14) with $i = L$ into Eq. (12) and considering the same reason as that giving rise to Eq. (16), we get

$$\begin{cases} \mathbf{D}_{uu}^{(L)} = 0 \\ \mathbf{D}_{tu}^{(L)} = \begin{bmatrix} \mathbf{G}_{(L)}^{(L)} \end{bmatrix}^{-1} \mathbf{H}_{(L)}^{(L)} \end{cases}. \tag{18}$$

In Eq. (18)$_2$, the matrix propagators $\mathbf{H}_{(L)}^{(L)}$ and $\mathbf{G}_{(L)}^{(L)}$ are of the dimension $N \times N$, which is different from the cases when $i < L$.

Similarly, for the other layers below the source, i.e., $i = L-1, L-2, \ldots, s+1$, we have

$$\begin{bmatrix} \mathbf{D}_{uu}^{(i)} \\ \mathbf{D}_{tu}^{(i)} \end{bmatrix} = -\begin{bmatrix} \mathbf{H}_{(i+1)}^{(i)} + \mathbf{G}_{(i+1)}^{(i)}\mathbf{D}_{tu}^{(i+1)} & -\mathbf{G}_{(i)}^{(i)} \end{bmatrix}^{-1} \mathbf{H}_{(i)}^{(i)}. \tag{19}$$

So far, the global matrix propagators are obtained for all the layers except for the source layer. But the problem has not been solved yet. In order to get the final solution of the problem, we need to form the system of equations in the source layer, which can be easily done by setting $i = s$ in Eq. (12)

$$\begin{bmatrix} \mathbf{H}_{(s)}^{(s)} & \mathbf{H}_{(s+1)}^{(s)} \end{bmatrix} \begin{bmatrix} \mathbf{u}_{(s)} \\ \mathbf{u}_{(s+1)} \end{bmatrix} = \begin{bmatrix} \mathbf{G}_{(s)}^{(s)} & -\mathbf{G}_{(s+1)}^{(s)} \end{bmatrix} \begin{bmatrix} \mathbf{t}_{(s)} \\ \mathbf{t}_{(s+1)} \end{bmatrix} + \mathbf{F}_s. \tag{20}$$

Here the subscript s in the $\mathbf{F}_s$ vector just indicates the source layer index. One more step needed to solve the above equations is to substitute $i = s-1$ and $s+1$ separately into Eqs. (17) and (19), which gives

$$\begin{cases} \mathbf{t}_{(s)} = \mathbf{D}_{tu}^{(s-1)} \mathbf{u}_{(s)} \\ \mathbf{t}_{(s+1)} = \mathbf{D}_{tu}^{(s+1)} \mathbf{u}_{(s+1)} \end{cases}. \tag{21}$$

Then Eq. (20) can be rewritten into by Eq. (21)

$$\begin{bmatrix} \mathbf{H}_{(s)}^{(s)} - \mathbf{G}_{(s)}^{(s)}\mathbf{D}_{tu}^{(s-1)} & \mathbf{H}_{(s+1)}^{(s)} + \mathbf{G}_{(s+1)}^{(s)}\mathbf{D}_{tu}^{(s+1)} \end{bmatrix} \begin{bmatrix} \mathbf{u}_{(s)} \\ \mathbf{u}_{(s+1)} \end{bmatrix} = \mathbf{F}_s, \tag{22}$$

which has the same number of unknowns and equations. Solving Eq. (22) by Gaussian elimination scheme, we can get the element displacements at the boundary of the source layer so that the element displacements and tractions at the other interfaces can be obtained by the following recursive equations

$$\begin{cases} \mathbf{u}_{(i)} = \mathbf{D}_{uu}^{(i)}\mathbf{D}_{uu}^{(i+1)}\cdots\mathbf{D}_{uu}^{(s-1)}\mathbf{u}_{(s)} \\ \mathbf{t}_{(i+1)} = \mathbf{D}_{tu}^{(i)}\mathbf{u}_{(i+1)} \end{cases}, \text{ for } i = 1, 2, \ldots, s-1, \tag{23}$$

$$\begin{cases} \mathbf{u}_{(i+1)} = \mathbf{D}_{uu}^{(i)}\mathbf{D}_{uu}^{(i+1)}\cdots\mathbf{D}_{uu}^{(s+1)}\mathbf{u}_{(s+1)} \\ \mathbf{t}_{(i)} = \mathbf{D}_{tu}^{(i)}\mathbf{u}_{(i)} \end{cases}, \text{ for } i = s+1, s+2, \ldots, L. \tag{24}$$

Then the displacement at any observation points in the whole model can be obtained by the integral equation in the domain

$$u(\mathbf{r}) + \int_{\Gamma} T^{*}(\mathbf{r},\mathbf{r}')u(\mathbf{r}')\mathrm{d}\Gamma(\mathbf{r}') = \int_{\Gamma} U^{*}(\mathbf{r},\mathbf{r}')t(\mathbf{r}')\mathrm{d}\Gamma(\mathbf{r}') + \delta_{si}U^{*}(\mathbf{r},\mathbf{s})F(\mathbf{r},\omega), \quad \mathbf{r} \in \Omega^{(i)}, \tag{25}$$

which can be transformed into a summation of the element displacements and tractions obtained just now without extra effort. Usually, the displacements at the free surface is of the most interest. By taking the inverse Fourier transform on the above frequency domain solution, we can finally get the time domain solution, i.e., the synthetic seismogram.

2.2. Numerical examples

In the following, we present the calculated synthetic seismograms for some typical models for which exiting results are available. Throughout the following examples, the source time function of the synthetic seismograms is a Ricker wavelet defined as

$$u(t) = \left(2\pi^2 f_0^2 t^2 - 1\right)\exp\left(-\pi^2 f_c^2 t^2\right), \tag{26}$$

where f_0 is the characteristic frequency of the wavelet and t is the time. Observation points are set along the free surface of all the models.

2.2.1. Semi-circular canyon model

The first model is a semi-circular canyon in an elastic homogenous half-space, as shown in Figure 2. Although the shape is too simple for a realistic canyon, the wave reflection and diffraction due to the topographical irregularity in time domain are quite complex, which can serve for an eligible test to the present method and our computation program. Here, we calculate the time-domain responses of the surface along the semi-circular canyon subject to incident SH waves. The shear velocity and mass density are taken as 1 km/s and 1 g/cm³ respectively in our calculation.

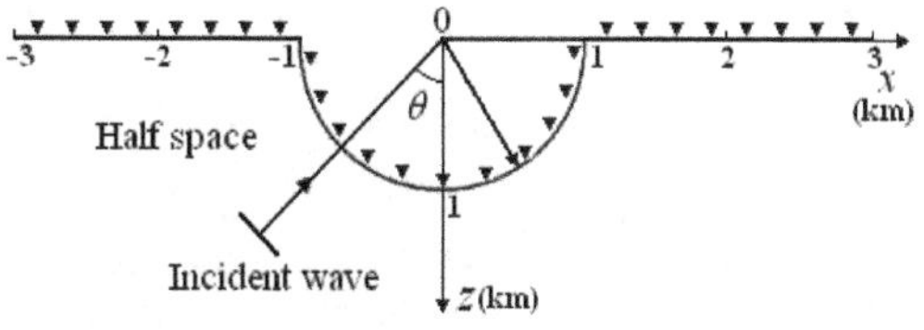

Figure 2. Semi-circular canyon half-space model

Figures 3 (a) and (b) separately show the antiplane responses of a canyon calculated by the present method due to incident *SH* waves with the angle of 0° and 30°. In the case of vertical incidence, the direct waves keep the same amplitude along the surface of the canyon except near the edges, where the destructive interference occurs. In the case of inclined incidence, the amplitudes of the direct waves inside the canyon decrease toward the rear-side edge. At the horizontal surface near the front-side edge, the peak amplitude becomes larger due to the constructive interference between the direct wave and the wave reflected at the canyon. The two seismograms were calculated by the discrete wavenumber method in [28]. The nice agreement can be seen between our results and those in [28].

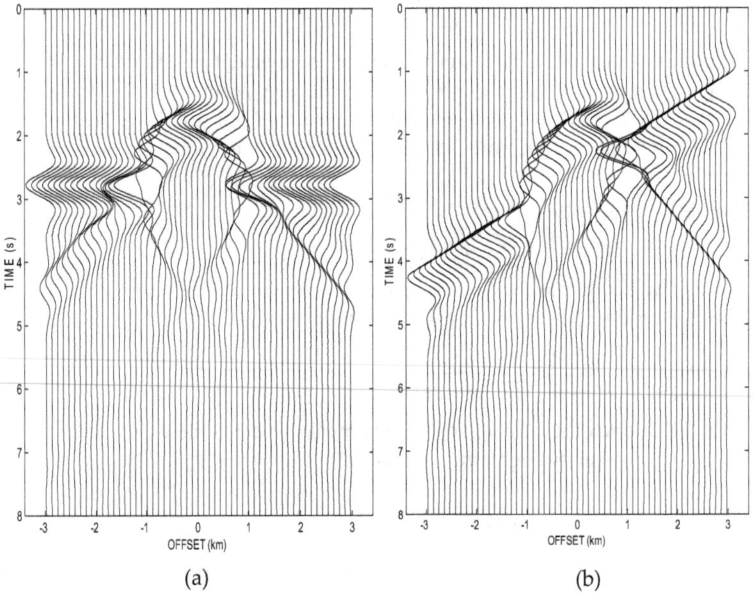

Figure 3. Time responses of the canyon model due to an *SH* wave incidence: (a) vertical incidence, (b) incidence with an angle of 30°. The characteristic frequency is 1.0 Hz.

2.2.2. Mountain model

Next, a mountain model with a point source is considered, as shown in Figure 4. The mountain shape topography is described by the equation

$$z(x) = \begin{cases} -h\left[1+\cos\left(\pi|x|\right)\right]/2, \text{for}|x| \le 1 \\ 0, \qquad\qquad\qquad\qquad \text{for}|x| \ge 1 \end{cases}, \tag{27}$$

where h = 1 km and the elastic constant and the mass density are 1 km/s and 1 g/cm³, respectively.

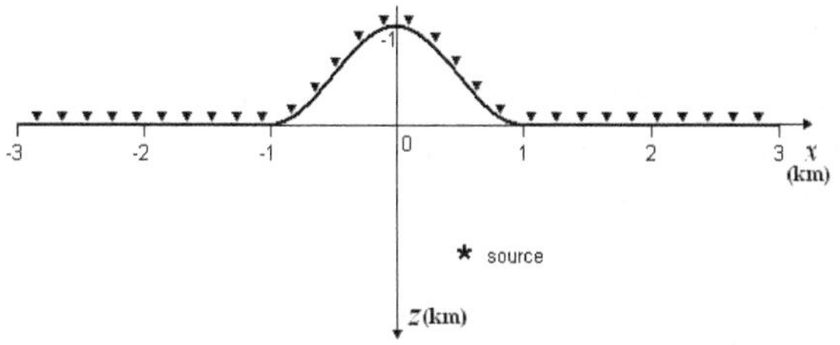

Figure 4. Mountain topography and the surrounding homogeneous half-space

Figure 5(a) shows the time responses along the mountain topography shown in Fig. 4 when the point source is plated right below the geometric center of the mountain; thus the synthetic seismogram has a symmetric feature about the center point. It can be seen from Figure 5(a) that the first arrival of seismic waves comes late at the locations on the mountain due to the higher altitude. We can also see strong diffracted waves produced by the mountain, i.e. the secondary phase. Figure 5(b) shows the time responses along the same topography as in Figure 5(a) but for the case when the point source is located at $x = 1$, $z = 2$ km. It can be seen from Figure 5(b) that although the first arrival of seismic waves has a nonsymmetric feature as expected, the diffracted waves are symmetric as that for the case shown in Figure 5(a). That indicates that the diffracted waves for both cases propagate along the free surface with the shear wave velocity of the media, and are produced by the mountain topography. The two seismograms were first calculated in [8] by using a global generalized reflection/transmission matrix method. Our calculated results show a good agreement with those in [8].

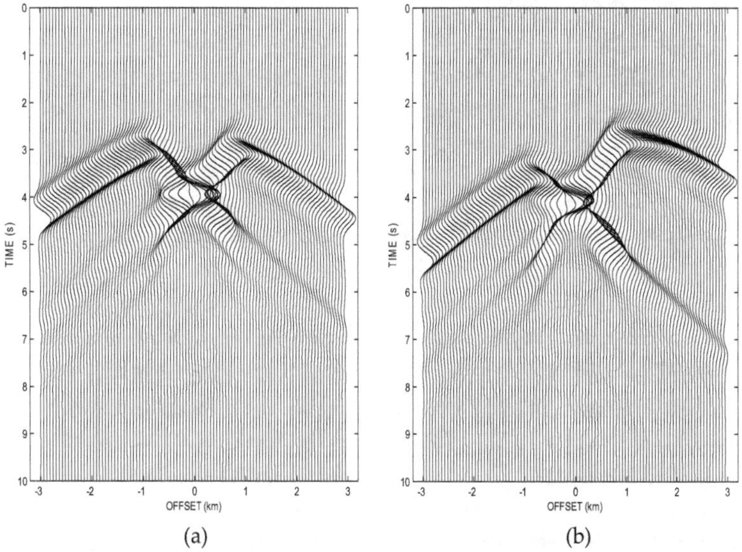

Figure 5. Synthetic seismograms for the mountain model in which the point source is placed at (a) $x = 0$, $z = 2$ km, (b) $x = 1$, $z = 2$ km. The characteristic frequency is 2.0 Hz.

2.2.3. *Deep basin model*

Consider the model configuration consisting of a deep basin structure, as shown in Figure 6. The basin has a trapezoidal shape with 1 km depth and 10 km width at the surface. The shear wave velocity of the basin-filling sediments and the half-space are 2.5 km/s and 1 km/s, respectively, and the ratio of mass density is set to 1:1.

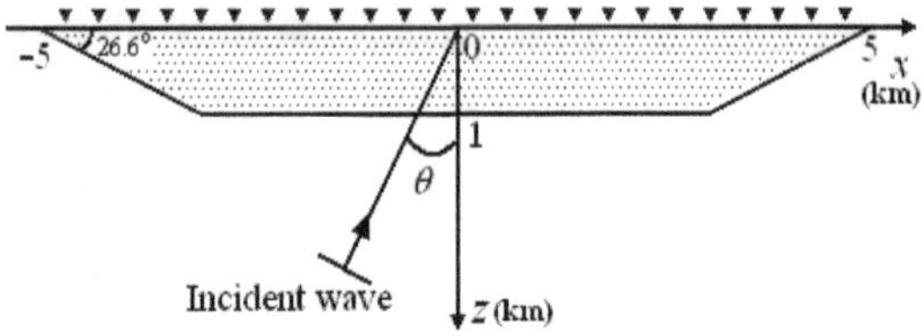

Figure 6. Deep basin topography and the surrounding homogeneous half-space

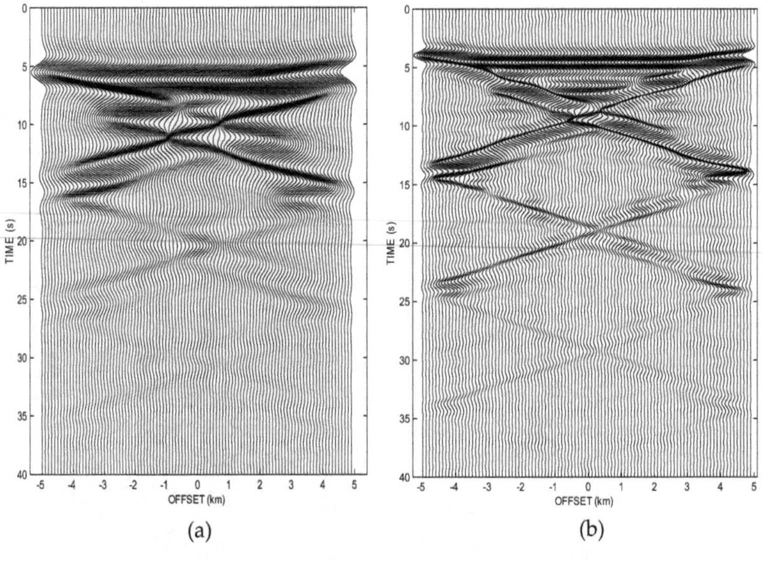

Figure 7. Time responses along the free surface of the basin in Figure 6 due to a vertically incident *SH*-wave: (a) characteristic frequency f_c is 0.25 Hz (4 sec), (b) f_c is 0.5 Hz (2 sec)

The time responses at the surface of the basin by a vertically incident *SH* wave are plotted in Figure 7(a). The characteristic frequency f_c is 0.25 Hz (4 s). This figure shows the horizontally propagating waves generated by the edges of the basin. The amplitude of Love waves (horizontally polarized shear surface waves) is smaller than that of the direct wave. Although these Love waves make the total duration longer, the time interval between the direct wave arrival and the Love wave arrival is less than 10 sec near the edges. Each arrival of reflected Love waves is well separated. Figure 7(b) shows the time responses for the same basin model as in Figure 7(a) but for a Ricker wavelet of characteristic frequency $f_c = 0.5$ Hz (2 s). Very similar phenomena to that shown in Figure 7(a) can be seen, but the amplitude of the Love wave is larger in this case. The X-shaped pattern of the surface-wave arrivals is more clearly shown for high-frequency incident waves, but the maximum amplitude of the Love wave decreases because energy splits into the fundamental mode and the other higher mode. The two seismograms were calculated first by the discrete wavenumber method in [29]. And nice agreement between our calculated and those in [29] can be observed.

2.2.4. Two-layer model with irregular interface

Finally, a two-layer model with an irregular interface is considered to have a point source, as shown in Figure 8. The irregular interface is described by

$$z(x) = h^{(1)} + \begin{cases} h, & \text{for } |x| \le d \\ \left\{1+\cos\left[\pi\left(|x|-d\right)/2d\right]\right\} h/2, & \text{for } d \le |x| \le 3d, \\ 0, & \text{for } |x| \ge 3d \end{cases} \tag{28}$$

where $h^{(1)} = 2.0$ km, $h = 0.5$ km, and $d = 0.5$ km. The material parameters are 1 km/s and 1 g/cm^3 for the top layer, and 1.5 km/s and 1 g/cm^3 for the bottom layer, respectively.

The synthetic seismograms for the two-layer model are shown in Figure 9. It can be seen from Figure 9 that all the seismic phases can be divided into two groups of hyperbolas, namely, a group of hyperbolas with a center point of $x = x_s$ and another group of hyperbolas with a center point of $x = 0$. The latter ones are due to the reflections by the flat free surface and the flat part of the interface. The former ones, however, are due to the diffractions by the irregular part of the interface. This model was first calculated by the global generalized reflection/transmission matrices method in [8]. Our result shows a nice agreement with that in [8].

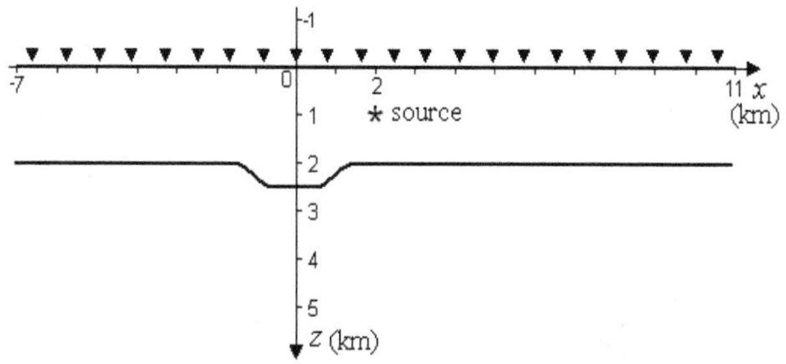

Figure 8. A two-layer model with an irregular interface: the source is at (2 km, 1 km)

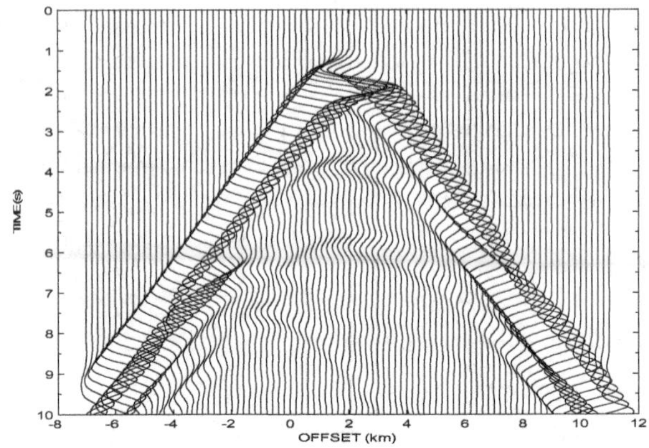

Figure 9. Synthetic seismograms for the model in Figure 8. The characteristic frequency is 1.0 Hz.

3. Seismoacoustic scattering modeling in mutilayered media

3.1. Methodology statement

The problem to be considered is illustrated in Figure 10. In this model, there are L homogeneous layers $\Omega^{(i)}$ (i=1, 2, …, L) over a half-space, among which the ith layer is bounded by two irregular interfaces $\Gamma_{(i-1)}$ and $\Gamma_{(i)}$ (Throughout the section, the superscripts with round brackets are used for layer index and the subscripts with round brackets for interface index). The uppermost layer is water which is assumed to be linear liquid without viscosity. An arbitrary seismic source is embedded in the sth layer. The properties of an arbitrary solid layer $\Omega^{(i)}$ are described by the elastic constants $\lambda^{(i)}$, $\mu^{(i)}$ and mass density $\rho^{(i)}$. This problem will be solved by a boundary element method with a global matrix propagator. In this Subsection, we will present the formulations of the method in the solid layers and the fluid layer respectively.

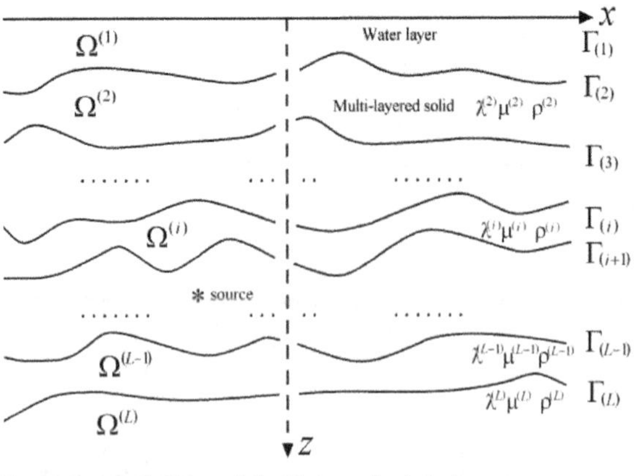

Figure 10. A multi-layered water/solid model with irregular interfaces

3.1.1. Formulations in solid layers

Consider the boundary element problem in a region Ω, where the elastic wave equation in frequency domain is given by

$$\mu\nabla^2\mathbf{u} + (\lambda+\mu)\nabla\nabla\cdot\mathbf{u} + \rho\omega^2\mathbf{u} = -\mathbf{f}, \tag{29}$$

in which **u** is the seismic displacement response in frequency domain, and **f** is the seismic source. Suppose the seismic source distribution consists of a simple point source at a position vector **s**. With the aid of the free-space Green's function, Eq. (29) can thus be transformed into the following boundary integral equation for the displacement $u_j(\mathbf{r})$ on the boundary Γ of the region Ω [30]

$$\Delta(\mathbf{r})u_k(\mathbf{r}) + \int_\Gamma u_j(\mathbf{r}')T_{jk}(\mathbf{r},\mathbf{r}')\,d\Gamma(\mathbf{r}') = \int_\Gamma t_j(\mathbf{r}')U_{jk}(\mathbf{r},\mathbf{r}')\,d\Gamma(\mathbf{r}') + f_jU_{jk}(\mathbf{r},\mathbf{s}), \tag{30}$$

where j and k can be 1, 2. **r** and **r**′ are the position vectors of "field point" and "source point" on the boundary Γ, respectively. The coefficient Δ(**r**) generally depends on the local geometry at **r**, $t_j(\mathbf{r}')$ are the traction components, and $U_{jk}(\mathbf{r}, \mathbf{r}')$ and $T_{jk}(\mathbf{r}, \mathbf{r}')$ are the fundamental solutions for displacements and tractions, respectively. The source exciting direction can be controlled arbitrarily by changing one of two components f_j. Application of Eq. (30) in domain $\Omega^{(i)}$ and discretization of interfaces $\Gamma_{(i)}$ and $\Gamma_{(i+1)}$ each into N elements (for simplicity, we use constant element) give rise to

$$\mathbf{H}^{(i)}\mathbf{u}^{(i)} = \mathbf{G}^{(i)}\mathbf{t}^{(i)} + \mathbf{F}\delta_{is}, \tag{31}$$

where $\mathbf{H}^{(i)}$ and $\mathbf{G}^{(i)}$ are the coefficient matrices obtained by integrating the fundamental solutions T_{jk} and U_{jk} over elements at both interfaces $\Gamma_{(i)}$ and $\Gamma_{(i+1)}$, and $\mathbf{u}^{(i)}$ and $\mathbf{t}^{(i)}$ are vectors containing element displacements and tractions at both interfaces $\Gamma_{(i)}$ and $\Gamma_{(i+1)}$, respectively. Obviously, $\mathbf{H}^{(i)}$ and $\mathbf{G}^{(i)}$ are of the dimension $4N\times4N$, and $\mathbf{u}^{(i)}$ and $\mathbf{t}^{(i)}$ are of the dimension $4N\times1$. In the multilayered medium (as shown in Figure 10), considering the continuity conditions of displacements and tractions at inner interfaces, Eq. (31) needs to be rewritten into

$$\begin{bmatrix}\mathbf{H}^{(i)}_{(i)} & \mathbf{H}^{(i)}_{(i+1)}\end{bmatrix}\begin{bmatrix}\mathbf{u}^{(i)}_{(i)}\\ \mathbf{u}^{(i)}_{(i+1)}\end{bmatrix} = \begin{bmatrix}\mathbf{G}^{(i)}_{(i)} & \mathbf{G}^{(i)}_{(i+1)}\end{bmatrix}\begin{bmatrix}\mathbf{t}^{(i)}_{(i)}\\ \mathbf{t}^{(i)}_{(i+1)}\end{bmatrix} + \mathbf{F}\delta_{si}, \tag{32}$$

where $\mathbf{u}^{(i)}$ and $\mathbf{t}^{(i)}$ with subscripts (i) and (i+1) corresponds separately to the element displacements and tractions at $\Gamma_{(i)}$ and $\Gamma_{(i+1)}$. Similarly, the subscripts (i) and (i+1) in $\mathbf{H}^{(i)}$ and $\mathbf{G}^{(i)}$ indicate the integration over elements at $\Gamma_{(i)}$ and $\Gamma_{(i+1)}$, respectively. Obviously, the matrices $\mathbf{H}^{(i)}$ and $\mathbf{G}^{(i)}$ with subscripts (i) or (i+1) are of the dimension $4N\times2N$.

For the same purpose as in Section 2, the upward direction is defined as the positive direction for tractions so that they have unique values at each interface [18]. Eq. (32) thus becomes

$$\mathbf{H}_{(i)}^{(i)}\mathbf{u}_{(i)} + \mathbf{H}_{(i+1)}^{(i)}\mathbf{u}_{(i+1)} = \mathbf{G}_{(i)}^{(i)}\mathbf{t}_{(i)} - \mathbf{G}_{(i+1)}^{(i)}\mathbf{t}_{(i+1)} + \mathbf{F}\delta_{si}, \tag{33}$$

where only one subscript index is needed to distinct element displacements and tractions at different interfaces, and the minus sign in front of $\mathbf{G}_{(i+1)}^{(i)}$ is due to the defined positive direction for tractions.

The global matrix propagators in the layers above the source are defined as

$$\begin{cases} \mathbf{u}_{(i)} = \mathbf{D}_{uu}^{(i)}\mathbf{u}_{(i+1)} \\ \mathbf{t}_{(i+1)} = \mathbf{D}_{tu}^{(i)}\mathbf{u}_{(i+1)} \end{cases}, \quad i = 2,3,\cdots,s-1, \tag{34}$$

in which one should note that the layer index i does not start from 1 but 2 since the first layer is a fluid layer in our model. Substitution of Eq. (34) into Eq. (33) yields

$$\mathbf{H}_{(i)}^{(i)}\mathbf{D}_{uu}^{(i)}\mathbf{u}_{(i+1)} + \mathbf{H}_{(i+1)}^{(i)}\mathbf{u}_{(i+1)} = \mathbf{G}_{(i)}^{(i)}\mathbf{D}_{tu}^{(i-1)}\mathbf{D}_{uu}^{(i)}\mathbf{u}_{(i+1)} - \mathbf{G}_{(i+1)}^{(i)}\mathbf{D}_{tu}^{(i)}\mathbf{u}_{(i+1)}. \tag{35}$$

Since $\mathbf{u}_{(i+1)}$ in Eq. (35) could be the displacement along the lower interface excited by an arbitrary source, the necessary condition of Eq. (35) is that its coefficient matrix equals zero, which leads to

$$\begin{pmatrix} \mathbf{D}_{uu}^{(i)} \\ \mathbf{D}_{tu}^{(i)} \end{pmatrix} = -\left[\mathbf{H}_{(i)}^{(i)} - \mathbf{G}_{(i)}^{(i)}\mathbf{D}_{tu}^{(i-1)}, \quad \mathbf{G}_{(i+1)}^{(i)}\right]^{-1}\mathbf{H}_{(i+1)}^{(i)}, \quad i = 2,3,\cdots,s-1. \tag{36}$$

Once the global matrix propagator $\mathbf{D}_{tu}^{(1)}$ in the fluid layer is obtained, those for the layers above the source can be obtained recursively from Eq. (36). For the moment, we have not known the expression of the global matrix propagator in the fluid layer yet.

Similarly, the global matrix propagators in the layers below the source are defined as

$$\begin{cases} \mathbf{u}_{(i+1)} = \mathbf{D}_{uu}^{(i)}\mathbf{u}_{(i)} \\ \mathbf{t}_{(i)} = \mathbf{D}_{tu}^{(i)}\mathbf{u}_{(i)} \end{cases}, \quad i = s+1, s+2, \cdots, L. \tag{37}$$

Substituting Eq. (37) into Eq. (33) and considering the same reason as in Eq. (36), we have

$$\begin{pmatrix} \mathbf{D}_{uu}^{(i)} \\ \mathbf{D}_{tu}^{(i)} \end{pmatrix} = -\left[\mathbf{H}_{(i+1)}^{(i)} + \mathbf{G}_{(i+1)}^{(i)}\mathbf{D}_{tu}^{(i+1)}, \quad -\mathbf{G}_{(i)}^{(i)}\right]^{-1}\mathbf{H}_{(i)}^{(i)}, \quad i = L-1, L-2, \cdots, s+1, \tag{38}$$

where $\mathbf{D}_{tu}^{(L)}$ is the global matrix propagator in the bottom layer which can be obtained by using the radiation condition at infinity, i.e. $\mathbf{u}^{(L+1)}=0$ and $\mathbf{t}^{(L+1)}=0$, as follows

$$\begin{cases} \mathbf{D}_{uu}^{(L)} = 0 \\ \mathbf{D}_{tu}^{(L)} = \left[\mathbf{G}_{(L)}^{(L)}\right]^{-1}\mathbf{H}_{(L)}^{(L)} \end{cases}. \tag{39}$$

So far, we have only defined the global matrix propagators for all the solid layers. The global matrix propagator in the fluid layer will be given next.

3.1.2. Formulations in fluid layer

The pressure (i.e. the infinitesimal pressure perturbation) is used as the variable to treat the interaction between the elastic waves and the acoustic waves in the fluid layer [20]. Thus the acoustic wave equation in frequency domain in a homogeneous isotropic fluid layer is given by

$$\nabla^2 p + k^2 p = 0, \tag{40}$$

where p is the fluid pressure and $k = \omega/V$ is the wavenumber with ω and V being the angular frequency and the acoustic wave velocity, respectively. Using the fundamental solution of Helmholtz equation, we can replace Eq. (40) by the following boundary integral equation for the fluid pressure $p(\mathbf{r})$ on the boundary Γ surrounding the fluid area

$$\mathrm{c}(\mathbf{r})p(\mathbf{r}) + \int_\Gamma p(\mathbf{r}')\frac{\partial G(\mathbf{r},\mathbf{r}')}{\partial n_{\mathbf{r}'}}d\Gamma(\mathbf{r}') = \int_\Gamma q(\mathbf{r}')\,G(\mathbf{r},\mathbf{r}')\,d\Gamma(\mathbf{r}') \tag{41}$$

where $\mathrm{c}(\mathbf{r})$ generally depends on the local geometry at a point $\mathbf{r}$, and $G(\mathbf{r},\mathbf{r}')$ is known as the Green's function of acoustics which physically represents the effect observed at the point $\mathbf{r}$ of a unit source at the point $\mathbf{r}'$ [31]. $q(\mathbf{r}') = \frac{\partial p(\mathbf{r}')}{\partial n_{\mathbf{r}'}}$, where the terminology $\frac{\partial *}{\partial n_{\mathbf{r}'}}$ represents the partial derivative of the function * with respect to the unit outward normal at the point $\mathbf{r}'$ on the boundary Γ. Applying Eq. (41) in the uppermost layer $\Omega^{(1)}$ and discretizing the interfaces $\Gamma_{(1)}$ and $\Gamma_{(2)}$ each into N elements (still constant element), we yield the matrix equation of the form

$$\mathbf{H}_{(2)}^{(1)}\mathbf{p}_{(2)} = \mathbf{G}_{(1)}^{(1)}\mathbf{q}_{(1)} + \mathbf{G}_{(2)}^{(1)}\mathbf{q}_{(2)}, \tag{42}$$

where $\mathbf{p}$ and $\mathbf{q}$ with subscripts $_{(1)}$ and $_{(2)}$ correspond separately to the element pressure and its normal derivative at $\Gamma_{(1)}$ and $\Gamma_{(2)}$ ($\mathbf{p}_{(1)}$ in Eq. (42) is suppressed since it's always equal to zero due to the free surface condition). Obviously, they are the vectors of the dimension $N\times1$. Follow the rule in the solid layers, we first define the global matrix propagators in the fluid layer as

$$\begin{cases}\mathbf{q}_{(1)} = \mathbf{D}_{qq}^{(1)}\mathbf{q}_{(2)} \\ \mathbf{p}_{(2)} = \mathbf{D}_{pq}^{(1)}\mathbf{q}_{(2)}\end{cases}, \tag{43}$$

where the first definition is actually unnecessary since we focus on the solution at the fluid-solid interface and ignore the normal derivative of the pressure at the free surface. Substitution of Eq. (43) into Eq. (42) yields

$$\mathbf{H}_{(2)}^{(1)}\mathbf{D}_{pq}^{(1)}\mathbf{q}_{(2)} = \mathbf{G}_{(1)}^{(1)}\mathbf{D}_{qq}^{(1)}\mathbf{q}_{(2)} + \mathbf{G}_{(2)}^{(1)}\mathbf{q}_{(2)}. \tag{44}$$

Since $\mathbf{q}_{(2)}$ in Eq. (44) could be the normal derivative of the pressure at the fluid-solid interface excited by an arbitrary source, the necessary condition of Eq. (44) is that its coefficient matrix equals zero, which leads to

$$\begin{pmatrix} \mathbf{D}_{qq}^{(1)} \\ \mathbf{D}_{pq}^{(1)} \end{pmatrix} = \left[-\mathbf{G}_{(1)}^{(1)}, \quad \mathbf{H}_{(2)}^{(1)}\right]^{-1} \mathbf{G}_{(2)}^{(1)}, \tag{45}$$

where one point noted here is that the global matrix propagators obtained in Eq. (45) have the dimension $N{\times}N$ which is different from those in Eqs. (34) and (37). Therefore, we need to extend the global matrix propagators defined for the fluid layer into the same dimension as those defined for the solid layers, i.e., from $N{\times}N$ to $2N{\times}2N$, by using the continuity conditions at the fluid-solid interface. They are : 1) the tangential traction (not stress, be careful with the terminology) is zero; 2) the normal traction is continuous; and 3) the normal displacement is continuous, which can be expressed in frequency domain as follows [20, 32]

$$\begin{aligned} \mathbf{t}\cdot\mathbf{s} &= 0 \\ \mathbf{t}\cdot\mathbf{n} &= -p \\ \mathbf{u}\cdot\mathbf{n} = -u_{\mathrm{f}} &= -\frac{1}{\rho_{\mathrm{f}}\omega^2}q' \end{aligned} \tag{46}$$

where $\mathbf{u}$ and $\mathbf{t}$ are the element displacements and tractions of the solid layer at the fluid-solid interface. And the $\mathbf{n}$ and $\mathbf{s}$ are the unit outward normal vector and the inplane tangential vector of the solid layer neighboring the fluid layer. ρ_f is the mass density of the fluid layer. The displacement of the fluid u_f is along the normal direction but points to the neighboring solid layer (i.e., the opposite direction of the normal vector $\mathbf{n}$), which leads to the minus sign in front of u_f.

Now, the global matrix propagators obtained in Eq. (45) can be extended into the same dimension as those defined in the solid layers, i.e. $2N{\times}2N$, by

$$\mathbf{D}_{tu}^{(1)} = \rho_{\mathrm{f}}\omega^2 \begin{bmatrix} \mathbf{D}_{pq}^{(1)}\cdot\mathbf{N}_1^{\mathrm{e}}\cdot(\mathbf{N}_1^{\mathrm{e}})^{\mathrm{T}} & \mathbf{D}_{pq}^{(1)}\cdot\mathbf{N}_2^{\mathrm{e}}\cdot(\mathbf{N}_1^{\mathrm{e}})^{\mathrm{T}} \\ \mathbf{D}_{pq}^{(1)}\cdot\mathbf{N}_1^{\mathrm{e}}\cdot(\mathbf{N}_2^{\mathrm{e}})^{\mathrm{T}} & \mathbf{D}_{pq}^{(1)}\cdot\mathbf{N}_2^{\mathrm{e}}\cdot(\mathbf{N}_2^{\mathrm{e}})^{\mathrm{T}} \end{bmatrix}, \tag{47}$$

where the matrices $\mathbf{N}_1^{\mathrm{e}}$ and $\mathbf{N}_2^{\mathrm{e}}$ are defined as follows:

$$\mathbf{N}_1^{\mathrm{e}} = \begin{bmatrix} n_1^{\mathrm{e}_1} & n_1^{\mathrm{e}_2} & \cdots & n_1^{\mathrm{e}_N} \\ n_1^{\mathrm{e}_1} & n_1^{\mathrm{e}_2} & \cdots & n_1^{\mathrm{e}_N} \\ \vdots & \vdots & \ddots & \vdots \\ n_1^{\mathrm{e}_1} & n_1^{\mathrm{e}_2} & \cdots & n_1^{\mathrm{e}_N} \end{bmatrix}, \mathbf{N}_2^{\mathrm{e}} = \begin{bmatrix} n_2^{\mathrm{e}_1} & n_2^{\mathrm{e}_2} & \cdots & n_2^{\mathrm{e}_N} \\ n_2^{\mathrm{e}_1} & n_2^{\mathrm{e}_2} & \cdots & n_2^{\mathrm{e}_N} \\ \vdots & \vdots & \ddots & \vdots \\ n_2^{\mathrm{e}_1} & n_2^{\mathrm{e}_2} & \cdots & n_2^{\mathrm{e}_N} \end{bmatrix}, \tag{48}$$

in which $n_1^{\mathrm{e}_i}$ and $n_2^{\mathrm{e}_i}$ are the two components of the unit outward normal vector of the neighboring solid layer at the element 'e_i', i = 1, 2, 3,..., N. Note that the multiplication between matrices in Eq. (47) is operated in the sense of 'element to element'.

3.1.3. *Solution in source layer*

We have clearly obtained the global matrix propagators for all the layers except for the source layer which is the place to finally solve the problem. Setting $i = s$ in Eq. (33), we have [19]

$$\mathbf{H}_{(s)}^{(s)}\mathbf{u}_{(s)} + \mathbf{H}_{(s+1)}^{(s)}\mathbf{u}_{(s+1)} = \mathbf{G}_{(s)}^{(s)}\mathbf{t}_{(s)} - \mathbf{G}_{(s+1)}^{(s)}\mathbf{t}_{(s+1)} + \mathbf{F}, \tag{49}$$

in which both the element displacements and tractions are unknowns, however, they cannot be solved from the present form of Eq. (49) yet. From the aforementioned definitions of the global matrix propagators, i.e., take $i = s-1$ in Eq. (34) and $i = s+1$ in Eq. (37), we can get the relationship between the traction and displacement at the sth and (s+1)th interfaces as [19]

$$\mathbf{t}_{(s)} = \mathbf{D}_{tu}^{(s-1)}\mathbf{u}_{(s)}, \mathbf{t}_{(s+1)} = \mathbf{D}_{tu}^{(s+1)}\mathbf{u}_{(s+1)}. \tag{50}$$

Substitution of Eq. (50) into Eq. (49) gives rise to [19]

$$\left[\mathbf{H}_{(s)}^{(s)} - \mathbf{G}_{(s)}^{(s)}\mathbf{D}_{tu}^{(s-1)}\right]\mathbf{u}_{(s)} + \left[\mathbf{H}_{(s+1)}^{(s)} + \mathbf{G}_{(s+1)}^{(s)}\mathbf{D}_{tu}^{(s+1)}\right]\mathbf{u}_{(s+1)} = \mathbf{F}, \tag{51}$$

which has the same number of unknowns and equations and can be solved without question.

Once the displacements at the interfaces enclosing the source layer are solved, the displacements at the fluid-solid interface can be obtained by

$$\mathbf{u}_{(2)} = \mathbf{D}_{uu}^{(2)}\mathbf{D}_{uu}^{(3)}\cdots\mathbf{D}_{uu}^{(s-1)}\mathbf{u}_{(s)}, \tag{52}$$

which is the displacement solution in frequency domain at the fluid-solid interface. By applying the inverse Fourier transform to Eq. (52), one can get the solution in time domain, i.e., the synthetic seismograms. A general procedure of applying the introduced approach to calculate a seismic ground motion in an irregularly multilayered media lain by a fluid layer is summarized as follows.

Step 1. Calculate the global matrix propagators in the uppermost layer, i.e. the fluid layer, by Eq. (45).

Step 2. Extend the global matrix propagators obtained in **Step** 1 by Eq. (47).

Step 3. Use the extended global matrix propagator in **Step** 2 to calculate the global matrix propagators in the solid layers above the source layer by Eq. (36).

Step 4. In case of a plane wave incidence problem or a seismic source located at the bottom layer, skip this step and jump to **Step** 5. Otherwise, calculate the global matrix propagators in the solid layers below the seismic source layer by Eqs. (38-39).

Step 5. Substitute all the global matrix propagators obtained in the above steps into the simultaneous matrix equation in the seismic source layer and solve the problem by Eq. (51).

Step 6. Obtain the displacement solutions at the fluid-solid interface by Eq. (52).

Step 7. Execute the Inverse Fourier Transform on the displacement solutions obtained in **Step 6** to get the seismic ground motion in time domain finally.

The above-mentioned procedure is for the case when the fluid layer is the top layer. In case that the fluid layer is not the uppermost layer of the model under consideration, such as a fluid layer surrounded by two solid layers, the treatment similar to Eq. (47) can be also applied at the two fluid-solid interfaces. The only difference of the solution procedure is the order of calculating the global matrix propagators in the fluid layer and the solid layer, the details of which are suppressed here.

Although the boundary element method is known as the best way to model wave propagation problems in unbounded media, it is still necessary to make some special treatment on the truncation edges of the model in order to avoid the appearance of some unphysical waves. For such purpose, many different techniques [33-40] applicable to the boundary element method have been developed.

3.2. Numerical examples

In this subsection, we conduct the numerical implementation of the fluid-solid formulation of the introduced method. Three examples are calculated for validity testing, considering both plane wave incidence and SV-wave source excitation. Although the models used in this subsection seem very simple, they are often used as examples for validity testing of some seismogram synthetic method.

We first calculate the case of a plane *P*-wave (Primary or pressure wave) vertically incident onto irregularly layered fluid-solid structures in the two models, as shown in Figure 11. The reason for the selection of the two models is because the synthetic results of seismoacoustic scattering due to the irregular fluid-solid interface are available in [20]. The existing results for the two models calculated by the reflection/transmission matrix method in [20] were validated in detail by the finite difference method [20], which concluded that the reflection/transmission matrix method could give more accurate results. Therefore, we will compare our results with those calculated by the reflection/ transmission matrix method.

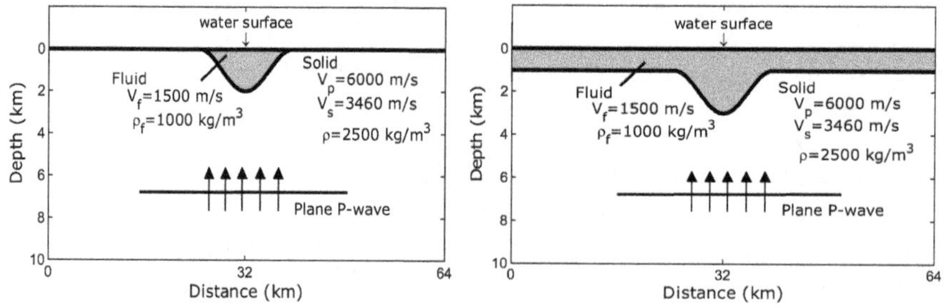

(a) (b)

Figure 11. The two models used to test the fluid-solid formulation: (a) Model 1; (b) Model 2 [20]

In order to make our comparison convincing, not only are the model parameters set to be the same as those used in [20] but also the time function of the input plane wave is selected as

$$f(t) = \begin{cases} \sqrt{\alpha/\pi}\exp(-\alpha t^2), -T_s/2 \le t \le T_s/2 \\ 0, \qquad t \ge T_s/2\,,\ t \le -T_s/2 \end{cases}, \tag{53}$$

where the pulse duration *Ts* is set to 4.2 s and α is set to 2.1. Eq. (53) is exactly the same as the Eq. (88) in [20] and its shape and spectrum are shown in Figure 12.

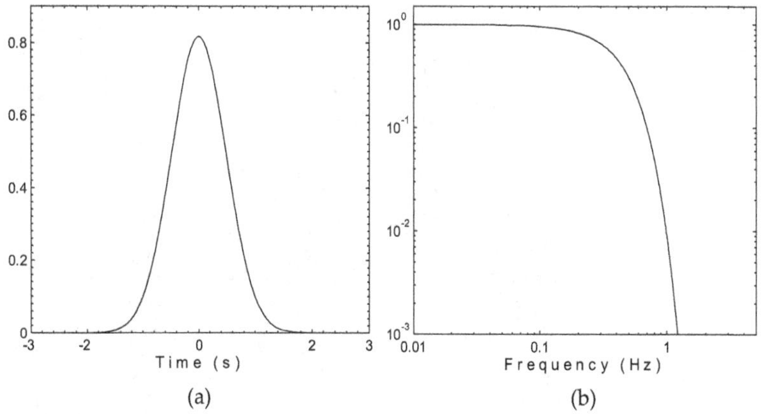

Figure 12. The time function (a) and its Fourier spectra (b) used for the two models in Figure 11

3.2.1. Model 1 subjected to a plane wave incidence

Figure 13 shows the displacements of both horizontal and vertical components along the fluid-solid interface of *Model* 1. The one on the left is calculated by the reflection/transmission matrix method in [20]. And the one on the right is calculated by the present method. Inside the irregular part of the model, i.e., the basin structure, the large-amplitude multiple reflections in the water basin are well observed in the synthetic waveforms calculated by both our method and the reflection/transmission matrix method. Both the scattered *P* wave (P) and *Rayleigh* wave (R), which are indicated in the left plot, can be well and clearly observed in our calculated seismograms.

3.2.2. Model 2 subjected to a plane wave incidence

Figure 14 shows the displacements of both horizontal and vertical components at the fluid-solid interface of *Model* 2. This model is a little bit more complicated than *Model* 1. The flat fluid layer on both sides of *Model* 2 will cause multiple reflections of the incident plane *P*-wave easily. The multiple reflections will be interfered by the waves scattered by the irregular fluid-solid interface, which leads to the complexity of the wave field. In this case, the 'observation points' are all located at the fluid layer bottom. The one on the left is calculated by the reflection/transmission matrix method in [20]. And the one on the right is calculated by the present method. Again, the synthetic seismograms calculated by both methods agree with each other very well. The periodic multiple reflections at the flat portion of the fluid layer are clearly observed and interfered by the scattered waves from the inside of the irregular part of the model, which are modeled very well by both methods.

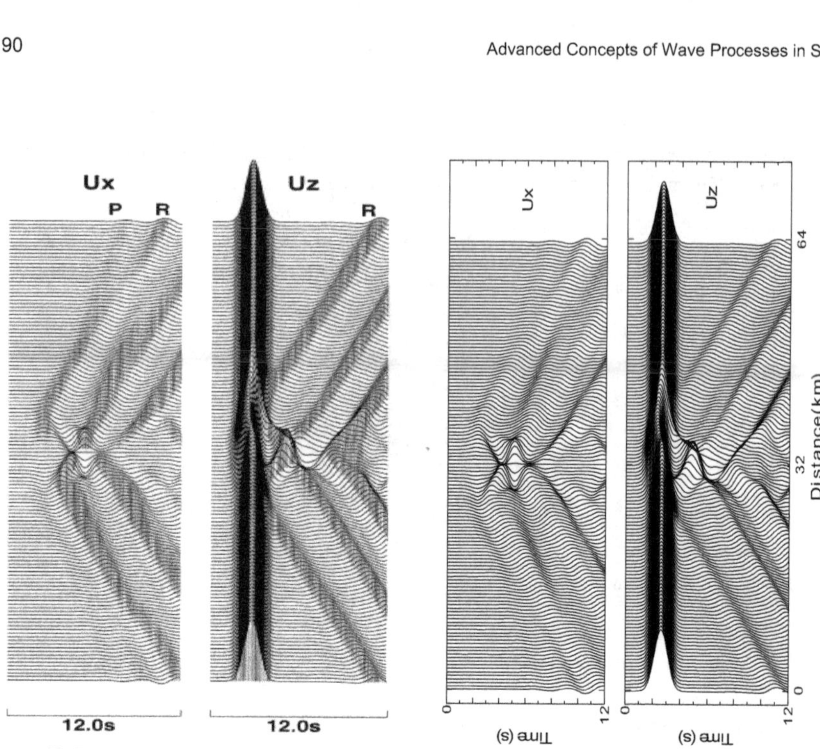

Figure 13. Time responses along the fluid-solid interface of *Model* 1: the one on the left calculated in [20], and the one on the right calculated by the present method

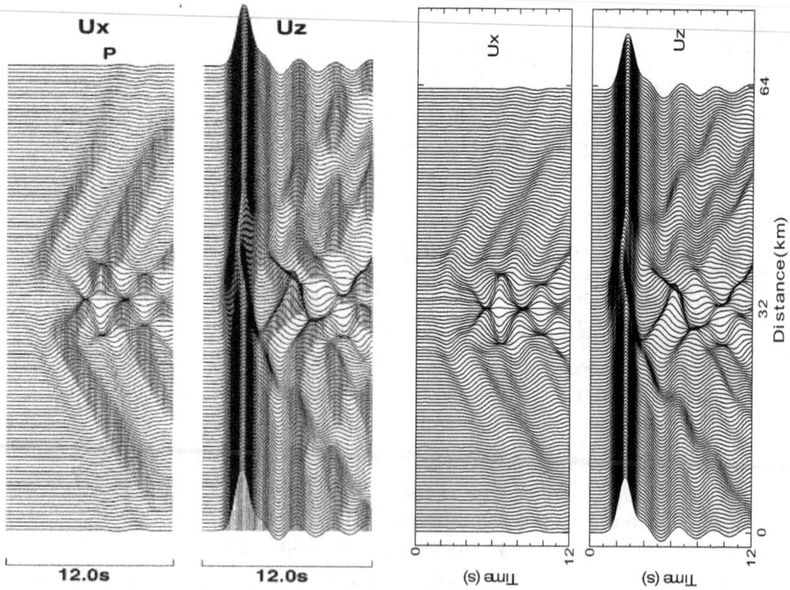

Figure 14. The same as Figure 13, but for *Model* 2

3.2.3. *Model 1 subjected to an SV-wave point source*

The above two testing calculations show the validity of the present method and the correctness of our computational program, respectively. However, the two numerical examples are only plane wave incidence cases. To validate the present method further, we show one more example using an SV-wave (Shear wave polarized in the vertical plane) excitation source as [23]

$$M_{xz} = -M_{zx} = -Mf(t)\delta(x)\delta(z), \tag{54}$$

where the source function *f*(*t*) is a Ricker wavelet with central frequency 0.25 Hz. The model used for the calculation is similar to *Model* 1 but with a width of 80 km in total, and the basin-like fluid-solid interface having a width of 60 km and thickness of 2.1 km. The material properties of the model is little different from *Model* 1, details of which can refer to [23]. The SV-wave source is located at (distance, depth) = (–0.1 km, 4.1 km). As for this case, the results calculated by the discrete wavenumber method are available in [23] so that we can make a full waveform comparison.

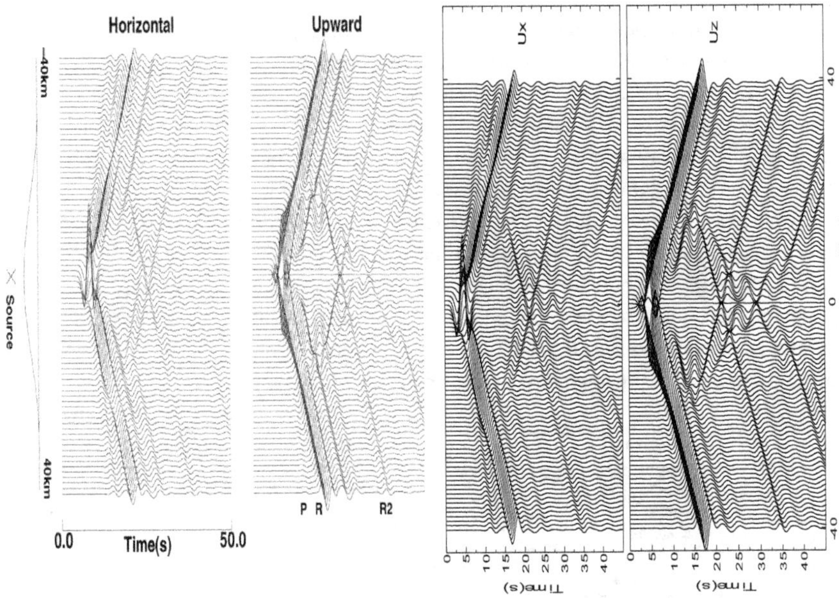

Figure 15. Time responses along the basin-like fluid-solid interface: the one on the left calculated in [23] with a time offset of 8 seconds, and the one on the right calculated by the present method.

Figure 15 shows the displacements of both horizontal and vertical components along the basin-like fluid-solid interface due to the SV-wave source presented in Eq. (54). The one on the left is calculated by the discrete wavenumber method in [23]. And the one on the right is calculated by the present method. It can be clearly seen that the large-amplitude multiple reflections in the fluid-solid basin part are well observed in the synthetic waveforms calculated by both the present method and the discrete wavenumber method. Furthermore, the first scattered *P*-waves, the first scattered *Rayleigh* waves, and the secondary scattered *Rayleigh* waves, which are separately indicated as *P*, *R* and *R*2 in the plot on the left, can be well and clearly observed in our calculated seismograms. Considering the time offset used in [23], the very good agreement between the results by the two methods further confirms the validity of the present method and the correctness of the fluid-solid formulation.

Up to now, we have validated the fluid-solid formulation by three examples, which results in the conclusion that the introduced approach can accurately cover the seismoacoustic scattering due to an irregular fluid-solid interface. In the next Section, we will show how to use the introduced approach to simulate the effects of a fluid layer on the synthetic seismograms and the water reverberation by three preliminary examples, respectively.

4. Water effects and water reverberation modeling

4.1. Water effects on seismogram synthesis

In this Subsection, we show the effect of a fluid layer on synthetic seismogram by using one of the testing models. In practice, seismologists or geophysicists are interested in the seismic ground motion at land in gulf areas where the fluid layer plays an important role in the recorded seismograms. Therefore, it is necessary and significant to simulate the seismoacoustic scattering in irregularly multilayered elastic media lain by a fluid layer due to some scenario earthquake event. Here, we select the first model used in Subsection 3.2 to show water layer effects due to the two reasons: a) this model is very simple and its calculated results are relatively easy to interpret; b) this model is kind of close to a practical gulf area although the solid layer is considered as an elastic half space. For the same model, we take into account two cases: the first case of a vertically incident plane *P* wave, as shown in Figure 16, the second case of an explosive source located in the solid half-space. We will show the calculated results of the two cases one by one in the following.

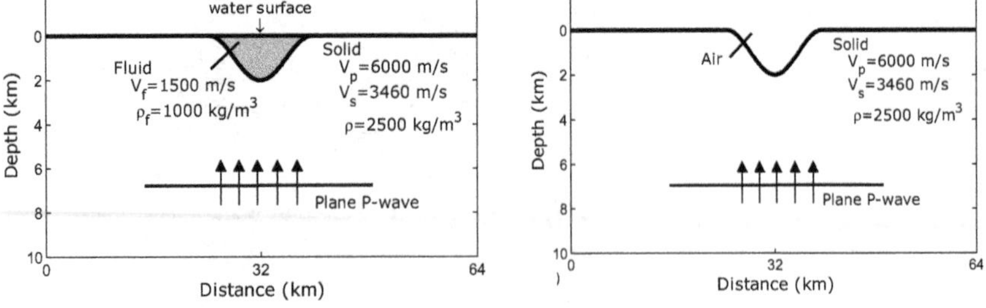

(a) (b)

Figure 16. The model used to show the effect of a fluid layer on the synthetic seismograms due to a vertically incident plane *P* wave: (a) with a fluid layer; (b) without a fluid layer.

4.1.1. Plane wave incidence

Figure 17 shows the time responses of the horizontal and vertical motions along the irregular solid interface due to a vertically incident plane *P* wave for the models with and without a fluid layer, respectively. It can be seen from Figure 17 that the direct waves appearing on the vertical component keep the same amplitude along the solid interface, regardless of the existence of the uppermost fluid layer. Outside the irregular part of the interface, later arrivals on the vertical component are mainly due to the wave reflection at the upper part of the irregular interface but they seem to contain some contribution of the diffracted waves since their amplitude changes very slowly. Judging from their particle motions and apparent velocities, we can say that they appear to become *Rayleigh* waves soon after the departure from the edges of the irregular part of the interface. Comparison between the results for the models with and without the fluid layer can be clearly seen in Figure 17. For the model shown in Figure 16(a), although the later arrivals on the vertical component inside the irregular part of the interface seem complicated, the multiple later arrivals outside the irregular part of the interface due to the multiple reflections caused by the fluid layer are clearly observed. In the case of the absence of the fluid layer, i.e. the model shown in Figure 16(b), there is only one later arrival on the vertical component, even the previous complicated later arrivals appearing inside the irregular part of the interface in the case of the presence of the fluid layer completely disappear, which appears to be very interesting results.

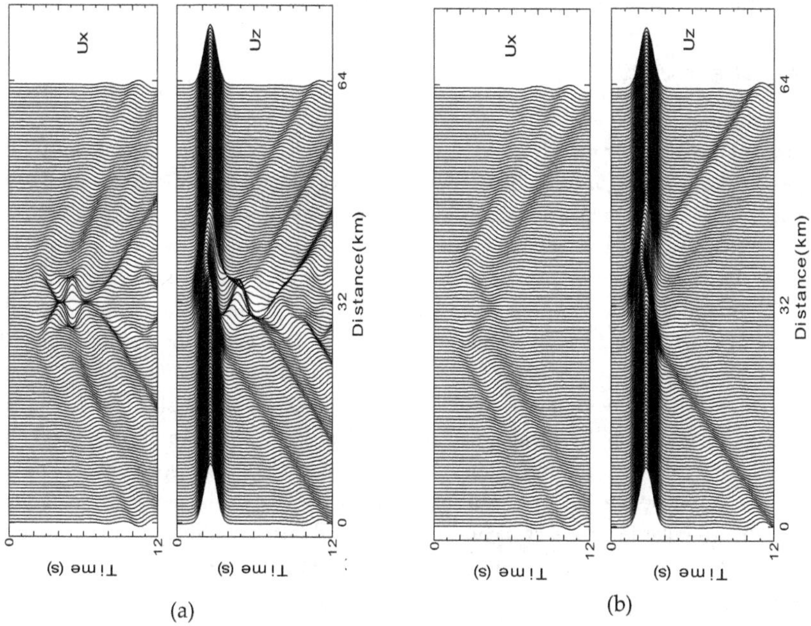

Figure 17. Time responses along the irregular solid interface in the models shown in Figure 16 due to a vertically incident plane *P* wave: (a) with a fluid layer; (b) without a fluid layer.

The earlier arrivals on the horizontal component generated by the impact and subsequent reflection of the incident plane *P* wave at the irregular part of the interface grows as the incident wave propagates upward and reaches its maximum at the edges, regardless of the existence of the uppermost fluid layer. Those diffracted waves gradually separate into *P* waves and *Rayleigh* waves with the increasing distance away from the irregular part edges. The difference between the results on the horizontal component for the models with and without the fluid layer can also be clearly seen, no matter inside the irregular part or outside the irregular part of the interface. Inside the irregular part of the interface the diffracted waves for the model with the fluid layer have much more complicated features than those for the model without the fluid layer. Different from the vertical component, multiple diffracted waves inside the irregular part of the model with the fluid layer can be clearly observed as well. For the model without the fluid layer, however, no multiple diffracted waves can be observed inside the irregular part of the surface. Outside the irregular part of the interface, the multiple arrivals of *P* waves and *Rayleigh* waves on the horizontal component can be seen for the model with the fluid layer, although it is difficult to distinguish them clearly. While for the model without the fluid layer, no more later arrivals present on the horizontal component, which is similar to the observation on the vertical component. But the amplitude of the first diffracted waves outside the irregular part of the surface of the model without the fluid layer (i.e. *P* waves) is a little bit larger than that of the model with the fluid layer, which is due to the energy dissipation by the fluid layer.

4.1.2. Explosive source

Besides the plane wave incidence, let us see what happens if we have an explosive source located in the solid layer. The source time function is a Ricker wavelet with central frequency 0.25 Hz. Figure 18 shows the time responses of the horizontal and vertical motions along the irregular solid interface due to an explosive source located at (distance, depth) = (32 km, 4 km) in the solid layer for *Model* 1 with and without a fluid layer, respectively. It can be seen from Figure 18 that the direct arrivals appearing along the irregular interface come from the explosive source directly and separate into two waves with the distance away from the center, regardless of the existence of the uppermost fluid layer. Those can be recognized as *P* and *Rayleigh* waves. The main difference of the results for the model with and without the fluid layer exists in the later multiple arrivals, which can be clearly observed from the comparison between Figure 18 (a) and (b). For the model with the fluid layer, the multiple reflections inside the basin-like fluid-solid interface can be clearly seen to generate the later multiple *Rayleigh* wave arrivals along the solid surface outside the irregular interface. While for the model without the fluid layer, no such multiple arrivals appear since it's just a half-space. Those phenomena resemble the case of plane wave incidence.

The above two examples clearly show the difference between the synthetic seismograms for the model with and without a fluid layer in the cases of plane wave incidence and explosive source excitation, from which we can say that the method presented in this Chapter can be used to study the effect of fluid layers on the seismic ground motions at land and ocean bottom seismic observations. Next, we will consider the case when we put the explosive source in a fluid layer and see how the introduced method can be used to simulate the water reverberation.

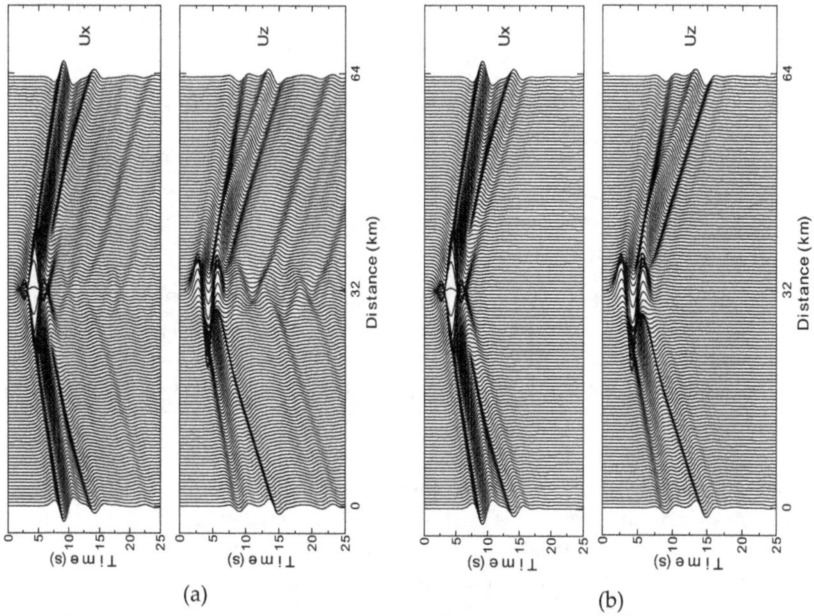

Figure 18. Time responses along the irregular solid interface in *Model* 1 due to an explosive source located at (distance, depth) = (32 km, 4 km): (a) with a fluid layer; (b) without a fluid layer.

4.2. Water reverberation modeling

4.2.1. Formulation when source is located in fluid layer

So far, we have given the formulation for seismic excitation in a multi-layered solid half-space overlain with a fluid layer. Readers may ask if the method can be applied to obtain strong ground motion at land when we have the source located in the fluid layer. The answer is definitely positive. For doing so, there are two key steps. The first one is to form the solution matrix equations in the fluid layer, which can be done from Eq. (42) as follows

$$\mathbf{H}^{(1)}_{(2)}\mathbf{p}_{(2)} = \mathbf{G}^{(1)}_{(1)}\mathbf{q}_{(1)} + \mathbf{G}^{(1)}_{(2)}\mathbf{q}_{(2)} + \mathbf{F}, \tag{55}$$

where $\mathbf{p}$ and $\mathbf{q}$ with subscripts $_{(2)}$ are connected by Eq. (43). In order to solve Eq. (55), we need to use the global matrix propagator $\mathbf{D}^{(1)}_{pq}$ defined in Eq. (43) to change Eq. (55) into

$$\begin{pmatrix} -\mathbf{G}_{(1)}^{(1)} & \mathbf{H}_{(2)}^{(1)}\mathbf{D}_{pq}^{(1)} - \mathbf{G}_{(2)}^{(1)} \end{pmatrix} \begin{pmatrix} \mathbf{q}_{(1)} \\ \mathbf{q}_{(2)} \end{pmatrix} = \mathbf{F}, \tag{56}$$

which has the same number of unknowns and equations and can be solved without problem. What is left is to get the global matrix propagator $\mathbf{D}_{pq}^{(1)}$, which is the other key step. We understand that the global matrix propagator $\mathbf{D}_{tu}^{(1)}$ in the solid layer right neighboring the fluid layer can be obtained recursively from the lowermost layer by using Eqs. (38-39). Then the continuity conditions at the fluid-solid interface Eq. (43) are used to obtain $\mathbf{D}_{pq}^{(1)}$ from $\mathbf{D}_{tu}^{(1)}$ as

$$\mathbf{D}_{pq}^{(1)} = \frac{1}{\rho_f\omega^2}\Sigma\mathbf{D}_{tu}^{(1)}/[\mathbf{N}_1^e \cdot (\mathbf{N}_1^e)^T + \mathbf{N}_2^e \cdot (\mathbf{N}_1^e)^T + \mathbf{N}_1^e \cdot (\mathbf{N}_2^e)^T + \mathbf{N}_2^e \cdot (\mathbf{N}_2^e)^T], \tag{57}$$

where the matrices $\mathbf{N}_1^e$ and $\mathbf{N}_2^e$ are defined as Eq. (48). The symbol Σ in Eq. (57) means the summation of the four quarters of $\mathbf{D}_{tu}^{(1)}$, and the multiplication and division of the matrices are operated in the sense of 'element to element'. The general solution steps can be summarized into the following.

Step 1. Calculate the global matrix propagators $\mathbf{D}_{tu}^{(1)}$ in the lowermost layer by Eq. (39).
Step 2. Calculate the global matrix propagators in all the solid layers above the bottom layer by Eqs. (38-39).
Step 3. Reduce the global matrix propagators obtained in the solid layer right neighboring the fluid layer into the one in the fluid layer by Eq. (57).
Step 4. Substitute the reduced global matrix propagator $\mathbf{D}_{pq}^{(1)}$ obtained in **Step 3** into the solution matrix equation (56) and solve for the normal derivative of pressure.
Step 5. Obtain the pressure solutions at the fluid-solid interface by Eq. (43). Then obtain the traction solutions at the fluid-solid interface by the first two in Eq. (46).
Step 6. Obtain the displacement solutions at the fluid-solid interface by $\mathbf{D}_{tu}^{(1)}$.
Step 7. Execute the Inverse Fourier Transform on the displacement solutions obtained in **Step 6** to get the ground motion in time domain finally.

4.2.2. Numerical example

First consider the same model as *Model* 1, with an explosive source located at (distance, depth) = (32 km, 0.5 km) and receivers along the fluid-solid interface. The time function of the source is a Ricker wavelet with central frequency 0.25 Hz. The calculated time responses of displacements are shown in Figure 19(a), from which we can clearly see the multiple reflections inside the water basin and the generated multiple arrivals along the solid surface ourside the basin part. The second calculation taken into account is also for *Model* 1 but with the water width enlarged to 30 km, results of which are shown in Figure 19(b) and the time responses of the displacements along the fluid-solid interface ressembles those in Figure 19(a). The main difference between the two calculated results exists in the time difference between the multiple arrivals, which is due to the enlarged width of the water basin area. This example implies that the present method can be used to simulate the water revibaration in the sea, which is of importantly practical significance to deep ocean acoustic experiments and so on.

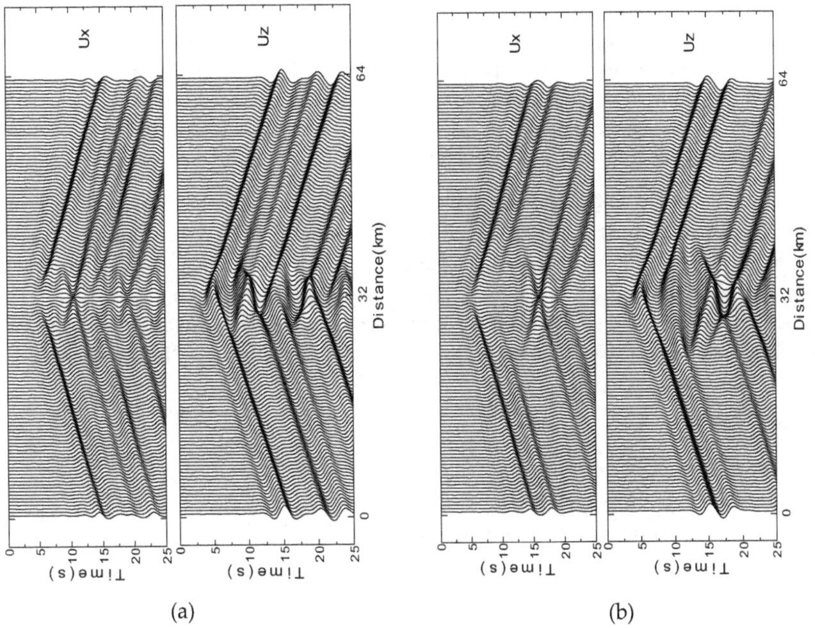

Figure 19. Time responses along the fluid-solid interface due to an explosive source located at (distance, depth) = (32 km, 0.5 km) in: (a) *Model* 1 and (b) *Model* 1 with a water width of 30 km.

5. Summary

In this Chapter, we presented an efficient approach based on the combination of the traditional boundary element method and the global matrix propagators for seismoacoutic scattering simulation in multilayered fluid-solid media. To make readers understand the introduced method easily, we first gave the mathematical formulation for SH-wave propagation simulation in multilayered solids, followed by some examples to test the validity of the formulation in solids. Then, we gradually went deep into the seismoacoustic scattering simulation due to an irregular fluid-solid interface, where the fluid pressure was used as the variable inside the fluid layer and the global matrix propagators defined in the fluid layer were extended successfully to connect the pressure wave with the elastic wave by using the standard continuity conditions at the fluid-solid interface (i.e. zero tangential traction, continuity of normal traction, and continuity of normal displacement). The synthetic waveforms in time domain for some selected models calculated by the present method agree well with those calculated by the reflection/transmission matrix method and those calculated by the discrete wavenumber method. The effects of a fluid layer on the synthetic seismograms can be exactly covered by the present method, and the water reverberation in the sea can also be simulated as well. The introduced method is especially suitable to simulate seismoacoustic scattering in a multilayered elastic structure overlain by a fluid layer. Since the global matrix propagators can be calculated recursively, the computer memory required for a multilayered model is the same as that for a two-layer model. In case of the application of dynamic allocation of matrices and saving the global matrix propagators on hard drive, the array size assigned for calculating a two-layer model is sufficient for a multilayered model. In that case, the advantages of the boundary element method can be preserved, which implies that seismoacoustic scattering synthesis due to a high-frequency excitation can be modelled with reduced computer resources.

Furthermore, if a calculated model is only partially modified, such as increasing the number of layers below the uppermost fluid layer in the case of a plane wave incidence or changing the free surface profile in the case of a point source excitation, not all the matrices need recalculating. That is due to the two merits of the present method: the global matrix propagators for the layers above the source are calculated downwards; and the global matrix propagators for the layers below the source are independent of the rest of the model.

Author details

Zheng-Hua Qian*
State Key Laboratory of Mechanics and Control of Mechanical Structures, Nanjing University of Aeronautics and Astronautics, Nanjing, China

Acknowledgement

The support from the State Key Laboratory of Mechanics and Control of Mechanical Structures, Nanjing University of Aeronautics and Astronautics, P. R. China is greatly appreciated. The financial support from the Global COE Program entitled "International Urban Earthquake Engineering Center for Mitigating Seismic Mega Risk", Tokyo Institute of Technology, Japan is gratefully acknowledged. The author (Qian) also appreciates very much the help of Prof. Chen X.F. from USTC and Prof. Ge Z. from Peking University for providing us with their program designed for multilayered solid media. And the author (Qian) thanks very much Prof. Fu L.Y. and Dr. Yu G.X. from the Institute of Geology and Geophysics, Chinese Academy of Science, P. R. China for their very much useful discussion and help on the application of BEM as well. The author (Qian) would also like to thank Prof. H. Takenaka from Kyushu University, Japan for giving the very constructive suggestions and comments on the research of seismoacoustic scattering simulation which help improve the introduced approach very much. The author (Qian) would like to appreciate the support from Prof. K. Kishimoto at Tokyo Institute of Technology, Japan. Finally, the author (Qian) would like to sincerely appreciate Prof. Sohichi Hirose from Tokyo Institute of Technolody, Japan for his so much inspiring discussion and comments on writing the programs of boundary element method.

6. References

[1] Kennett BLN (1983) Seismic wave propagation in stratified media. New York: Cambridge University Press. 497p.

* Corresponding Author

[2] Miklowitz J, Achenbach JD (1978) Modern Problems in Elastic Wave Propagation. New York: John Wiley & Sons. 561p.
[3] Giovine P, Oliveri F (1995) Dynamics and Wave Propagation in Dilatant Granular Materials. Meccanica 30: 341–357.
[4] Godano C, Oliveri F (1999) Nonlinear seismic waves: a model for site effects. Int. J. Non-linear Mech. 34:457–468.
[5] Zhao CG, Dong J, Gao FP, Jeng DS (2006) Seismic responses of a hemispherical alluvial valley to SV waves: a three-dimensional analytical approximation. Acta Mechanica Sinica 22: 547–557.
[6] Cerveny V (2001) Seismic Ray Theory. Cambridge: Cambridge University Press. 722p.
[7] Chen XF (1990) Seismogram synthesis for multi-layered media with irregular interfaces by global generalized reflection/transmission matrices method-I. Theory of two-dimensional SH case. Bull. Seism. Soc. Am. 80: 1696–1724.
[8] Chen XF (1995) Seismogram synthesis for multi-layered media with irregular interfaces by global generalized reflection/transmission matrices method–II. Applications for 2D SH case. Bull. Seism. Soc. Am. 85: 1094–1106.
[9] Chen, XF (1996) Seismogram synthesis for multi-layered media with irregular interfaces by global generalized reflection/transmission matrices method–III. Theory of 2D P-SV case. Bull. Seism. Soc. Am. 86: 389–405.
[10] Fu LY, Bouchon M (2004) Discrete wavenumber solutions to numerical wave propagation in piecewise heterogeneous media–I. Theory of two-dimensional SH case. Geophys. J. Int. 157: 481–498.
[11] Zhang W, Chen XF (2006) Traction image method for irregular free surface boundaries in finite difference seismic wave simulation. Geophys. J. Int. 167: 337–353.
[12] Koketsu K, Fujiwara H, Ikegami Y (2004) Finite-element simulation of seismic ground motion with a voxel mesh. Pure Appl. Geophys. 161: 2183–2198.
[13] Wang YB, Takenaka H, Furumura T (2001) Modelling seismic wave propagation in a two-dimensional cylindrical whole-earth model using the pseudospectral method. Geophys. J. Int. 145: 689–708.
[14] Komititsch D, Tromp J (2002) Spectral-element simulations of global seismic wave propagation-I. Validation. Geophys. J. Int. 149: 390–412.
[15] Fu LY, Mu YG (1994) Boundary element method for elastic wave forward modeling. Acta Geophys. Sinica 37: 521–529.
[16] Fu LY (2002) Seismogram synthesis for piecewise heterogeneous media. Geophys. J. Int. 150: 800–808.
[17] Bouchon M, Schultz CA, Toksoz MN (1995) A fast implementation of boundary integral equation methods to calculate the propagation of seismic waves in laterally varying layered media. Bull. Seism. Soc. Am. 85: 1679–1687.
[18] Ge ZX, Chen XF (2007) Wave propagation in irregularly layered elastic models: a boundary element approach with a global reflection/transmission matrix propagators. Bull. Seism. Soc. Am. 97: 1025–1031.
[19] Ge ZX, Chen XF (2008) An efficient approach for simulating wave propagation with the boundary element method in multilayered media with irregular interfaces. Bull. Seism. Soc. Am. 98: 3007–3016.
[20] Okamoto T, Takenaka H (1999) A reflection/transmission matrix formulation for seismoacoustic scattering by an irregular fluid-solid interface. Geophys. J. Int. 139: 531–546.

[21] Lay T, Wallace TC (1995) Modern Global Seismology. San Diego: Academic Press. 521p.
[22] Stephen RA, (1988) A review of finite difference methods for seismo-acoustics problems at the seafloor. Rev. Geophys. 26: 445–458.
[23] Okamoto T, Takenaka H. (2005) Fluid-solid boundary implementation in the velocity-stress finite-difference method. Zisin. 57: 355–364. (in Japanese with English abstract)
[24] Murphy JE, Chin-Bing SA (1989) A finite-element model for ocean acoustic propagation and scattering. J. Acoust. Soc. Am. 86: 1478–1483.
[25] Aki K, Larner KL (1970) Surface motion of a layered media having an irregular interface due to incident plane SH waves. J. Geophys. Res. 75: 933–954.
[26] Fokkema JT (1981) Reflection and transmission of acoustic waves by the periodic interface between a solid and a fluid. Wave Motion 3: 145–157.
[27] Achenbach JD (1973) Wave propagation in elastic solids. London: North-Holland Publishing Company. 55-59p.
[28] Kawase H (1988) Time-domain response of a semi-circular canyon for incident SV, P, and Rayleigh waves calculated by the discrete wavenumber boundary element method. Bull. Seism. Soc. Am. 78: 1415–1437.
[29] Kawase H, Aki K (1989) A study on the response of a soft basin for incident S, P, and Rayleigh waves with special reference to the long duration observed in Mexico City. Bull. Seism. Soc. Am. 79: 1361–1382.
[30] Balas J, Sladek J, Sladek V (1990) Stress Analysis by Boundary Element Methods. New York: Elsevier Science Ltd. 702p.
[31] Kirkup SM (2007) The Boundary Element Method in Acoustics. 2nd Edition, Available: http://www.kirkup.info/papers. Accessed 2012 March 28.
[32] Medina F, Dominguez J (1989) Boundary elements for the analysis of the seismic response of dams including dam-water-foundation interaction effects I. Engineering Analysis with Boundary Elements 6: 153–163.
[33] Fujiwara H, Takenaka H (1994) Calculation of surface waves for a thin basin structure using a direct boundary element method with normal modes. Geophys. J. Int. 117: 69–91.
[34] Yokoi T, Takenaka H (1995) Treatment of an infinitely extended free surface for indirect formulation of the boundary element method. J. Phys. Earth 43: 79–103.
[35] Yokoi T, Sanchez-Sesma FJ (1998) A hybrid calculation technique of the indirect boundary element method and the analytical solutions for three-dimensional problems of topography. Geophys. J. Int. 133: 121–139.
[36] Takenaka H, Kennett BLN, Fujiwara H (1996) Effect of 2-D topography on the 3-D seismic wavefield using a 2.5-D discrete wavenumber-boundary integral equation method. Geophys. J. Int. 124: 741–755.
[37] Liu E, Zhang Z, Yue J, Dobson A (2008) Boundary integral modelling of elastic wave propagation in multi-layered 2D media with irregular interfaces. Commun. Comput. Phys. 3: 52–62.
[38] Fu LY, Wu RS (2000) Infinite boundary element absorbing boundary for wave propagation simulations. Geophysics 65: 596–602.
[39] Davies TG, Bu S (1996) Infinite boundary elements for the analysis of halfspace problems. Computers and Geotechnics 19: 137–151.
[40] Bu S (1997) Infinite boundary elements for the dynamic analysis of machine foundations. Int. J. Numer. Meth. Engng. 40: 3901–3917.

Chapter 8

Velocity-Stress Equations for Wave Propagation in Anisotropic Elastic Media

Sheng-Tao John Yu, Yung-Yu Chen and Lixiang Yang

Additional information is available at the end of the chapter

1. Introduction

Conventionally, the second-order elastodynamics equation has been employed to model waves in elastic solids:

$$\rho \frac{\partial^2 \mathbf{w}}{\partial t^2} = \nabla \cdot \left(c^{[4]} \nabla \mathbf{w} \right), \tag{1.1}$$

where ρ is the density of the medium, $\mathbf{w}$ the displacement, and $c^{[4]}$ the fourth-order stiffness tensor [2]. Equation (1.1) has been derived based on the equation of motion in conjunction with the elastic constitutive equation. Equation (1.1) has been solved by the finite-difference methods, e.g., [21], and the time-domain finite-element methods, e.g., [33], for propagating waves, and the frequency-domain finite-element methods, e.g., [6], for normal mode analysis of standing waves.

Alternatively, one can use a modern numerical method [11, 12] to solve the first-order velocity-stress equations to obtain the transient solution of wave propagation in elastic solids. For example, Virieux [25] modeled waves in earth crust by using a finite-difference method to solve the velocity-stress equations. LeVeque [12, 13] simulated stress waves in isotropic, elastic solids by using a upwind scheme. Shorr [22] developed a specialized formulation of the finite-element method to simulate propagating waves. Käser and Dumbser [10] used the discontinuous Galerkin method [20] to model waves in anisotropic earth crust. Yu et al. [32] simulated non-linear stress-wave propagation in hypo-elastic media by using the space-time Conservation Element and Solution Element (CESE) method [4]. In the setting of the second-order wave equation, Eq. (1.1), ample theoretical analyses about anisotropic elasticity can be found in the literature, e.g., [2, 24]. In general, mathematical properties of the first-order velocity-stress equations must be similar to that of the second-order wave equation Eq. (1.1). However, only limited discussions, e.g., Puente et al. [18] and Yang et al. [30], about the eigen structure of the velocity-stress equations can be found in the literature.

The objective of the present chapter is to analyze the mathematical properties of the first-order velocity-stress equations. In particular, their connection to the conventional second-order wave equations, Eq. (1.1) is of interest. To this end, this paper will be focused on analyzing the eigen structure of the velocity-stress equations. The approach employed is identical to that demonstrated in Warming et al. [29]. To be general, the derivation will be based on the properties of a triclinic solid, which have 21 independent stiffness constants and no symmetry. As such, the result will be applicable to all anisotropic, elastic solids. Moreover, in an one-dimensional space, the Jacobian matrix of the velocity-stress equations is shown to be equivalent to the classical Christoffel matrix of the second-order wave equations. In contrast to the conventional approach, however, a harmonic solution of plane waves is not assumed. To demonstrate the capabilities of the new formulation, the CESE method is applied to solve the velocity-stress equations for modeling wave propagation in a block of beryl, a solid with hexagonal symmetry. Wave expansion from a point source is demonstrated. The calculated group velocity profile compared well with the analytical solution. Additional result of waves interacting with interfaces separating regions with different lattice orientations are also reported. All results show salient features of wave propagation.

In the remainder of this section of Introduction, the equations of elastodynamics are reviewed in order to stage the further derivation in the following sections. First, the equation of motion without the body force is considered:

$$\frac{\partial \mathbf{v}}{\partial t} = \frac{1}{\rho}\nabla \cdot T, \tag{1.2}$$

where $\mathbf{v}$ is the velocity vector and T is the Cauchy stress tensor. T can be related to the strain tensor S by the elastic relation:

$$T = c^{[4]}S, \tag{1.3}$$

where $c^{[4]}$ is the fourth-order stiffness tensor. To proceed, Eq. (1.3) is differentiated by time t to yield:

$$\frac{\partial T}{\partial t} = \frac{1}{2}c^{[4]}\left(\nabla \mathbf{v} + (\nabla \mathbf{v})^t\right), \tag{1.4}$$

where the superscript t denotes transpose. The model equations include Eqs. (1.2) and (1.4). In an three-dimensional space, there are nine independent equations with the velocity and stress components as the unknowns. Next, the Voigt notation [2] is applied and the second-order tensors in Eq. (1.4) is changed to be 6-component vectors:

$$\begin{gathered}\begin{pmatrix} T_{11} & T_{12} & T_{13} \\ T_{12} & T_{22} & T_{23} \\ T_{13} & T_{23} & T_{33} \end{pmatrix} \rightarrow \begin{pmatrix} T_1 & T_6 & T_5 \\ T_6 & T_2 & T_4 \\ T_5 & T_4 & T_3 \end{pmatrix}, \\ \begin{pmatrix} S_{11} & S_{12} & S_{13} \\ S_{12} & S_{22} & S_{23} \\ S_{13} & S_{23} & S_{33} \end{pmatrix} \rightarrow \begin{pmatrix} S_1 & S_6/2 & S_5/2 \\ S_6/2 & S_2 & S_4/2 \\ S_5/2 & S_4/2 & S_3 \end{pmatrix}. \end{gathered} \tag{1.5}$$

In the following, boldfaced $\mathbf{T} = (T_1, T_2, T_3, T_4, T_5, T_6)^t$ and $\mathbf{S} = (S_1, S_2, S_3, S_4, S_5, S_6)^t$ denote the 6-component stress and strain vectors. The non-boldfaced T and S in Eqs. (1.2) (1.3), and (1.4) denote the second-order stress and strain tensors. Aided by the Voigt notation, the elastic constitutive relation is rewritten as $\mathbf{T} = C\mathbf{S}$, where C is the second-order 6×6 stiffness matrix.

The rest of this chapter is organized as in the following. Section 2 shows the first-order velocity-stress equations in a vector-matrix form. Section 3 proves that the velocity-stress equations are hyperbolic. Section 4 shows that the Jacobian matrix of the one-dimensional velocity-stress equations is equivalent to the classic Christoffel matrix. As a concrete example, Section 5 illustrates a special form of the velocity-stress equations for elastodynamics in beryl, a solid of hexagonal symmetry. Section 6 reports the numerical solution of wave propagation in a block of beryl. The solution is obtained by applying the CESE method to solve the velocity-stress equations. In Section 7 we provide the concluding remarks. At the end of the chapter, a list of cited references and several appendices are attached.

2. The model equation

To proceed, a Cartesian coordinate system is employed to formulate the velocity-stress equations Eqs. (1.2) and (1.4):

$$\begin{aligned} \frac{\partial \mathbf{v}}{\partial t} - \frac{1}{\rho}\sum_{\mu=1}^{3} K^{(\mu)} \frac{\partial \mathbf{T}}{\partial x_\mu} &= 0, \\ \frac{\partial \mathbf{T}}{\partial t} - C\sum_{\mu=1}^{3} K^{(\mu)^t} \frac{\partial \mathbf{v}}{\partial x_\mu} &= 0, \end{aligned} \tag{2.1}$$

where $K^{(\mu)}$ with $\mu = 1,2,3$ are 3×6 matrices:

$$K^{(\mu)} \stackrel{\text{def}}{=} \begin{pmatrix} \delta_{\mu 1} & 0 & 0 & 0 & \delta_{\mu 3} & \delta_{\mu 2} \\ 0 & \delta_{\mu 2} & 0 & \delta_{\mu 3} & 0 & \delta_{\mu 1} \\ 0 & 0 & \delta_{\mu 3} & \delta_{\mu 2} & \delta_{\mu 1} & 0 \end{pmatrix}. \tag{2.2}$$

In Eq. (2.2), δ_{ij} is the Kronecker delta. The rank, i.e., the dimension of the column and row spaces [23], of $K^{(\mu)}$ with $\mu = 1,2,3$ is three. In the above equations, the value of each entry in C (or $c^{[4]}$) must be measured based on the same Cartesian coordinates employed in formulating Eq. (2.1). This coordinate system could be arbitrary. However, the entries of C reported in the literature, e.g., Auld [2], are usually tabulated based on a chosen coordinate system, which is aligned with the lattice orientation of the solid of interest.

To proceed, the velocity-stress equations, Eq. (2.1), is recast into a vector-matrix form:

$$\frac{\partial \mathbf{u}}{\partial t} + \sum_{\mu=1}^{3} A^{(\mu)} \frac{\partial \mathbf{u}}{\partial x_\mu} = 0. \tag{2.3}$$

where the unknown vector $\mathbf{u} \stackrel{\text{def}}{=} (v_1, v_2, v_3, T_1, T_2, T_3, T_4, T_5, T_6)^t$. Matrices $A^{(1)}$, $A^{(2)}$, and $A^{(3)}$ are the three Jacobian matrices:

$$A^{(\mu)} \stackrel{\text{def}}{=} \begin{pmatrix} 0_{3\times 3} & -\frac{1}{\rho}K^{(\mu)} \\ -CK^{(\mu)^t} & 0_{6\times 6} \end{pmatrix}, \quad \mu = 1,2,3, \tag{2.4}$$

where $0_{m\times n}$ denotes a $m \times n$ null matrix. Since the rank of matrices $K^{(1)}$, $K^{(2)}$, and $K^{(3)}$ is 3, matrices $A^{(1)}$, $A^{(2)}$, and $A^{(3)}$ can be easily reduced to their Echelon forms [23] and the result shows that the rank of each of the three 9×9 matrices $A^{(1)}$, $A^{(2)}$, and $A^{(3)}$ is 6.

3. Hyperbolicity of the velocity-stress equations

In this section, the hyperbolicity of the velocity-stress equations Eq. (2.3) is proved based on analyzing a linear combination of the three Jacobian matrices:

$$B \stackrel{\text{def}}{=} \sum_{\mu=1}^{3} h_\mu A^{(\mu)} = h_1 A^{(1)} + h_2 A^{(2)} + h_3 A^{(3)}, \tag{3.1}$$

where h_1, h_2, and h_3 are arbitrary real numbers. Without losing any generality, we let h_1, h_2, and h_3 be the three components of a unit vector $\mathbf{h}$, i.e., $|\mathbf{h}| = 1$. Equation (2.3) is hyperbolic if all eigenvalues of B are real, and B can be diagonalized by its eigenvector matrix [11].

Physical meaning of the hyperbolicity of the velocity-stress equations can be illustrated by considering a plane wave propagating in the direction of $\mathbf{h}$, along which Eq. (2.3) is reduced to be one-dimensional. In this uni-axial direction, a new coordinate is defined as $y \stackrel{\text{def}}{=} \sum_{\mu=1}^{3} h_\mu x_\mu$. Aided by the new coordinate y, the three spatial derivative terms in Eq. (2.3) can be combined:

$$\sum_{\mu=1}^{3} A^{(\mu)} \frac{\partial \mathbf{u}}{\partial x_\mu} = \sum_{\mu=1}^{3} A^{(\mu)} \frac{\partial \mathbf{u}}{\partial y} \frac{\partial y}{\partial x_\mu} = \sum_{\mu=1}^{3} h_\mu A^{(\mu)} \frac{\partial \mathbf{u}}{\partial y} = B \frac{\partial \mathbf{u}}{\partial y}. \tag{3.2}$$

Aided by Eq. (3.2), a one-dimensional version of Eq. (2.3) for modeling wave propagating in the direction of $\mathbf{h}$ becomes

$$\frac{\partial \mathbf{u}}{\partial t} + B \frac{\partial \mathbf{u}}{\partial y} = 0. \tag{3.3}$$

Therefore, hyperbolicity of Eq. (2.3) implies that the eigenvalues of the Jacobian matrix B of Eq. (3.3) are all real and they represent different wave speeds. Moreover, Eq. (3.3) can be diagonalized so that Eq. (3.3) can be decoupled into 9 independent scalar convection equations, each of which would propagate a constant profile, i.e., a Riemann invariant, in the direction $\mathbf{h}$ by its own wave speed, i.e., the corresponding eigenvalue.

In the following, we will show that all eigenvalues of B are real. Then we will show that B is diagonalizable by its eigenvector matrix. To proceed, we introduce the following definition:

$$G \stackrel{\text{def}}{=} \sum_{\mu=1}^{3} h_\mu K^{(\mu)} = \begin{pmatrix} h_1 & 0 & 0 & 0 & h_3 & h_2 \\ 0 & h_2 & 0 & h_3 & 0 & h_1 \\ 0 & 0 & h_3 & h_2 & h_1 & 0 \end{pmatrix}. \tag{3.4}$$

Recall that $|\mathbf{h}| = 1$. Since h_1, h_2, and h_3 are not all-zero, the rank of matrix G is 3. Aided by this definition of G, matrix B can be written as

$$B = \begin{pmatrix} 0_{3\times3} & -\frac{1}{\rho} G \\ -CG^t & 0_{6\times6} \end{pmatrix}. \tag{3.5}$$

Because the rank of G is 3, the rank of the 6×3 matrix CG^t must be less than or equal to 3, such that $\text{rank}(B) \le 6$. The eigenvalues of B can be obtained by solving its characteristic equation:

$$\det(B - \lambda I_9) = 0, \tag{3.6}$$

where I_n is a $n \times n$ identity matrix and λ the eigenvalue. We assume that eigenvalues of B are either zero or non-zero. If all non-zero eigenvalues of B are real, all eigenvalues of B are real. The existence of zero eigenvalues of B is self-evident because $\text{rank}(B) \leq 6$ [23]. For non-zero eigenvalues, aided by the Schur complements, Eq. (3.6) can be transformed to be

$$\det(\Gamma - \rho\lambda^2 I_3) = 0, \tag{3.7}$$

where

$$\Gamma \stackrel{\text{def}}{=} GCG^t \tag{3.8}$$

is a 3×3 matrix. For completeness, the Schur complements are provided in Appendix 8. To proceed, we recall that the stiffness matrix C is symmetric and positive-definite [17]. Therefore, matrix Γ must be symmetric and positive-definite. As a result, the eigenvalues of Γ, i.e., $\rho\lambda^2$, are real and positive. Since the density ρ must be positive, λ^2 is positive. Therefore, the non-zero eigenvalues of B must be real and in positive-negative pairs, i.e., for a positive eigenvalue λ_+, there must be a negative counterpart $\lambda_- = -\lambda_+$.

To recap, the spectrum of B includes the null eigenvalue and three non-zero, positive-negative pairs of eigenvalues. Eigenvalues in pairs could be repeated such that there could be at least one and at most three pairs of distinct non-zero eigenvalues. For isotropic solids, B has two pairs of distinct non-zero eigenvalues. The above proof of the real spectrum of B is underpinned by the property of matrix C, which is symmetry and positive-definiteness. As will be shown in the following, this property of C is equally important in constructing the eigenvector matrix of B.

To proceed, we show that matrix B can be diagonalized by its eigenvector matrix. Essentially, we need to find nine linearly independent eigenvectors for B. First, we show that there exist six linearly independent eigenvectors associated with the non-zero eigenvalues. We then show the existence of three linearly independent eigenvectors associated with the null eigenvalue. To proceed, we consider the eigen problem of matrix B:

$$B\mathbf{q}_+ = \lambda_+\mathbf{q}_+, \tag{3.9}$$

where λ_+ is a positive eigenvalue of B and

$$\mathbf{q}_+ \stackrel{\text{def}}{=} \begin{pmatrix} \mathbf{q}_v \\ \mathbf{q}_T \end{pmatrix} \tag{3.10}$$

the associated eigenvector containing a 3-component vector $\mathbf{q}_v$ and a 6-component vector $\mathbf{q}_T$. Aided by the definition of matrix B, Eq. (3.5), the eigen problem, Eq. (3.9), can be recast to

$$\begin{pmatrix} -\frac{1}{\rho}G\mathbf{q}_T \\ -CG^t\mathbf{q}_v \end{pmatrix} = \lambda_+ \begin{pmatrix} \mathbf{q}_v \\ \mathbf{q}_T \end{pmatrix},$$

which gives:

$$\Gamma\mathbf{q}_v = \rho\lambda_+^2\mathbf{q}_v, \tag{3.11}$$

$$\mathbf{q}_T = -\frac{1}{\lambda_+}CG^t\mathbf{q}_v. \tag{3.12}$$

By arbitrarily specifying a value to an entry of $\mathbf{q}_v$, $\mathbf{q}_v$ can be obtained by solving Eq. (3.11). $\mathbf{q}_T$ can then be obtained by substituting $\mathbf{q}_v$ into Eq. (3.12). As such, the eigenvector $\mathbf{q}_+$ can be fully constructed.

Essentially, the eigen problem of the 9×9 matrix B in Eq. (3.9), is reduced to a eigen problem for the 3×3 matrix Γ in Eq. (3.11). Since Γ is a 3×3, symmetric, positive-definite matrix, the spectral theorem [23] of linear algebra asserts that Γ always has three linearly independent eigenvectors, regardless whether Γ has degenerate eigenvalues or not. These three eigenvectors of Γ are denoted by $\mathbf{q}_{v,1}$, $\mathbf{q}_{v,2}$, and $\mathbf{q}_{v,3}$. The three associated eigenvectors of B with positive eigenvalues are denoted by $\mathbf{q}_{+,1}$, $\mathbf{q}_{+,2}$, and $\mathbf{q}_{+,3}$. Aided by Eq. (3.12), $\mathbf{q}_{+,1}$, $\mathbf{q}_{+,2}$, and $\mathbf{q}_{+,3}$ are readily determined by $\mathbf{q}_{v,1}$, $\mathbf{q}_{v,2}$, and $\mathbf{q}_{v,3}$. Because $\mathbf{q}_{v,1}$, $\mathbf{q}_{v,2}$, and $\mathbf{q}_{v,3}$ are linearly independent, $\mathbf{q}_{+,1}$, $\mathbf{q}_{+,2}$, and $\mathbf{q}_{+,3}$ must also be linearly independent. By following exactly the same procedure, one can derive the three independent eigenvectors $\mathbf{q}_{-,1}$, $\mathbf{q}_{-,2}$, and $\mathbf{q}_{-,3}$ associated with the three negative eigenvalues $\lambda_- = -\lambda_+$. To proceed, since $\lambda_- = -\lambda_+ \neq \lambda_+$,

$$\mathbf{q}_{\pm,i} \stackrel{\text{def}}{=} \begin{pmatrix} \mathbf{q}_{v,i} \\ \pm\mathbf{q}_{T,i} \end{pmatrix}, \quad i = 1,2,3, \tag{3.13}$$

are the six linearly independent eigenvectors of B. The corresponding non-zero eigenvalues of B can be written as $\lambda_{\pm,i}$, $i = 1,2,3$, with $\lambda_{-,i} = -\lambda_{+,i}$, $i = 1,2,3$, and $\lambda_{+,1} \leq \lambda_{+,2} \leq \lambda_{+,3}$. We note that it is possible for B to have degenerate non-zero eigenvalues. Nevertheless, all six eigenvectors associated with the non-zero eigenvalues are linearly independent. This concludes the discussions about the eigenvectors of B associated with the non-zero eigenvalues.

The remaining three eigenvectors of B are associated with the null eigenvalue. Recall that B is a 9×9 matrix with $\text{rank}(B) \leq 6$. Thus, the nullity of B must be greater than or equal to 3, i.e., there are at least three linearly independent vectors spanning the null space of B [23]. Aided by the definition of null space and eigenvector, the basis of the null space of B consists of the eigenvectors associated with the null eigenvalue of B. As such, there must be exactly three eigenvectors associated with the null eigenvalue of B, because (i) there are at most nine linearly independent eigenvectors of B, (ii) there are at least three linearly independent eigenvectors associated with the null eigenvalues of B, and (iii) there are exactly six linearly independent eigenvectors associated with the non-zero eigenvalues of B. These three eigenvectors are denoted by $\mathbf{q}_{0,1}$, $\mathbf{q}_{0,2}$, and $\mathbf{q}_{0,3}$. This concludes the discussion of three linearly independent eigenvectors associated with the null eigenvalue of B.

Together, there are nine linearly independent eigenvectors of B: $\mathbf{q}_{\pm,i}$, $\mathbf{q}_{0,i}$, $i = 1,2,3$. The diagonalizing eigenvector matrix of B can be constructed as:

$$Q \stackrel{\text{def}}{=} \left(\mathbf{q}_{-,3}\ \mathbf{q}_{-,2}\ \mathbf{q}_{-,1}\ \mathbf{q}_{0,1}\ \mathbf{q}_{0,2}\ \mathbf{q}_{0,3}\ \mathbf{q}_{+,1}\ \mathbf{q}_{+,2}\ \mathbf{q}_{+,3} \right), \tag{3.14}$$

where Q is a 9×9 matrix composed of nine column vectors. We then perform similarity transformation $\Lambda = Q^{-1}BQ$ to diagonalize B, where the eigenvalue matrix Λ is composed of $\lambda_{-,3}$, $\lambda_{-,2}$, $\lambda_{-,1}$, λ_0, λ_0, λ_0, $\lambda_{+,1}$, $\lambda_{+,2}$, and $\lambda_{+,3}$ with $\lambda_0 = 0$. This concludes the discussions of diagonalizability of B by the above nine linearly independent eigenvectors.

As an aside, each of Jacobian matrices $A^{(\mu)}$, $\mu = 1,2,3$, is a special case of B with $h_\nu = \delta_{\mu\nu}$, $\nu = 1,2,3$. For these special cases, we let $\mathbf{q}_{\pm,i}^{(\mu)}$, $\mathbf{q}_{0,i}^{(\mu)}$, $i = 1,2,3$, be the eigenvectors of $A^{(\mu)}$,

$\mu = 1,2,3$. If we let $\mu = 1$, it can be straightforwardly shown that the fifth, sixth, and seventh columns of $A^{(1)}$ are all-zero. For $\mu = 2$, the fourth, sixth, and eighth columns of $A^{(2)}$ are all-zero. For $\mu = 3$, the fourth, fifth, and ninth columns of $A^{(3)}$ are all-zero. For completeness, the eigenvectors associated with the trivial eigenvalue for $A^{(\mu)}$, $\mu = 1,2,3$, are tabulated in the following:

$$\mathbf{q}_{0,1}^{(1)} = (0,0,0,0,1,0,0,0,0)^t,\ \mathbf{q}_{0,1}^{(2)} = (0,0,0,1,0,0,0,0,0)^t,\ \mathbf{q}_{0,1}^{(3)} = (0,0,0,1,0,0,0,0,0)^t,$$
$$\mathbf{q}_{0,2}^{(1)} = (0,0,0,0,0,1,0,0,0)^t,\ \mathbf{q}_{0,2}^{(2)} = (0,0,0,0,0,1,0,0,0)^t,\ \mathbf{q}_{0,2}^{(3)} = (0,0,0,0,1,0,0,0,0)^t,$$
$$\mathbf{q}_{0,3}^{(1)} = (0,0,0,0,0,0,1,0,0)^t,\ \mathbf{q}_{0,3}^{(2)} = (0,0,0,0,0,0,0,1,0)^t,\ \mathbf{q}_{0,3}^{(3)} = (0,0,0,0,0,0,0,0,1)^t. \tag{3.15}$$

Since Eq. (2.3) is hyperbolic, we proceed to show the characteristic form of Eq. (3.3). We pre-multiply Eq. (3.3) with Q^{-1} to have

$$\frac{\partial Q^{-1}\mathbf{u}}{\partial t} + Q^{-1}BQ\frac{\partial Q^{-1}\mathbf{u}}{\partial y} = 0. \tag{3.16}$$

Note that a 9×9 identity matrix $I_9 = QQ^{-1}$ is inserted in the second term of Eq. (3.16). Since Q^{-1} is a constant matrix, it can be moved into the partial differentiation operator. Aided by the eigenvalue matrix Λ and the characteristic variables

$$\hat{\mathbf{u}} \stackrel{\text{def}}{=} Q^{-1}\mathbf{u},$$

Eq. (3.16) can be rewritten in the characteristic form as

$$\frac{\partial \hat{\mathbf{u}}}{\partial t} + \Lambda\frac{\partial \hat{\mathbf{u}}}{\partial y} = 0, \tag{3.17}$$

where Λ is a diagonal matrix consisting of the eigenvalues of B. Equation (3.17) contains nine decoupled equations:

$$\frac{\partial \hat{u}_i}{\partial t} + \lambda_i\frac{\partial \hat{u}_i}{\partial y} = 0, \quad i = 1,\ldots,9, \tag{3.18}$$

where λ_i is the ith eigenvalue of B and $\hat{u}_i$ the ith characteristic variable. If Eq. (2.3) is hyperbolic, each scalar equation in Eq. (3.18) describes convection of a characteristic variable $\hat{u}_i$ with a real wave speed λ_i.

4. Coordinate transformation and the Christoffel matrix

In this section, we show how matrix B is connected to the classic Christoffel matrix. Essentially, we will show that matrix Γ in Eq. (3.8) is in fact the Christoffel matrix for plane waves propagating in the direction of $\mathbf{h}$. Recall that Γ is obtained by solving the eigen problem for non-zero eigenvalues of B. To proceed, two Cartesian coordinate systems are considered: (i) The global coordinate system (x_1-x_2-x_3), in which the governing equations Eqs. (2.3) and

(3.3) are formulated, and (ii) the local coordinate system $(\bar{x}_1\text{-}\bar{x}_2\text{-}\bar{x}_3)$, in which the values of the stiffness coefficients are tabulated according to the lattice orientation of the solids, e.g., [2]. Because the values of the stiffness constants depend on the coordinate system employed, the stiffness coefficients reported in the literature need to be transformed into the global coordinate system where the model equations are formulated. One coordinate system can be obtained from the other by coordinate rotation. For completeness, the transformation rules are provided in Appendix 8. Vectors, tensors, and 6-component vectors formulated in the local coordinate system are denoted by an over-bar. The entities formulated in the global coordinate system have no bar. For completeness, Appendix 8 shows the coordinate transformation of the model equations. It turns out that the forms of the governing equations in two different coordinate systems are identical.

To connect the present formulation to the classical Christoffel matrices tabulated in the literature, we proceed to transform matrix Γ formulated in global coordinates to be $\bar{\Gamma}$ in local coordinates. We first define the following matrices associated with the Jacobian matrices $A^{(\mu)}, \mu = 1,2,3$:

$$\Gamma^{(\mu)} \stackrel{\text{def}}{=} K^{(\mu)} C K^{(\mu)^t}, \quad \mu = 1,2,3, \tag{4.1}$$

which are special cases of Γ with $h_\nu = \delta_{\mu\nu}, \nu = 1,2,3$. $\Gamma^{(1)}, \Gamma^{(2)}$, and $\Gamma^{(3)}$ are tensors in $\mathbb{E}^3$ and they can be transformed to local coordinates by using the rotation matrix R shown in Eq. (8.1) in Appendix 8:

$$\bar{\Gamma}^{(\mu)} = R^t \Gamma^{(\mu)} R. \tag{4.2}$$

$\bar{\Gamma}^{(\mu)}$ has the three eigenvectors $\bar{\mathbf{q}}_{v,i}^{(\mu)}, i = 1,2,3$. $\mathbf{q}_{v,i}^{(\mu)}$ and $\bar{\mathbf{q}}_{v,i}^{(\mu)}$ are the same vector defined in the global and local coordinate systems, respectively. The algebraic relation between the $\Gamma^{(\mu)}$ and $\bar{\Gamma}^{(\mu)}$ in Eq. (4.2) needs to be further elaborated. To proceed, Bond's matrix [2] is used to transform the stiffness matrix from the local coordinates to the global coordinates:

$$C = M \bar{C} M^t. \tag{4.3}$$

Details of the transformation are provided in Appendix 8. For a wide range of media, entries of $\bar{C}$ in the local coordinate system are widely available in the literature, e.g., [2]. Substituting Eqs. (4.1) and (4.3) into Eq. (4.2) gives:

$$\bar{\Gamma}^{(\mu)} = L^{(\mu)} \bar{C} L^{(\mu)^t}, \quad \mu = 1,2,3, \tag{4.4}$$

where

$$L^{(\mu)} \stackrel{\text{def}}{=} R^t K^{(\mu)} M \tag{4.5}$$

is a 3×6 matrix with its entries as the components of the direction cosine corresponding to x_μ-axis. To proceed, Appendix 8 shows that:

$$K^{(\mu)} M = R \sum_{\nu=1}^{3} r_{\mu\nu} K^{(\nu)}. \tag{4.6}$$

The algebra involved is lengthy but straightforward. Aided by Eq. (4.6) and the orthogonality of matrix R, Eq. (4.5) becomes

$$L^{(\mu)} = \sum_{\nu=1}^{3} r_{\mu\nu} K^{(\nu)} = \begin{pmatrix} r_{\mu 1} & 0 & 0 & 0 & r_{\mu 3} & r_{\mu 2} \\ 0 & r_{\mu 2} & 0 & r_{\mu 3} & 0 & r_{\mu 1} \\ 0 & 0 & r_{\mu 3} & r_{\mu 2} & r_{\mu 1} & 0 \end{pmatrix}. \tag{4.7}$$

To proceed, the relation between Γ and $\bar{\Gamma}$ is derived in the following. Let L be the linear combination of $L^{(1)}$, $L^{(2)}$, and $L^{(3)}$:

$$L \overset{\text{def}}{=} \sum_{\mu=1}^{3} h_\mu L^{(\mu)}. \tag{4.8}$$

Aided by Eq. (4.5), Eq. (4.8) can be expanded and becomes

$$L = \sum_{\mu=1}^{3} h_\mu R^t K^{(\mu)} M = R^t \left(\sum_{\mu=1}^{3} h_\mu K^{(\mu)} \right) M. \tag{4.9}$$

Aided by Eqs. (3.8), (3.4), and (4.9), Γ can be readily transformed to $\bar{\Gamma}$:

$$\bar{\Gamma} = R^t \Gamma R = R^t G C G^t R = R^t \left(\sum_{\mu=1}^{3} h_\mu K^{(\mu)} \right) M \bar{C} M^t \left(\sum_{\mu=1}^{3} h_\mu K^{(\mu)^t} \right) R = L \bar{C} L^t. \tag{4.10}$$

Next, the unit vector $\mathbf{h}$ is also transformed into the local coordinates:

$$\bar{\mathbf{h}} = R^t \mathbf{h}, \tag{4.11}$$

where $\bar{\mathbf{h}} = (\bar{h}_1, \bar{h}_2, \bar{h}_3)$ is the unit vector of the wave propagation direction defined in the local coordinates and $\bar{h}_1$, $\bar{h}_2$, and $\bar{h}_3$ are the direction cosines in the local coordination system. Aided by Eqs. (4.7) and (4.11), matrix L defined in Eq. (4.8) can be expanded as

$$L = \begin{pmatrix} \bar{h}_1 & 0 & 0 & 0 & \bar{h}_3 & \bar{h}_2 \\ 0 & \bar{h}_2 & 0 & \bar{h}_3 & 0 & \bar{h}_1 \\ 0 & 0 & \bar{h}_3 & \bar{h}_2 & \bar{h}_1 & 0 \end{pmatrix}. \tag{4.12}$$

By comparing Eqs. (4.10) and (4.12) to the Christoffel matrix tabulated in the literature, e.g., Auld [2], it can be seen that $\bar{\Gamma}$ in Eq. (4.10) is indeed the Christoffel matrix. Special cases of $\bar{\Gamma}$ in the directions along the axes of the local coordinate system, i.e., $\bar{x}_1$-, $\bar{x}_2$-, and $\bar{x}_3$-axis, are $\bar{\Gamma}^{(\mu)}$, $\mu = 1, 2, 3$, respectively.

5. Elastic solids with hexagonal symmetry

In this section, we illustrate how the above formulation is applied to model waves in elastic solids of hexagonal symmetry. A solid of hexagonal symmetry is shown in Figure 1. The Cartesian coordinates are chosen such that the x_3-axis is aligned with the hexagonal cylinder of the medium, and the x_2-axis is aligned with the crystal lattice. In this coordinate system,

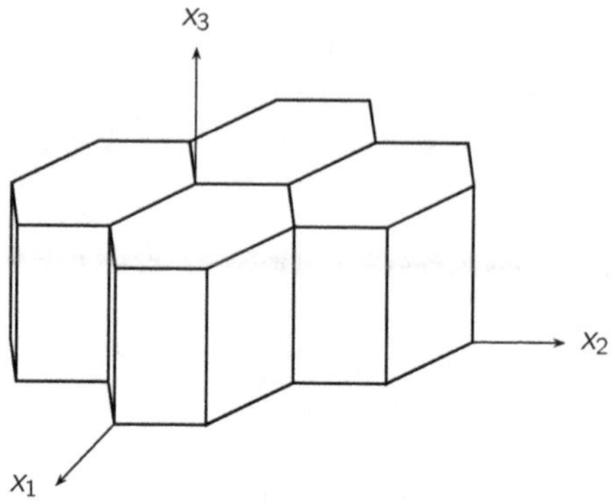

Figure 1. The Cartesian coordinates and the lattice structure of solids of hexagonal symmetry.

9 of the 21 stiffness constants are non-zero and 5 of them, i.e., c_{11}, c_{12}, c_{13}, c_{33} and c_{44}, are independent. The stiffness matrix is

$$C = \begin{pmatrix} c_{11} & c_{12} & c_{13} & 0 & 0 & 0 \\ c_{12} & c_{11} & c_{13} & 0 & 0 & 0 \\ c_{13} & c_{13} & c_{33} & 0 & 0 & 0 \\ 0 & 0 & 0 & c_{44} & 0 & 0 \\ 0 & 0 & 0 & 0 & c_{44} & 0 \\ 0 & 0 & 0 & 0 & 0 & c_{66} \end{pmatrix}. \tag{5.1}$$

where $c_{66} = (c_{11} - c_{12})/2$. Aided by Eq. (5.1), the Jacobian matrix $A^{(1)}$, i.e., Eq. (2.4) in the model equation Eq. (2.3) become

$$A^{(1)} = \begin{pmatrix} 0_{3\times3} & -\frac{1}{\rho}K^{(1)} \\ -CK^{(1)^t} & 0_{6\times6} \end{pmatrix}, \quad \text{where} \quad -CK^{(1)^t} = \begin{pmatrix} -c_{11} & 0 & 0 \\ -c_{12} & 0 & 0 \\ -c_{13} & 0 & 0 \\ 0 & 0 & 0 \\ 0 & 0 & -c_{44} \\ 0 & -c_{66} & 0 \end{pmatrix}, \tag{5.2}$$

Similarly, for $A^{(2)}$ and $A^{(3)}$, the related matrices are

$$-CK^{(2)^t} = \begin{pmatrix} 0 & -c_{12} & 0 \\ 0 & -c_{11} & 0 \\ 0 & -c_{13} & 0 \\ 0 & 0 & -c_{44} \\ 0 & 0 & 0 \\ -c_{66} & 0 & 0 \end{pmatrix} \quad \text{and} \quad -CK^{(3)^t} = \begin{pmatrix} 0 & 0 & -c_{13} \\ 0 & 0 & -c_{13} \\ 0 & 0 & -c_{33} \\ 0 & -c_{44} & 0 \\ -c_{44} & 0 & 0 \\ 0 & 0 & 0 \end{pmatrix}. \tag{5.3}$$

The above Jacobian matrices in the model equations are valid only if the local coordinates ($\bar{x}_1$, $\bar{x}_2$, $\bar{x}_3$) are employed, in which the entries of the stiffness matrix such as that shown in Eq. (5.1) can be directly taken from the literature. To proceed, we transform the model equations from the local coordinates $(\bar{x}_1, \bar{x}_2, \bar{x}_3)$ to the global coordinates (x_1, x_2, x_3). In general, the rotation can be defined by the three Euler angles [1]: α, β, and γ. The coordinate transformation matrix R, Eq. (8.1), can be readily specified:

$$R(\alpha,\beta,\gamma) = \begin{pmatrix} \cos\gamma\cos\beta\cos\alpha - \sin\gamma\sin\alpha & \cos\gamma\cos\beta\sin\alpha + \sin\gamma\cos\alpha & -\cos\gamma\sin\beta \\ -\sin\gamma\cos\beta\cos\alpha - \cos\gamma\sin\alpha & -\sin\gamma\cos\beta\sin\alpha + \cos\gamma\cos\alpha & \sin\gamma\sin\beta \\ \sin\beta\cos\alpha & \sin\beta\sin\alpha & \cos\beta \end{pmatrix}.$$

With $r_{\mu\nu}$ denoting the (μ,ν) entry of R, we have $\partial\bar{x}_\mu/\partial x_\nu = r_{\mu\nu}$ with $\mu,\nu = 1,2,3$. Aided by the chain rule, the model equation Eq. (2.3) becomes

$$\frac{\partial \mathbf{u}}{\partial t} + \sum_{\mu=1}^{3} \bar{A}^{(\mu)} \frac{\partial \mathbf{u}}{\partial \bar{x}_\mu} = 0, \quad \text{or} \quad \frac{\partial u_i}{\partial t} + \sum_{\mu=1}^{3}\sum_{j=1}^{9} \bar{a}_{ij}^{(\mu)} \frac{\partial u_j}{\partial \bar{x}_\mu} = 0, \quad i = 1,2,\ldots,9, \tag{5.4}$$

where $\bar{a}_{ij}^{(\mu)} = \sum_{\nu=1}^{3} r_{\mu\nu} a_{ij}^{(\nu)}$. To connect the present formulation to the Christoffel matrices for solids of hexagonal symmetry, Eq. (5.4) is reduced to be one-dimensional for modeling wave propagation along the $\bar{x}_1$ axis:

$$\frac{\partial \mathbf{u}}{\partial t} + \bar{A}^{(1)} \frac{\partial \mathbf{u}}{\partial \bar{x}_1} = 0, \quad \text{or} \quad \frac{\partial u_i}{\partial t} + \sum_{j=1}^{9} \bar{a}_{ij}^{(1)} \frac{\partial u_j}{\partial \bar{x}_1} = 0, \quad i = 1,2,\ldots,9, \tag{5.5}$$

As shown in Fig. 2, two angles are needed to specify a particular direction in the three-dimensional space. The Euler angles are chosen to be $(\theta, -\phi, 0)$, where $0 \le \theta \le 2\pi$ and $-\pi/2 \le \phi \le \pi/2$. The transformation matrix R of a given pair of angles $(\theta, -\phi)$ is

$$R(\theta,-\phi,0) = \begin{pmatrix} \cos\phi\cos\theta & \cos\phi\sin\theta & \sin\phi \\ -\sin\theta & \cos\theta & 0 \\ -\sin\phi\cos\theta & -\sin\phi\sin\theta & \cos\phi \end{pmatrix}. \tag{5.6}$$

Aided by the above coordinate transformation, the Jacobian matrix $\bar{A}^{(1)}$ in Eq. (5.5) becomes

$$\bar{A}^{(1)} = r_{11}A^{(1)} + r_{12}A^{(2)} + r_{13}A^{(3)} = \left(\begin{array}{c|c} 0_{3\times3} & \bar{A}_v \\ \hline \bar{A}_T & 0_{6\times6} \end{array}\right), \tag{5.7}$$

$$\bar{A}_v = \frac{1}{\rho}\begin{pmatrix} -\cos\phi\cos\theta & 0 & 0 & 0 & -\sin\phi & -\cos\phi\sin\theta \\ 0 & -\cos\phi\sin\theta & 0 & -\sin\phi & 0 & -\cos\phi\cos\theta \\ 0 & 0 & -\sin\phi & -\cos\phi\sin\theta & -\cos\phi\cos\theta & 0 \end{pmatrix},$$

$$\bar{A}_T = \begin{pmatrix} -c_{11}\cos\phi\cos\theta & -c_{12}\cos\phi\sin\theta & -c_{13}\sin\phi \\ -c_{12}\cos\phi\cos\theta & -c_{11}\cos\phi\sin\theta & -c_{13}\sin\phi \\ -c_{13}\cos\phi\cos\theta & -c_{13}\cos\phi\sin\theta & -c_{33}\sin\phi \\ 0 & -c_{44}\sin\phi & -c_{44}\cos\phi\sin\theta \\ -c_{44}\sin\phi & 0 & -c_{44}\cos\phi\cos\theta \\ -c_{66}\cos\phi\sin\theta & -c_{66}\cos\phi\cos\theta & 0 \end{pmatrix},$$

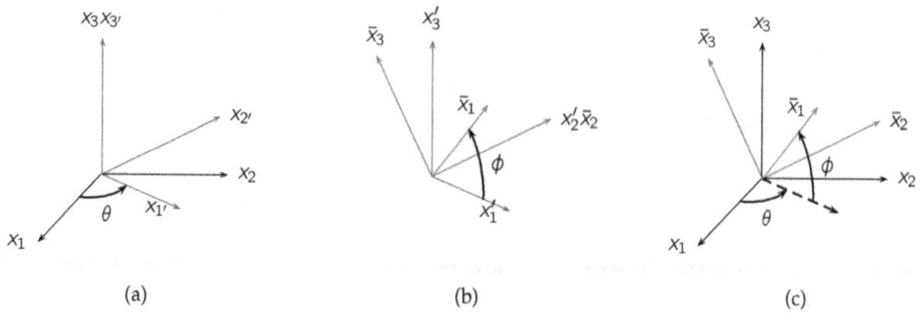

Figure 2. Rotations of the Cartesian coordinate system. (a) Rotation with respect to the x_3 axis. (b) Rotation with respect to the $\bar{x}_2$ axis. (c) Overall rotation.

Matrix $\bar{A}^{(1)}$ is a 9×9 matrix, there are nine eigenvalues, which are calculated by solving the polynomial:

$$\det(\bar{A}^{(1)} - \lambda I_9) = \det \begin{pmatrix} -\lambda I_3 & \bar{A}_v \\ \bar{A}_T & -\lambda I_6 \end{pmatrix} = 0,$$

where I_n is a rank n identity matrix. Based on the Schur complement [15, pp. 475], the non-trivial eigenvalues of $\bar{A}^{(1)}$ can be obtained by solving a simpler equation

$$\det(\bar{A}_v \bar{A}_T - \kappa^2 I_3) = 0, \tag{5.8}$$

where

$$\bar{A}_v \bar{A}_T = \frac{1}{\rho} \begin{pmatrix} \Gamma_{11} & \Gamma_{12} & \Gamma_{13} \\ \Gamma_{21} & \Gamma_{22} & \Gamma_{23} \\ \Gamma_{31} & \Gamma_{32} & \Gamma_{33} \end{pmatrix},$$

and

$$\begin{aligned}
\Gamma_{11} &= c_{11} \cos^2 \phi \cos^2 \theta + c_{44} \sin^2 \phi + c_{66} \cos^2 \phi \sin^2 \theta, \\
\Gamma_{22} &= c_{11} \cos^2 \phi \sin^2 \theta + c_{44} \sin^2 \phi + c_{66} \cos^2 \phi \cos^2 \theta, \\
\Gamma_{33} &= c_{33} \sin^2 \phi + c_{44} \cos^2 \phi \sin^2 \theta + c_{44} \cos^2 \phi \cos^2 \theta. \\
\Gamma_{12} = \Gamma_{21} &= c_{12} \cos^2 \phi \cos \theta \sin \theta + c_{66} \cos^2 \phi \cos \theta \sin \theta, \\
\Gamma_{13} = \Gamma_{31} &= c_{13} \cos \phi \sin \phi \cos \theta + c_{44} \cos \phi \sin \phi \cos \theta, \\
\Gamma_{23} = \Gamma_{32} &= c_{13} \cos \phi \sin \phi \sin \theta + c_{44} \cos \phi \sin \phi \sin \theta.
\end{aligned} \tag{5.9}$$

To proceed, according to Eq. (5.6), we let $\bar{\mathbf{h}} = (\bar{h}_1, \bar{h}_2, \bar{h}_3)$ be the unit vector parallel to the $\bar{x}_1$ axis in the local coordinate system, i.e., $\bar{h}_1 = r_{11} = \cos\phi \cos\theta$, $\bar{h}_2 = r_{12} = \cos\phi \sin\theta$, and $\bar{h}_3 = r_{13} = \sin\phi$. Aided by the definition of $\bar{\mathbf{h}}$, Eq. (5.9) can be rewritten as

$$\begin{aligned}
\Gamma_{11} &= c_{11} \bar{h}_1^2 + c_{44} \bar{h}_3^2 + c_{66} \bar{h}_2^2, \\
\Gamma_{22} &= c_{11} \bar{h}_2^2 + c_{44} \bar{h}_3^2 + c_{66} \bar{h}_1^2, \\
\Gamma_{33} &= c_{44} \bar{h}_1^2 + c_{44} \bar{h}_2^2 + c_{33} \bar{h}_3^2, \\
\Gamma_{12} = \Gamma_{21} &= (c_{12} + c_{66}) \bar{h}_1 \bar{h}_2, \\
\Gamma_{13} = \Gamma_{31} &= (c_{13} + c_{44}) \bar{h}_1 \bar{h}_3, \\
\Gamma_{23} = \Gamma_{32} &= (c_{13} + c_{44}) \bar{h}_2 \bar{h}_3.
\end{aligned} \tag{5.10}$$

Equation (5.10) are identical to the Christoffel equations for hexagonal solids as shown in the literature, e.g., Auld [2].

Shown in Eq. (5.8), matrix $\bar{A}_v\bar{A}_T$ is real and symmetric. Thus, its three eigenvalues κ_1^2, κ_2^2, and κ_3^2 are positive and real. As such, the non-trivial eigenvalues of $\bar{A}^{(1)}$ are $\pm\kappa_i$ with $i = 1,2,$ and 3. Moreover, κ_1, κ_2, and κ_3 are the three wave speeds in the hexagonal solids. Simple manipulation also shows that the other three trivial eigenvalues of $\bar{A}^{(1)}$ are null. Therefore, all nine eigenvalues are real.

6. Modeling wave propagation in a block of beryl

In this section, we apply the space-time Conservation Element and Solution Element (CESE) method [4] to solve the velocity-stress equations for modeling wave propagation in a block of beryl, a elastic solid of hexagonal symmetry. The CESE method is a modern numerical method for time accurate solutions of linear and non-linear hyperbolic partial differential equations. The CESE method treats space and time in a unified way in calculating the space-time flux. The method is simple, robust, and its operational count is comparable to that of a second-order central difference scheme. The backbone of the CESE method is the a scheme [4], which is neutrally stable without artificial dissipation for solving a scalar convection equation with a constant coefficient. For more complex wave equations, the a scheme is extended with added artificial damping, including the a-α-ϵ scheme [4] for shock capturing and the c-τ scheme [5] for Courant-Friedrichs-Lewy (CFL) number insensitive calculations. To model multi-dimensional problems, the CESE method uses unstructured meshes [27] with triangles and tetrahedra as the basic elements in two- and three-dimensional spaces. Extensions to use quadrilaterals and hexahedra are also available [36]. In the past, the CESE method has been successfully applied to solve a wide range of various fluid dynamics problems, including aero-acoustics [14], cavitations [19], complex shock waves [8], detonations [26], magnetohydrodynamics (MHD)[34, 35], etc. Moreover, the method has been applied to model electromagnetic waves [28] and nonlinear stress waves in isotropic solids [3, 31, 32]. In the present chapter, the following results are obtained by using the a-α-ϵ scheme.

The two-dimensional velocity-stress equations are employed to model wave propagation in a block of beryl, which is a solid of hexagonal symmetry. The material properties are taken from [16]: $\rho = 2.7\,\mathrm{g/cm^3}$, $c_{11} = 26.94 \times 10^{11}\,\mathrm{dynes/cm^2}$, $c_{12} = 9.61 \times 10^{11}\,\mathrm{dynes/cm^2}$, $c_{13} = 6.61 \times 10^{11}\,\mathrm{dynes/cm^2}$, $c_{33} = 23.63 \times 10^{11}\,\mathrm{dynes/cm^2}$, and $c_{44} = 6.53 \times 10^{11}\,\mathrm{dynes/cm^2}$. The computational domain is $-1\,\mathrm{m} \le x_{(1)} \le 1\,\mathrm{m}$ and $-1\,\mathrm{m} \le x_{(2)} \le 1\,\mathrm{m}$. A unstructured mesh composed of triangular elements is used. The square domain is divided into 2.2 millions of triangular elements. The maximum length of the element edge is $2.68 \times 10^{-3}\,\mathrm{m}$. The initial condition is a two-dimensional step function $\mathbf{v} = \mathbf{a}\mathcal{H}(\sigma - |\mathbf{x}|)$, where $\mathbf{a}$ represents a constant initial velocity to initiate the expanding wave from the origin. The constant σ in the initial condition is chosen to be $5 \times 10^{-3}\,\mathrm{m}$.

The two-dimensional code is parallelized by domain decomposition. First, a graph of element connectivity is built based on the unstructured mesh. The connectivity graph is processed by METIS [7] to partition the domain. As shown in Fig. 3, the overall spatial domain is decomposed into 16 sub-domains for parallel computing. Numerical calculations for elements in each sub-domain is distributed to a workstation as a part of a networked cluster. In all cases, $\Delta t = 65 \times 10^{-9}\,\mathrm{s}$ and the maximum CFL number ν is about 0.95. Three cases of wave propagation are calculated: (1) calculation of the group velocity profile to assess

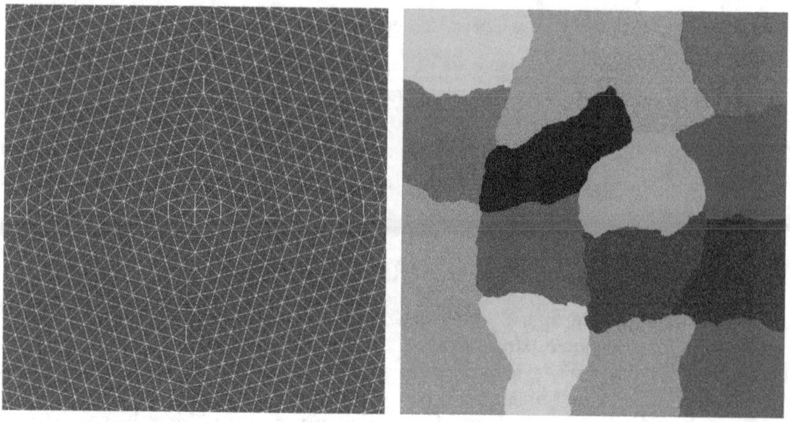

(a) (b)

Figure 3. Two-dimensional mesh (2.2 million triangular elements). (a) close look at the mesh around origin. (b) decomposed 16 sub-domains.

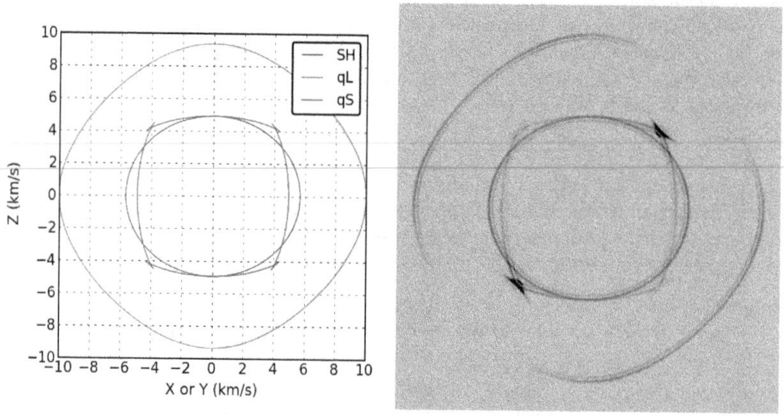

(a) (b)

Figure 4. Comparison between the analytical solutions of the group velocities (in SH, qS, and qL polarization) and the calculated energy profiles for beryl at $t = 91$ μs, including: (a) The analytical solution, and (b) The calculated energy density profiles.

numerical accuracy, (2) waves propagation in a plane of an arbitrary orientation to assess the non-reflective boundary condition, and (3) waves propagation in a plane composed of three blocks of beryl in different lattice orientations.

In Case 1, two-dimensional numerical simulations are performed to calculate the density of the total energy, which is the summation of the kinetic energy and the strain energy normalized by area. For elastic solids, the group velocity is identical to the energy velocity [2]. Here, we consider wave propagation in 100 (x_2-x_3) or 010 (x_1-x_3) plane of the hexagonal

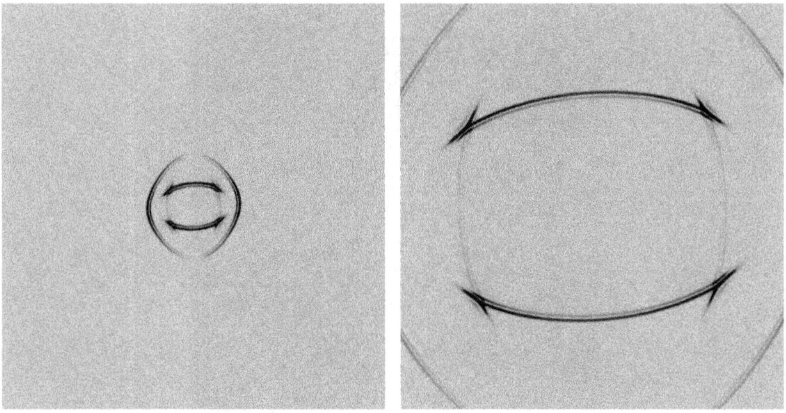

(a) (b)

Figure 5. Plots for the normalized total energy density ($e/\max(e)$) at two times: (a) $t = 26\,\mu$s, (b) $t = 130\,\mu$s.

solid. The orientation of the solid is set to be $\theta = 0^\circ$ and $\phi = 90^\circ$, to have $x_{(1)}$ aligned to x_3, and $x_{(2)}$ aligned to x_2. The group velocities have the analytical solution. The group velocities of pure-shear (SH), quasi-longitudinal (qL), and quasi-shear (qS) waves in a block of beryl are plotted in Fig. 4(a) [16]. The simulation result, shown in Fig. 4(b), compares well with the analytical solution.

In Case 2, the orientation of the computational plane is in the direction of $\theta = 20^\circ$ and $\phi = 70^\circ$. Similar to that in Case 1, the initial condition is a velocity source at the origin with $\sigma = 0.00$ m. Figure 5 shows two snapshots of the calculated energy profiles at t = 26 and 130 μs. With the non-reflecting boundary condition implemented on the four boundaries of the computational domain, waves propagate out of the domain without spurious reflection.

For Case 3, we calculate wave propagation in a domain composed of three blocks of beryl with three different lattice orientations. The partition of the domain is shown in Fig. 6. Shown in Fig. 7(a), an initial, cylindrical wave is generated at the origin. This is because of the lattice orientation of the central block such that an isotropic wave expansion occurs. This wave expands and interacts with the interfaces separating the neighbouring blocks of beryl with different lattice orientations. Figure 7 shows four snapshots of the calculated energy profiles at different times. Due to complex wave/interface interactions, all wave modes, including SH, qL and qS are excited at the interface. As a results, refracted waves at different speeds can be seen in Figs. 7(b) and 7(c). The original circular wave front is fractured after passing through the interfaces, as shown in Fig. 7(d). Complex wave features have been successfully captured by the numerical results.

7. Concluding remarks

In this chapter, we have presented the first-order velocity-stress equations as an alternative to the conventional second-order wave equations for modeling wave propagation in anisotropic

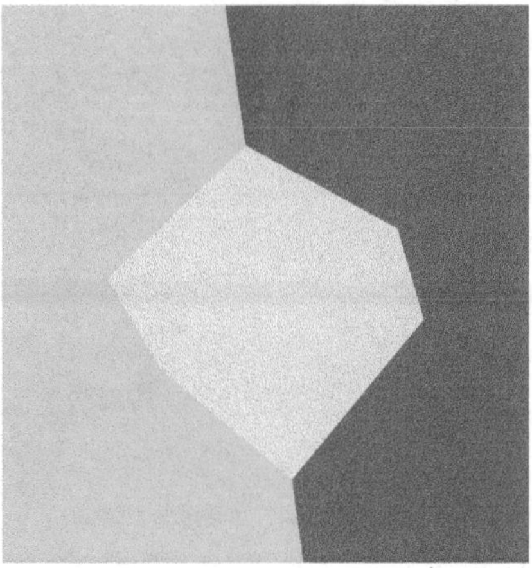

Figure 6. The computational domain is divided into three regions. The orientations of the solids in the central, left and right regions are $\theta = 0°$ and $\phi = 0°$, $\theta = 0°$ and $\phi = 60°$, and $\theta = 0°$ and $\phi = 30°$, respectively.

elastic solids. The eigen structure of the equations has been thoroughly studied by analyzing the composed Jacobian matrix of the first-order equations, i.e., $B = \sum_{\mu=1}^{3} h_\mu A^{(\mu)}$. The hyperbolicity of the equations has been rigorously proved by showing the real spectrum of matrix B and diagonalizability of B. B has a degenerate null eigenvalue. For non-zero eigenvalues, B has at least one and at most three positive-negative pairs of real eigenvalues. Moreover, B has nine linearly independent eigenvectors, which can be used to diagonalize B. The eigenvalues represent the wave speeds, and the eigenvectors are related to the polarization of the waves. Since the first-order velocity-stress equations are hyperbolic, the Cauchy problem is well-posed [9]. For numerical simulation, the proven hyperbolicity justifies the use of modern finite-volume methods, originally developed for solving conservation laws, to solve the velocity-stress equations. For upwind methods [12] and discontinuous Galerkin methods [18], information about eigenvalues and eigenvectors of matrix B are critically important. The derived characteristic relations of the governing equations can be directly used to derive the Riemann solver and the flux function as the building blocks of the methods. Moreover, by clearly defining the local and global coordinate systems, we also show that matrix B is directly connected to the classic Christoffel matrix. Therefore, nine coupled first-order velocity-stress equations are shown to be directly related to the classic second-order elastodynamic equations. To demonstrate the application of the velocity-stress equations, we apply the space-time CESE method to solve the equations for modeling wave propagation in a block of beryl. The calculated energy profiles compare well with the analytical solution. We also simulate wave propagation in a heterogeneous solid composed of three blocks of beryl with different lattice orientations. Numerical results show complex wave/interface interactions.

Author details

Sheng-Tao John Yu, Yung-Yu Chen and Lixiang Yang
The Department of Mechanical Engineering, The Ohio State University, Columbus, OH 43210, USA

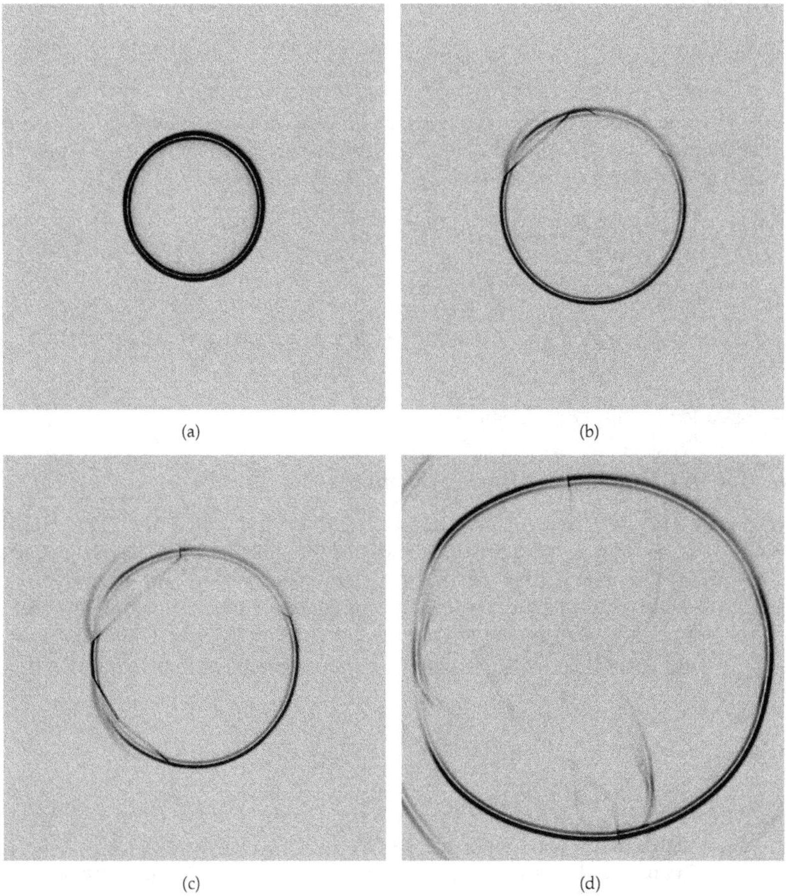

Figure 7. Waves propagation in a heterogeneous domain. The total energy density $(e/\max(e))$ are plotted at different times: (a) $t = 72\,\mu s$, (b) $t = 96\,\mu s$, (c) $t = 108\,\mu s$, (d) $t = 180\,\mu s$. $\Delta t = 60\,ns$, and CFL number $= 0.95$.

8. Appendix

A: The Schur complements

Consider matrix M, which can be divided into 4 sub matrices:

$$M = \begin{pmatrix} A & B \\ C & D \end{pmatrix},$$

where $A, B, C,$ and D are $p \times p$, $p \times q$, $q \times p$, and $q \times q$ matrices, respectively. If A^{-1} and D^{-1} exist, then matrices $D - CA^{-1}B$ and $A - BD^{-1}C$ are the Schur complements of A and D [15]. In this case, matrix M can be factored out as

$$\begin{pmatrix} A & B \\ C & D \end{pmatrix} = \begin{pmatrix} I_p & 0_{p\times q} \\ CA^{-1} & I_q \end{pmatrix} \begin{pmatrix} A & B \\ 0_{q\times p} & D - CA^{-1}B \end{pmatrix},$$

or

$$\begin{pmatrix} A & B \\ C & D \end{pmatrix} = \begin{pmatrix} A - BD^{-1}C & B \\ 0_{q\times p} & D \end{pmatrix} \begin{pmatrix} I_p & 0_{p\times q} \\ D^{-1}C & I_q \end{pmatrix}.$$

This factorization is useful to calculate $\det(M)$.

B: Rotation of Cartesian coordinate system

This appendix provides the transformation relations for the global and local coordinate systems. Let $\mathbf{e}_1, \mathbf{e}_2,$ and $\mathbf{e}_3$ be the natural basis of the global coordinate system, and let $\bar{\mathbf{e}}_1, \bar{\mathbf{e}}_2,$ and $\bar{\mathbf{e}}_3$ be the natural basis of the local coordinate system. An arbitrary vector $\mathbf{b}$ can be represented by using either the global coordinates or the local coordinates as: $\mathbf{b} = \sum_{\mu=1}^{3} b_\mu \mathbf{e}_\mu = \sum_{\mu=1}^{3} \bar{b}_\mu \bar{\mathbf{e}}_\mu$. Let $\mathbf{b}$ denote the vector defined in the global coordinates and $\bar{\mathbf{b}}$ defined in the local coordinates. The transformation between $\mathbf{b}$ and $\bar{\mathbf{b}}$ is written as: $\mathbf{b} = R\bar{\mathbf{b}}$, where

$$R = \begin{pmatrix} r_{11} & r_{12} & r_{13} \\ r_{21} & r_{22} & r_{23} \\ r_{31} & r_{32} & r_{33} \end{pmatrix} \tag{8.1}$$

is the rotation matrix and it is orthonormal, i.e., $R^{-1} = R^t$ or $\sum_{k=1}^{3} r_{ik} r_{jk} = \sum_{k=1}^{3} r_{ki} r_{kj} = \delta_{ij}$, $i, j = 1, 2, 3$. The natural bases of the global and local coordinate systems are also related through R defined in Eq. (8.1):

$$\mathbf{e}_i = \sum_{j=1}^{3} r_{ij} \bar{\mathbf{e}}_j, \quad \bar{\mathbf{e}}_i = \sum_{j=1}^{3} r_{ji} \mathbf{e}_j, \quad i = 1, 2, 3. \tag{8.2}$$

To proceed, consider the transformation of a tensor E:

$$E\mathbf{b} = R(\bar{E}\bar{\mathbf{b}}) = R(\bar{E}R^t R\bar{\mathbf{b}}) = (R\bar{E}R^t)(R\bar{\mathbf{b}}) = (R\bar{E}R^t)\mathbf{b} \Rightarrow E = R\bar{E}R^t, \tag{8.3}$$

where E is defined in the global coordinates, and $\bar{E}$ in the local coordinates.

To transform the 6-component stress vector **T**, we use the Bond matrix [2]. Aided by matrix R, the Bond transformation matrix M is

$$M \stackrel{\text{def}}{=} \begin{pmatrix} r_{11}^2 & r_{12}^2 & r_{13}^2 & 2r_{12}r_{13} & 2r_{11}r_{13} & 2r_{11}r_{12} \\ r_{21}^2 & r_{22}^2 & r_{23}^2 & 2r_{22}r_{23} & 2r_{21}r_{23} & 2r_{21}r_{22} \\ r_{31}^2 & r_{32}^2 & r_{33}^2 & 2r_{32}r_{33} & 2r_{31}r_{33} & 2r_{31}r_{32} \\ r_{21}r_{31} & r_{22}r_{32} & r_{23}r_{33} & r_{22}r_{33}+r_{32}r_{23} & r_{21}r_{33}+r_{23}r_{31} & r_{21}r_{32}+r_{22}r_{31} \\ r_{11}r_{31} & r_{12}r_{32} & r_{13}r_{33} & r_{12}r_{33}+r_{13}r_{32} & r_{11}r_{33}+r_{13}r_{31} & r_{11}r_{32}+r_{12}r_{31} \\ r_{11}r_{21} & r_{12}r_{22} & r_{13}r_{23} & r_{12}r_{23}+r_{13}r_{22} & r_{11}r_{23}+r_{13}r_{21} & r_{11}r_{22}+r_{12}r_{21} \end{pmatrix}. \tag{8.4}$$

The application of M to the stress vector is $\mathbf{T} = M\bar{\mathbf{T}}$, where **T** is the 6-component stress vector defined in the global coordinates and $\bar{\mathbf{T}}$ in local coordinates. Similarly, the Bond matrix for strain N [2] is used to transform the 6-component strain vector **S** as $\mathbf{S} = N\bar{\mathbf{S}}$. Simple manipulation shows that $N^{-1} = M^t$. Finally, we verify Eq. (4.3). The elastic constitutive equations in the global and local coordinate systems are stated as $\mathbf{T} = C\mathbf{S}$ and $\bar{\mathbf{T}} = \bar{C}\bar{\mathbf{S}}$. Aided by the Bond's matrices M and N, $\bar{C}$ can be related to C in the following equation:

$$\mathbf{T} = M\bar{\mathbf{T}} = M(\bar{C}\bar{\mathbf{S}}) = M\bar{C}(N^{-1}\mathbf{S}) = (M\bar{C}M^t)\mathbf{S} = C\mathbf{S}.$$

C: Rotation of governing equations

This appendix shows that the velocity-stress equations, Eq. (1.4), remain in the same form when formulated in different Cartesian coordinate systems. To proceed, we consider an arbitrary tensor E. In the following, we show that $\nabla \cdot E = \bar{\nabla} \cdot (ER)$, where $\bar{\nabla}\cdot$ is the divergence operator defined in the local coordinate system, and R is the rotation matrix R defined in Appendix 8. Consider the ith component of $\nabla \cdot E$ with $i = 1, 2, 3$:

$$\begin{aligned}(\nabla \cdot E)_i &= \sum_{j=1}^{3} \frac{\partial e_{ij}}{\partial x_j} = \sum_{j=1}^{3}\sum_{k=1}^{3} \frac{\partial e_{ij}}{\partial \bar{x}_k}\frac{\partial \bar{x}_k}{\partial x_j} = \sum_{j=1}^{3}\sum_{k=1}^{3} \frac{\partial e_{ij}}{\partial \bar{x}_k} r_{jk} = \sum_{k=1}^{3} \frac{\partial}{\partial \bar{x}_k}\left(\sum_{j=1}^{3} e_{ij} r_{jk}\right) \\ &= \left(\bar{\nabla} \cdot (ER)\right)_i .\end{aligned} \tag{8.5}$$

The above equation shows that $\nabla \cdot E = \bar{\nabla} \cdot (ER)$ because R is a constant spatial tensor.

Next, apply coordinate transformation to the equation of motion, Eq. (1.2):

$$R^t \frac{\partial \mathbf{v}}{\partial t} = \frac{1}{\rho} R^t \nabla \cdot T. \tag{8.6}$$

Aided by Eq. (8.5), Eq. (8.6) can be rewritten as:

$$\frac{\partial \bar{\mathbf{v}}}{\partial t} = \frac{1}{\rho} R^t \bar{\nabla} \cdot (TR) = \frac{1}{\rho} \bar{\nabla} \cdot (R^t TR) = \frac{1}{\rho} \bar{\nabla} \cdot \bar{T}.$$

Then perform the coordinate transformation for constitutive equations, Eq. (1.3):

$$R^t TR = R^t c^{[4]} SR. \tag{8.7}$$

Recast $c^{[4]}$ in Eq. (8.7) into the index form:

$$\begin{aligned}\bar{T}_{mn} &= r_{im}T_{ij}r_{jn} = r_{im}c_{ijkl}S_{kl}r_{jn} = r_{im}c_{ijkl}r_{ko}\bar{S}_{op}r_{lp}r_{jn} \\ &= r_{im}r_{jn}r_{ko}r_{lp}c_{ijkl}\bar{S}_{op} = \bar{c}_{mnop}\bar{S}_{op},\end{aligned} \tag{8.8}$$

where c_{ijkl} and $\bar{c}_{mnop}$ are the components of the fourth-order stiffness tensor $c^{[4]}$ and $\bar{c}^{[4]}$, respectively. For conciseness, Eq. (8.8) uses Einstein summation convention. By using the direct notation, Eq. (8.8) can be recast as $\bar{T} = \bar{c}^{[4]}\bar{S}$.

D: Supplemental algebraic relations for the direction cosine matrices

This appendix provides the algebraic relations to facilitate the derivation of Eq. (4.6). Aided by Eqs. (2.2) and (8.4), we expand $K^{(\mu)}M$, $\mu = 1, 2, 3$, as:

$$K^{(1)}M = \begin{pmatrix} r_{11}^2 & r_{12}^2 & r_{13}^2 & 2r_{12}r_{13} & 2r_{11}r_{13} & 2r_{11}r_{12} \\ r_{11}r_{21} & r_{12}r_{22} & r_{13}r_{23} & r_{12}r_{23}+r_{13}r_{22} & r_{11}r_{23}+r_{13}r_{21} & r_{11}r_{22}+r_{12}r_{21} \\ r_{11}r_{31} & r_{12}r_{32} & r_{13}r_{33} & r_{12}r_{33}+r_{13}r_{32} & r_{11}r_{33}+r_{13}r_{31} & r_{11}r_{32}+r_{12}r_{31} \end{pmatrix}, \tag{8.9}$$

$$K^{(2)}M = \begin{pmatrix} r_{11}r_{21} & r_{12}r_{22} & r_{13}r_{23} & r_{12}r_{23}+r_{13}r_{22} & r_{11}r_{23}+r_{13}r_{21} & r_{11}r_{22}+r_{12}r_{21} \\ r_{21}^2 & r_{22}^2 & r_{23}^2 & 2r_{22}r_{23} & 2r_{21}r_{23} & 2r_{21}r_{22} \\ r_{21}r_{31} & r_{22}r_{32} & r_{23}r_{33} & r_{22}r_{33}+r_{32}r_{23} & r_{21}r_{33}+r_{23}r_{31} & r_{21}r_{32}+r_{22}r_{31} \end{pmatrix}, \tag{8.10}$$

$$K^{(3)}M = \begin{pmatrix} r_{11}r_{31} & r_{12}r_{32} & r_{13}r_{33} & r_{12}r_{33}+r_{13}r_{32} & r_{11}r_{33}+r_{13}r_{31} & r_{11}r_{32}+r_{12}r_{31} \\ r_{21}r_{31} & r_{22}r_{32} & r_{23}r_{33} & r_{22}r_{33}+r_{32}r_{23} & r_{21}r_{33}+r_{23}r_{31} & r_{21}r_{32}+r_{22}r_{31} \\ r_{31}^2 & r_{32}^2 & r_{33}^2 & 2r_{32}r_{33} & 2r_{31}r_{33} & 2r_{31}r_{32} \end{pmatrix}. \tag{8.11}$$

By comparing Eqs. (8.9), (8.10), and (8.11) with Eq. (4.7), Eq. (4.6) is reached.

9. References

[1] Arfken, G. [2005]. *Mathematical methods for physicists*, 6th ed. / edn, Elsevier, Boston.

[2] Auld, B. A. [1990]. *Acoustic Fields and Waves in Solids*, 2nd edn, R.E. Krieger.

[3] Cai, M., Yu, S. J. & Zhang, M. [2006]. Theoretical and numerical solutions of linear and nonlinear elastic waves in a thin rod, *AIAA 2006-4778, 42nd AIAA/ASME/SAE/ASEE Joint Propulsion Conference and Exhibit, Sacramento, Califonia, July 9-12, 2006*, Sacramento, Califonia.

[4] Chang, S. [1995]. The method of space-time conservation element and solution element - a new approach for solving the Navier-Stokes and euler equations, *J. Comput. Phys.* 119(2): 295–324.

[5] Chang, S. & Wang, X. [2003]. Multi-Dimensional courant number insensitive CE/SE euler solvers for applications involving highly nonuniform meshes, *39th AIAA/ASME/SAE/ASEE Joint Propulsion Conference and Exhibit*, Huntsville, Alabama.

[6] Kacimi, A. E. & Laghrouche, O. [2009]. Numerical modelling of elastic wave scattering in frequency domain by the partition of unity finite element method, *International Journal for Numerical Methods in Engineering* 77(12): 1646–1669.
[7] Karypis, G. & Kumar, V. [1998]. A fast and high quality multilevel scheme for partitioning irregular graphs, *SIAM Journal on Scientific Computing* 20(1): 359–392.
[8] Kim, C., Yu, S. J. & Zhang, Z. [2004]. Cavity flow in scramjet engine by Space-Time conservation and solution element method, *AIAA Journal* 42(5): 912–919.
[9] Kreiss, H. O. [1973]. *Methods for the approximate solution of time dependent problems*, Global Atmospheric Research Programme - WMO-ICSU Joint Organizing Committee.
[10] Käser, M. & Dumbser, M. [2006]. An arbitrary high-order discontinuous galerkin method for elastic waves on unstructured meshes; i. the two-dimensional isotropic case with external source terms, *Geophysical Journal International* 166(2): 855–877.
[11] Kulikovskii, A., Pogorelov, N. & Semenov, A. Y. [2000]. *Mathematical Aspects of Numerical Solution of Hyperbolic Systems*, 1 edn, Chapman and Hall/CRC.
[12] LeVeque, R. J. [2002a]. *Finite-Volume Methods for Hyperbolic Problems*, Cambridge texts in applied mathematics, Cambridge University Press, Cambridge.
[13] LeVeque, R. J. [2002b]. Finite-volume methods for non-linear elasticity in heterogeneous media, *International Journal for Numerical Methods in Fluids* 40(1-2): 93–104.
[14] Loh, C., Hultgren, L. & Chang, S. [2001]. Wave computation in compressible flow using space-time conservation element and solution element method, *AIAA Journal* 39(5): 794–801.
[15] Meyer, C. D. [2000]. *Matrix Analysis and Applied Linear Algebra*, Society for Industrial and Applied Mathematics, Philadelphia.
[16] Musgrave, M. J. P. [1954]. On the propagation of elastic waves in aeolotropic media. II. media of hexagonal symmetry, *Proceedings of the Royal Society of London. Series A, Mathematical and Physical Sciences* 226(1166): 356–366.
[17] Musgrave, M. J. P. [1970]. *Crystal Acoustics; Introduction to the Study of Elastic Waves and Vibrations in Crystals*, Holden-day, San Francisco.
[18] Puente, J. d. l., Käser, M., Dumbser, M. & Igel, H. [2007]. An arbitrary high-order discontinuous galerkin method for elastic waves on unstructured meshes; IV. anisotropy, *Geophysical Journal International* 169(3): 1210–1228.
[19] Qin, J. R., Yu, S. J. & Lai, M. [2001]. Direct calculations of cavitating flows in fuel delivery pipe by the Space-Time CESE method, *Journal of Fuels and Lubricants, SAE Transaction* 108: 1720–1725.
[20] Reed, W. H. & Hill, T. R. [1973]. Triangular mesh methods for the neutron transport equation, *Technical Report LA-UR-73-479*, Los Alamos Scientific Laboratory.
[21] Saenger, E. H., Gold, N. & Shapiro, S. A. [2000]. Modeling the propagation of elastic waves using a modified finite-difference grid, *Wave Motion* 31(1): 77–92.
[22] Shorr, B. F. [2004]. *The Wave Finite Element Method*, 1 edn, Springer.
[23] Strang, G. [2006]. *Linear Algebra and Its Applications*, 4th ed edn, Thomson, Brooks/Cole, Belmont, Calif.
[24] Ting, T. C. [1996]. *Anisotropic Elasticity: Theory and Applications*, Oxford University Press, New York.
[25] Virieux, J. [1986]. P-SV wave propagation in heterogeneous media: Velocity-stress finite-difference method, *Geophysics* 51(4): 889–901.
[26] Wang, B., He, H. & Yu, S. J. [2005]. Direct calculation of wave implosion for detonation initiation, *AIAA Journal* 43(10): 2157–2169.

[27] Wang, X. & Chang, S. [1999]. A 2D Non-Splitting unstructured triangular mesh euler solver based on the Space-Time conservation element and solution element method, *Computational Fluid Dynamics Journal* 8(2): 309–325.

[28] Wang, X., Chen, C. & Liu, Y. [2002]. The space-time CE/SE method for solving maxwell's equations in time-domain, *Antennas and Propagation Society International Symposium, 2002. IEEE*, Vol. 1, pp. 164–167 vol.1.

[29] Warming, R. F., Beam, R. M. & Hyett, B. J. [1975]. Diagonalization and simultaneous symmetrization of the Gas-Dynamic matrices, *Mathematics of Computation* 29(132): 1037–1045.

[30] Yang, L., Chen, Y. & Yu, S. J. [2011]. Velocity-Stress equations for waves in solids of hexagonal symmetry solved by the Space-Time CESE method, *ASME Journal of Vibration and Acoustics* 133(2): 021001.

[31] Yang, L., Lowe, R. L., Yu, S. J. & Bechtel, S. E. [2010]. Numerical solution by the CESE method of a First-Order hyperbolic form of the equations of dynamic nonlinear elasticity, *ASME Journal of Vibrations and Acoustics* 132(5): 051003.

[32] Yu, S. J., Yang, L., Lowe, R. L. & Bechtel, S. E. [2010]. Numerical simulation of linear and nonlinear waves in hypoelastic solids by the CESE method, *Wave Motion* 47(3): 168–182.

[33] Zhang, J. & Gao, H. [2009]. Elastic wave modelling in 3-D fractured media: an explicit approach, *Geophysical Journal International* 117(3): 1233–1241.

[34] Zhang, M., Yu, S. J., Lin, S., Chang, S. & Blankson, I. [2004]. Solving magnetohydrodynamic equations without special treatment for Divergence-Free magnetic field, *AIAA Journal* 42(12): 2605–2608.

[35] Zhang, M., Yu, S. J., Lin, S. H., Chang, S. & Blankson, I. [2006]. Solving the MHD equations by the space-time conservation element and solution element method, *Journal of Computational Physics* 214(2): 599–617.

[36] Zhang, Z., Yu, S. T. J. & Chang, S. [2002]. A Space-Time conservation element and solution element method for solving the two- and Three-Dimensional unsteady euler equations using quadrilateral and hexahedral meshes, *Journal of Computational Physics* 175(1): 168–199.

Permissions

The contributors of this book come from diverse backgrounds, making this book a truly international effort. This book will bring forth new frontiers with its revolutionizing research information and detailed analysis of the nascent developments around the world.

We would like to thank Pasquale Giovine, for lending his expertise to make the book truly unique. He has played a crucial role in the development of this book. Without his invaluable contribution this book wouldn't have been possible. He has made vital efforts to compile up to date information on the varied aspects of this subject to make this book a valuable addition to the collection of many professionals and students.

This book was conceptualized with the vision of imparting up-to-date information and advanced data in this field. To ensure the same, a matchless editorial board was set up. Every individual on the board went through rigorous rounds of assessment to prove their worth. After which they invested a large part of their time researching and compiling the most relevant data for our readers. Conferences and sessions were held from time to time between the editorial board and the contributing authors to present the data in the most comprehensible form. The editorial team has worked tirelessly to provide valuable and valid information to help people across the globe.

Every chapter published in this book has been scrutinized by our experts. Their significance has been extensively debated. The topics covered herein carry significant findings which will fuel the growth of the discipline. They may even be implemented as practical applications or may be referred to as a beginning point for another development. Chapters in this book were first published by InTech; hereby published with permission under the Creative Commons Attribution License or equivalent.

The editorial board has been involved in producing this book since its inception. They have spent rigorous hours researching and exploring the diverse topics which have resulted in the successful publishing of this book. They have passed on their knowledge of decades through this book. To expedite this challenging task, the publisher supported the team at every step. A small team of assistant editors was also appointed to further simplify the editing procedure and attain best results for the readers.

Our editorial team has been hand-picked from every corner of the world. Their multi-ethnicity adds dynamic inputs to the discussions which result in innovative outcomes. These outcomes are then further discussed with the researchers and contributors who give their valuable feedback and opinion regarding the same. The feedback is then collaborated with the researches and they are edited in a comprehensive manner to aid the understanding of the subject.

Apart from the editorial board, the designing team has also invested a significant amount of their time in understanding the subject and creating the most relevant covers. They scrutinized every image to scout for the most suitable representation of the subject and create an appropriate cover for the book.

The publishing team has been involved in this book since its early stages. They were actively engaged in every process, be it collecting the data, connecting with the contributors or procuring relevant information. The team has been an ardent support to the editorial, designing and production team. Their endless efforts to recruit the best for this project, has resulted in the accomplishment of this book. They are a veteran in the field of academics and their pool of knowledge is as vast as their experience in printing. Their expertise and guidance has proved useful at every step. Their uncompromising quality standards have made this book an exceptional effort. Their encouragement from time to time has been an inspiration for everyone.

The publisher and the editorial board hope that this book will prove to be a valuable piece of knowledge for researchers, students, practitioners and scholars across the globe.

List of Contributors

Jing Ba and Hong Cao
Geophysical Department, Research Institute of Petroleum Exploration and Development, PetroChina, Beijing, China
Key Laboratory of Geophysics, PetroChina, Beijing, China

Qizhen Du
School of Geosciences, China University of Petroleum (East China), Qingdao, China

Pasquale Giovine
Dipartimento di Meccanica e Materiali - Università degli Studi "Mediterranea", Via Graziella 1, Località Feo di Vito, I-89122 Reggio Calabria, Italy

K. Daneshjou
Department of Mechanical Engineering, Iran University of Science and Technology, Tehran, Iran

H. Ramezani
Young Researchers Club, Islamic Azad University - Shahrerey branch, Tehran, Iran

R. Talebitooti
Department of Automotive Engineering, Iran University of Science and Technology, Tehran, Iran

Yu-Cheng Liu, Yun-Fan Hwang and Jin-Huang Huang
Feng Chia University, Electroacoustic Graduate Program, Taiwan, R.O.C.

Zhengde Dai
School of Mathematics and Physics, Yunnan University, Kunming 650091, P.R.China

Jun Liu and Gui Mu
College of Mathematics and Information Science, Qujing Normal University, Qujing 655000, P.R.China

Murong Jiang
School of Information Sciences and Engineering, Yunnan University, Kunming 650091, P.R.China

Auckpath Sawangsuriya
Bureau of Road Research and Development, Department of Highways, Thailand

Zheng-Hua Qian
State Key Laboratory of Mechanics and Control of Mechanical Structures, Nanjing University of Aeronautics and Astronautics, Nanjing, China

Sheng-Tao John Yu, Yung-Yu Chen and Lixiang Yang
The Department of Mechanical Engineering, The Ohio State University, Columbus, OH 43210, USA

Printed in the USA
CPSIA information can be obtained
at www.ICGtesting.com
JSHW011420221024
72173JS00004B/602

9 781632 400185